Remote Sensing and GIS Accuracy Assessment

Remote Sensing and GIS Accuracy Assessment

Edited by

Ross S. Lunetta
John G. Lyon

CRC Press
Taylor & Francis Group
Boca Raton London New York

CRC Press is an imprint of the
Taylor & Francis Group, an **informa** business

CRC Press
Taylor & Francis Group
6000 Broken Sound Parkway NW, Suite 300
Boca Raton, FL 33487-2742

First issued in paperback 2019

© Taylor & Francis Group, LLC
CRC Press is an imprint of Taylor & Francis Group, an Informa business

No claim to original U.S. Government works

ISBN-13: 978-1-56670-443-4 (hbk)
ISBN-13: 978-0-367-39407-3 (pbk)
Library of Congress Card Number 2004045728

Library of Congress Cataloging-in-Publication Data

Remote sensing and GIS accuracy assessment / edited by Ross S. Lunetta, John G. Lyon.
     p. cm.
    Includes bibliographical references (p.).
    ISBN 1-56670-443-X
     1. Remote sensing—Congresses. 2. Geographic information systems—Congresses. 3. Spatial analysis (Statistics)—Congresses. I. Lunetta, Ross S. II. Lyon, J. G. (John G.)

G70.39.R45 2004
621.36'78—dc22

2004045728

**Visit the Taylor & Francis Web site at
http://www.taylorandfrancis.com**

**and the CRC Press Web site at
http://www.crcpress.com**

Preface

The development of robust accuracy assessment methods for the validation of spatial data represents a difficult challenge for the geospatial science community. The importance and timeliness of this issue are related directly to the dramatic escalation in the development and application of spatial data throughout the latter 20th century. This trend, which is expected to continue, will become increasingly pervasive and continue to revolutionize future decision-making processes. However, our current ability to validate large-area spatial data sets represents a major impediment to many future applications. Problems associated with assessing spatial data accuracy are primarily related to their valued characteristic of being continuous data and to the associated geometric or positional errors implicit with all spatial data. Continuous data typically suffer from the problem of spatial autocorrelation, which violates the important statistical assumption of "independent" data. Positional errors tend to introduce anomalous errors with the combining of multiple data sets or layers. The majority of large-area spatial data coverages are derived from remote sensor data and subsequently analyzed in a GIS to provide baseline information for data-driven assessments to facilitate the decision-making process.

This important topic was the focus of a special symposium sponsored by the U.S. Environmental Protection Agency (EPA) on "Remote Sensing and GIS Accuracy Assessment" on December 11–13, 2001, in Las Vegas, Nevada. The symposium evaluated the important scientific elements relevant to the performance of accuracy assessments for remote sensing-derived data and GIS data analysis and integration products. A keynote address was delivered by Russell G. Congalton that provided attendees with an historical accuracy assessment overview and that identified current technical gaps and established important issues that were the subject of intense debates throughout the symposium. A total of 27 technical papers were presented by an international group of scientists representing federal, state, and local governments, academia, and nongovernmental organizations. Specific technical presentations examined sampling issues, reference data collection, edge and boundary effects, error matrix and fuzzy assessments, error budget analysis, and special issues related to change detection accuracy assessment.

Abstracts submitted for presentation were evaluated for technical merit and assigned to technical sessions by the program committee members. Members then served as technical session chairs, thus maintaining responsibility for session content. Subsequent to the symposium, presenters were invited to submit manuscripts for consideration as chapters. This book contains 20 chapters that represent the important symposium outcomes. All chapters have undergone peer review and were determined to be suitable for publication. The editors have arranged the book into a series of complementary scientific topics to provide the reader with a detailed treatise on spatial data accuracy assessment issues.

The symposium chairs would like to thank the program committee members for their organization of individual technical sessions and participation as session chairs and presenters.

Ross S. Lunetta and John G. Lyon
U.S. Environmental Protection Agency

Acknowledgments

Symposium Sponsor
U.S. Environmental Protection Agency
Office of Research and Development

Symposium Chairs and Book Editors
Ross S. Lunetta (U.S. Environmental Protection Agency)
John G. Lyon (U.S. Environmental Protection Agency)

Program Committee and Session Chairs
Gregory S. Biging (University of California at Berkeley)
Russell G. Congalton (University of New Hampshire)
Christopher D. Elvidge (National Oceanographic and Atmospheric Administration)
John S. Iiames (U.S. Environmental Protection Agency)
S. Taylor Jarnagin (U.S. Environmental Protection Agency)
Michael Jennings (U.S. Geological Survey)
Bruce K. Jones (U.S. Environmental Protection Agency)
Siamak Khorram (North Carolina State University)
Thomas R. Loveland (U.S. Geological Survey)
Thomas H. Mace (National Aeronautics and Space Administration)
Anthony R. Olsen (U.S. Environmental Protection Agency)
Elijah Ramsey III (U.S. Geological Survey)
Terrence E. Slonecker (U.S. Environmental Protection Agency)
Stephen V. Stehman (State University of New York)
James D. Wickham (U.S. Environmental Protection Agency)
L. Dorsey Worthy (U.S. Environmental Protection Agency)

Acknowledgments

Contributors

Elisabeth A. Addink
University of Michigan
Ann Arbor, Michigan

Frank Baarnes
Oak Ridge Associated Universities
Oak Ridge, Tennessee

Michele Barson
Bureau of Rural Sciences
Canberra, ACT, Australia

Latha Baskaran
Pennsylvania State University
University Park, Pennsylvania

Kimberly E. Baugh
Cooperative Institute for Research in
  Environmental Sciences
University of Colorado
Boulder, Colorado

Vivienne Bordas
Bureau of Rural Science
Kingston, ACT, Australia

Mark A. Bowersox
Town of Pittsford
Pittsford, New York

Daniel G. Brown
University of Michigan
Ann Arbor, Michigan

Halil I. Cakir
North Carolina State University
Raleigh, North Carolina

Russell G. Congalton
University of New Hampshire
Durham, New Hampshire

Stephen D. DeGloria
Cornell University
Ithaca, New York

John B. Dietz
Cooperative Institute for Research on the
  Atmosphere
University of Colorado
Boulder, Colorado

Samuel E. Drake
University of Arizona
Tucson, Arizona

Jiunn-Der Dub
University of Michigan
Ann Arbor, Michigan

Donald W. Ebert
U.S. Environmental Protection Agency
National Exposure Research Laboratory
Las Vegas, Nevada

Curtis M. Edmonds
U.S. Environmental Protection Agency
National Exposure Research Laboratory
Las Vegas, Nevada

Christopher D. Elvidge
NOAA National Geophysical Data Center
Boulder, Colorado

Sarah R. Falzarano
National Park Service
Flagstaff, Arizona

Donald Garofalo
U.S. Environmental Protection Agency
National Exposure Research Laboratory
Reston, Virginia

Michael F. Goodchild
University of California
Santa Barbara, California

Kass Green
Space Imaging
Emeryville, California

Bert Guindon
Canada Centre for Remote Sensing
Ottawa, Canada

Daniel T. Heggem
U.S. Environmental Protection Agency
National Exposure Research Laboratory
Las Vegas, Nevada

Vinita Ruth Hobson
Cooperative Institute for Research in
  Environmental Sciences
University of Colorado
Boulder, Colorado

John S. Iiames
Environmental Protection Agency
National Exposure Research Laboratory
  (E243-05)
Research Triangle Park, North Carolina

S. Taylor Jarnagin
U.S. Environmental Protection Agency
National Exposure Research Laboratory
Reston, Virginia

Eugene Jaworski
Eastern Michigan University
Ypsilanti, Michigan

David B. Jennings
U.S. Environmental Protection Agency
National Exposure Research Laboratory
Reston, Virginia

K. Bruce Jones
U.S. Environmental Protection Agency
National Exposure Research Laboratory
Las Vegas, Nevada

Christopher O. Justice
University of Maryland
College Station, Maryland

William G. Kepner
U.S. Environmental Protection Agency
Las Vegas, Nevada

Siamak Khorram
North Carolina State University
Raleigh, North Carolina

Gregory Knight
Pennsylvania State University
University Park, Pennsylvania

Joseph F. Knight
U.S. Environmental Protection Agency
National Exposure Research Laboratory
Research Triangle Park, North Carolina

Phaedon C. Kyriakidis
University of California
Santa Barbara, California

Timothy E. Lewis
U.S. Environmental Protection Agency
National Center for Environmental Assessment
  (B243-01)
Research Triangle Park, North Carolina

Ricardo D. Lopez
U.S. Environmental Protection Agency
National Exposure Research Laboratory
Las Vegas, Nevada

Kim Lowell
Université Laval
Québec, Canada

Xiaohang Liu
San Francisco State University
San Francisco, California

John G. Lyon
U.S. Environmental Protection Agency
National Exposure Research Laboratory
Las Vegas, Nevada

John K. Maingi
Morehead State University
Morehead, Kentucky

Kim Malafant
Complexia
Belconnen, ACT, Australia

Stuart E. Marsh
University of Arizona
Tucson, Arizona

Philippe Mayaux
Institute for Environment and Sustainability
Ispra, Italy

Jeffrey T. Morisette
NASA Goddard Space Flight Center
Greenbelt, Maryland

Anne C. Neale
U.S. Environmental Protection Agency
National Exposure Research Laboratory
Las Vegas, Nevada

Gene Nelson
National Wetlands Research Center
Lafayette, Louisiana

Ingrid L. Nelson
Cooperative Institute for Research in
  Environmental Sciences
University of Colorado
Boulder, Colorado

Andrea Wright Parmenter
Cornell University
Ithaca, New York

Andrew N. Pilant
U.S. Environmental Protection Agency
National Exposure Research Laboratory
  (E243-05)
Research Triangle Park, North Carolina

R. Gil Pontius
Clark University
Worcester, Massachusetts

Jeffrey L. Privette
NASA Goddard Space Flight Center
Greenbelt, Maryland

Elijah Ramsey III
National Wetlands Research Center
Lafayette, Louisiana

Jeffery Safran
Cooperative Institute for Research in
  Environmental Sciences
University of Colorado
Boulder, Colorado

Guofan Shao
Purdue University
West Lafayette, Indiana

Susan M. Skirvin
University of Arizona
Tucson, Arizona

Terrance Slonecker
U.S. Environmental Protection Agency
National Exposure Research Laboratory
Reston, Virginia

Ruth Spell
Ducks Unlimited, Inc.
Rancho Cordova, California

Stephen V. Stehman
State University of New York
Syracuse, New York

Alan Strahler
Boston University
Boston, Massachusetts

Beth Suedmeyer
Clark University
Worcester, Massachusetts

John Sydenstricker-Neto
Cornell University
Ithaca, New York

Kathryn A. Thomas
USGS Southwest Biological Science Center
Flagstaff, Arizona

Liem T. Tran
Pennsylvania State University
University Park, Pennsylvania

Benjamin T. Tuttle
Cooperative Institute for Research in
  Environmental Sciences
University of Colorado
Boulder, Colorado

Christopher J. Watts
Universidad de Sonora
Hermosillo, Sonora, Mexico

David Williams
U.S. Environmental Protection Agency
National Exposure Research Laboratory
Reston, Virginia

David R. Williams
Lockheed Martin Environmental Services
Las Vegas, Nevada

Wenchum Wu
Purdue University
West Lafayette, Indiana

Contents

CHAPTER 1

Putting the Map Back in Map Accuracy Assessment

Russell G. Congalton

CONTENTS

1.1 INTRODUCTION

The need for assessing the accuracy of a map generated from any remotely sensed data has become universally recognized as an integral project component. In the last few years, most projects have required that a certain level of accuracy be achieved for the project and map to be deemed a success. With the widespread application of geographic information systems (GIS) employing remotely sensed data as layers, the need for such an assessment has become even more critical. There are a number of reasons why this assessment is so important, including:

- The need to perform a self-evaluation and to learn from your mistakes
- The ability to compare method/algorithms/analysts quantitatively
- The desire to use the resulting maps/spatial information in some decision-making process

There are many examples in the literature as well as an overwhelming selection of anecdotal evidence to demonstrate the need for accuracy assessment. Many different groups have mapped and/or quantified the amount of tropical deforestation occurring in South America or Southeast Asia. Estimates have ranged by almost an order of magnitude. Which estimate is correct? Without a valid accuracy assessment we may never know. Several federal, state, and local agencies have created maps of wetlands in a county on the eastern shore of Maryland. Techniques used to make these maps included satellite imagery, aerial photography (various scales and film types), and ground sampling. Comparing the various maps yielded little agreement about where wetlands actually existed. Without a valid accuracy assessment we may never know which of these maps to use.

It is no longer sufficient just to make a map using remotely sensed or other spatial data. It is absolutely necessary to take some steps toward assessing the accuracy or validity of that map. There are a number of ways to investigate the accuracy/error in spatial data including, but not limited to, visual inspection, nonsite-specific analysis, generating difference images, error budget analysis, and quantitative accuracy assessment.

The goal of this chapter is to review the current knowledge of accuracy assessment methods and stimulate the reader to further the progression of diagnostic techniques and information to support the appropriate application of spatial data. The ultimate objective is to motivate everyone to conduct or demand an appropriate accuracy assessment or validation and make certain it is included as an essential metadata element.

1.2 ACCURACY ASSESSMENT OVERVIEW

1.2.1 Historical Review

The history of accuracy assessment of digital, remotely sensed data is relatively short, beginning in about 1975. Before 1975 maps derived from analog, remotely sensed data (i.e., photo interpretation) were rarely subjected to any kind of quantitative accuracy assessment. Field checking was typically performed as part of the interpretation process, but no overall map accuracy or other quantitative measures of quality were typically incorporated into the analysis. Only after photo interpretation began being used as reference data to compare maps derived from digital, remote sensor data did issues concerning the accuracy of the photo interpretation arise. All the accuracy assessment techniques mentioned in this chapter can be applied to assessing the accuracy of both analog and digital remotely sensed data (Congalton and Mead, 1983; Congalton et al., 1983).

The history of accuracy assessment can be effectively divided into four developmental epochs or ages. In the beginning, no real accuracy assessment was performed; rather, an "it-looks-good" mentality prevailed. This approach is typical of many new, emerging technologies. Despite the maturing of the technology over the last 25 years, some remote sensing analysts are still stuck in this mentality. Of course, the map must "look good" before any further analysis should be performed. Why assess a map that is obviously poor? However, while "looking good" is a required characteristic, it is not sufficient for a valid assessment.

The second age of accuracy assessment could be called the epoch of nonsite-specific assessment. During this period, overall acreages were compared between ground estimates and the map without regard for location (Meyer et al., 1975). In some instances, such as imagery with very large pixels (e.g., AVHRR imagery), a nonsite-specific assessment may be the best and/or only choice for validation. For most imagery, the age of nonsite-specific assessment quickly gave way to the age of the site-specific assessment (third age). In a site-specific assessment, actual places on the ground (i.e., locations) were compared to the same place on the map and a measure of overall accuracy (i.e., percentage correct) was computed.

Finally, the fourth and current age of accuracy assessment could be called the age of the error matrix. This epoch includes a significant number of analysis techniques, most importantly the Kappa analysis. A brief review of the techniques and considerations of the error matrix age can be found below and is described in more detail in Congalton and Green (1999).

1.2.2 Established Techniques and Considerations

Since the mid-1980s the error matrix has been accepted as the standard descriptive reporting tool for accuracy assessment of remotely sensed data. The use of the error matrix has significantly improved our ability to conduct accuracy assessments. In addition, analysis tools including discrete multivariate techniques have facilitated the comparison and development of various methodologies, algorithms, and approaches. Many factors affect the compilation of the error matrix and must be considered when designing any accuracy assessment. The current state of knowledge concerning the error matrix, analysis techniques, and some considerations are briefly reviewed here.

1.2.3 The Error Matrix

An error matrix is a square array of numbers organized in rows and columns that express the number of sample units (i.e., pixels, clusters of pixels, or polygons) assigned to a particular category relative to the actual category as indicated by the reference data (Table 1.1). The columns typically represent the reference data and the rows indicate the map generated from the remotely sensed data. Reference data are assumed correct and can be collected from a variety of sources, including photographic interpretation, ground or field observation, and ground or field measurement

The error matrix, once correctly generated, can be used as a starting point for a series of descriptive and analytical statistical techniques. The most common and simplest descriptive statistic is overall accuracy, which is computed by dividing the total correct (i.e., the sum of the major diagonal) by the total number of sample units in the error matrix. In addition, individual category accuracies can be computed in a similar manner. Traditionally, the total number of correct sample

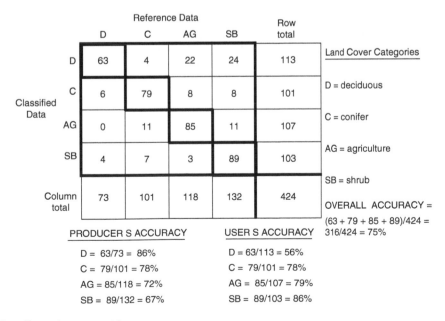

Table 1.1   Example error matrix.

units in a category is divided by the total number of sample units of that category from the reference data (i.e., the column total). This accuracy measure relates to the probability of a reference sample unit's being correctly classified and is really a measure of omission error. This accuracy measure is often called the *producer's accuracy* because the producer of the classification is interested in how well a certain area can be classified. On the other hand, if the total number of correct sample units in a category is divided by the total number of sample units that were classified into that category on the map (i.e., the row total), then this result is a measure of commission error. This measure is called *user's accuracy* or reliability and is indicative of the probability that a sample unit classified on the map actually represents that category on the ground (Story and Congalton, 1986).

1.2.4 Discrete Multivariate Analysis

In addition to these descriptive techniques, an error matrix is an appropriate beginning for many analytical statistical techniques, especially discrete multivariate techniques. Starting with Congalton et al. (1983), discrete multivariate techniques have been used for performing statistical tests on the classification accuracy of digital, remotely sensed data. Since that time many others have adopted these techniques as the standard accuracy assessment tools (Rosenfield and Fitzpatrick-Lins, 1986; Campbell, 1987; Hudson and Ramm, 1987; Lillesand and Kiefer, 1994).

One analytical step to perform once the error matrix has been built is to "normalize" or standardize the matrix using a technique known as "MARGFIT" (Congalton et al., 1983). This technique uses an iterative proportional fitting procedure that forces each row and column in the matrix to sum to one. The rows and column totals are called marginals, hence the technique's name, MARGFIT. In this way, differences in sample sizes used to generate the matrices are eliminated and, therefore, individual cell values within the matrix are directly comparable. Also, because the iterative process totals the rows and columns, the resulting normalized matrix is more indicative of the off-diagonal cell values (i.e., the errors of omission and commission) than is the original matrix. The major diagonal of the normalized matrix can be summed and divided by the total of the entire matrix to compute a normalized overall accuracy.

A second discrete multivariate technique of use in accuracy assessment is called Kappa (Cohen, 1960). Kappa can be used as another measure of agreement or accuracy. Kappa values can range from +1 to −1. However, since there should be a positive correlation between the remotely sensed classification and the reference data, positive values are expected. Landis and Koch (1977) lumped the possible ranges for Kappa into three groups: a value greater than 0.80 (i.e., 80%) represents strong agreement; a value between 0.40 and 0.80 (i.e., 40%–80%) represents moderate agreement; and a value below 0.40 (i.e., 40%) represents poor agreement.

The equations for computing Kappa can be found in Congalton et al. (1983), Rosenfield and Fitzpatrick-Lins (1986), Hudson and Ramm (1987), and Congalton and Green (1999), to list just a few. It should be noted that the Kappa equation assumes a multinomial sampling model and that the variance is derived using the Delta method (Bishop et al., 1975).

The power of the Kappa analysis is that it provides two statistical tests of significance. Using this technique, it is possible to test whether an individual land-cover (LC) map generated from remotely sensed data is significantly better than a map generated by randomly assigning labels to areas. The second test allows for the comparison of any two matrices to see whether they are statistically, significantly different. In this way, it is possible to determine that one method/algorithm/analyst is different from another one and, based on a chosen accuracy measure (e.g., overall accuracy), to conclude which is better.

1.2.5 Sampling Size and Scheme

Sample size is another important consideration when assessing the accuracy of remotely sensed data. Each sample point collected is expensive. Therefore, sample size must be kept to a minimum, yet it is critical to maintain a large enough sample size so that any analysis performed is statistically valid. Many researchers, notably Hord and Brooner (1976), van Genderen and Lock (1977), Tortora (1978), Hay (1979), Rosenfield et al. (1982), and Congalton (1988a), have published equations and guidelines for choosing the appropriate sample size. The majority of researchers have used an equation based on the binomial distribution or the normal approximation to the binomial distribution to compute the required sample size. These techniques are statistically sound for computing the sample size needed to compute the overall accuracy of a classification or the overall accuracy of a single category. The equations are based on the proportion of correctly classified samples (pixels, clusters, or polygons) and on some allowable error. However, these techniques were not designed to choose a sample size for creating an error matrix. In the case of an error matrix, it is not simply a matter of correct or incorrect. Given an error matrix with n land-cover categories, for a given category there is one correct answer and $n - 1$ incorrect answers. Sufficient samples must be acquired to be able to adequately represent this confusion. Therefore, the use of these techniques for determining the sample size for an error matrix is not inappropriate. Instead, the use of the multinomial distribution is recommended (Tortora, 1978).

Traditional thinking about sampling does not often apply because of the large number of pixels in a remotely sensed image. For example, a 0.5% sample of a single Landsat Thematic Mapper (TM) scene can be over 300,000 pixels. Most, if not all, assessments should *not* be performed on a per-pixel basis because of problems with exact single pixel location. Practical considerations more often dictate the sample size selection. A balance between what is statistically sound and what is practically attainable must be found. A generally accepted rule of thumb is to use a minimum of 50 samples for each LC category in the error matrix. This rule also tends to agree with the results of computing sample size using the multinomial distribution (Tortora, 1978). If the area is especially large or the classification has a large number of LC categories (i.e., more than 12 categories), the minimum number of samples should be increased to 75 to 100 samples per category.

The number of samples for each category can also be weighted based on the relative importance of that category within the objectives of the mapping or on the inherent variability within each of the categories. Sometimes it is better to concentrate the sampling on the categories of interest and increase their number of samples while reducing the number of samples taken in the less important categories. Also, it may be useful to take fewer samples in categories that show little variability, such as water or forest plantations, and increase the sampling in the categories that are more variable, such as uneven-aged forests or riparian areas. In summary, the goal is to balance the statistical recommendations to obtain an adequate sample from which to generate an appropriate error matrix within the objectives, time, cost, and practical limitations of the mapping project.

Along with sample size, sampling scheme is an important part of any accuracy assessment. Selection of the proper scheme is absolutely critical to generating an error matrix that is representative of the entire classified image. Poor choice in sampling scheme can result in significant biases being introduced into the error matrix that may over- or underestimate the true accuracy. In addition, the use of the proper sampling scheme may be essential depending on the analysis techniques to be applied to the error matrix.

Many researchers have expressed opinions about the proper sampling scheme to use, including everything from simple random sampling to stratified, systematic, unaligned sampling. Despite all these opinions, very little work has actually been performed in this area. Congalton (1988a) performed sampling simulations on three spatially diverse areas (forest, agriculture, and rangeland) and concluded that in all cases simple random sampling without replacement and stratified random sampling provided satisfactory results. Despite the desirable statistical properties of simple random sampling, this sampling scheme is not always very practical to apply. Simple random sampling tends to undersample small but possibly very important areas unless the sample size is significantly increased. For this reason, stratified random sampling is recommended where a minimum number of samples are selected from each strata (i.e., category). Even stratified random sampling can be somewhat impractical because of having to collect ground information for the accuracy assessment at random locations on the ground.

Two difficult problems arise when using random locations: (1) the location can be very difficult to access and (2) they can only be selected after the classification has been performed. This second condition limits the accuracy assessment data to being collected late in the project instead of in conjunction with the training data collection, thereby increasing the costs of the project. In addition, in some projects the time between the project beginning and the accuracy assessment may be so long as to cause temporal problems in collecting reference data.

1.2.6 Spatial Autocorrelation

Spatial autocorrelation is said to occur when the presence, absence, or degree of a certain characteristic affects the presence, absence, or degree of the same characteristic in neighboring units (Cliff and Ord, 1973). This condition is particularly important in accuracy assessment if an error at a certain location can be found to influence errors at surrounding locations positively or negatively (Campbell, 1981). Work by Congalton (1988b) on Landsat MSS data from three areas of varying spatial diversity (agriculture, range, and forest) showed a positive influence as much as 30 pixels (1.8 km) away. More recent work by Pugh and Congalton (2001) using Landsat TM data in a forested environment showed similar issues with spatial autocorrelation. These results affect the choice of sample size and, especially, the sampling scheme used in the accuracy assessment.

1.3 CURRENT ISSUES AND NEEDS

1.3.1 Sampling Issues

The major sampling issue of importance today is the choice of the sample unit. Historically, a single pixel has often been chosen as the sample unit. However, it is extremely difficult to know exactly where that pixel is on the reference data, especially when the reference data are generated on the ground (using field work). Despite recent advances in Global Positioning System (GPS) technology, it is very rare to achieve adequate location information for a single pixel. Many times the GPS unit is used under dense forest canopy and the GPS signals are weak or absent. Location becomes even more problematic with the new high-spatial resolution sensors such as Space Imaging IKONOS or Digital Globe imagery with pixels as small as 1 m. Also, it is nearly impossible to match the corners of a pixel on an image to the ground despite our best registration algorithms. Therefore, using a single pixel as the sampling unit can cause much of the error represented in the error matrix to be positional error rather than thematic error. Since the goal of the error matrix is to measure thematic error, it is best to take steps to avoid including positional error. Single pixels should not be used for the sample unit. Instead, some cluster of pixels or a polygon should be chosen.

1.3.2 Edge and Boundary Effects

Traditionally, accuracy assessment has been performed to avoid the boundaries between different LC classes by taking samples near the center of each polygon, or at least away from the edges. Avoiding the edges also helps to minimize the locational error as discussed in the last section. Where exactly to draw the line between different cover types on the ground is very subjective. Most LC or vegetation maps divide a rather continuous environment called Earth into a number of discrete categories. The number of categories varies with the objective of the mapping, and our ability to separate different categories depends on the variability within and between each category.

All this information should be represented in a well-defined, mutually exclusive, and totally exhaustive classification scheme. However, in many instances, it would be useful to know more about the boundaries or edges of different LC types. For example, when performing change detection (i.e., looking for changes over time), it is important to know whether real change exists and that change is going to occur along the boundaries between cover types. Therefore, it is important that more research and study be undertaken to better understand this boundary and edge issue.

1.3.3 Reference Data Collection

Reference data are typically assumed to be correct and are used to evaluate the results of the LC mapping. If the reference data are wrong, then the LC map will be unfairly judged. If the reference data are inefficiently collected, then the project may suffer from unnecessarily high costs or an insufficient number of samples to properly evaluate the results. Reference data are a critical, very expensive, and yet often overlooked component of any spatial analysis. For example, aerial photographic interpretation is often used as reference data for assessing a LC map generated from digital satellite imagery. The photographic interpretation is assumed correct because it often has greater spatial resolution than the satellite imagery and because photogrammetry has become a time-honored skill that is accepted as accurate. However, photographic interpretation is subjective and can be significantly wrong. If the interpretation is wrong, then the results of the accuracy assessment could indicate that the satellite-based map is of poor accuracy when actually it is the reference data that are inappropriate.

There are numerous examples in the literature documenting problems with collecting improper or inadequate reference data. One especially insidious problem with reference data collection is the size of the sample area in which the reference data are collected. Clearly, it is important to collect reference information that is representative of the mapped area. In other words, if the map is generated with remotely sensed data that have 30- × 30-m pixels, it does not make sense to collect reference data for a 5-m² area. A current example of this situation is the use of the Forest Inventory and Analysis (FIA) plots collected by the U.S. Forest Service across the country. It is important that these inventory plots be large enough to provide valid reference data.

The opposite situation must also be carefully monitored. For example, it is not appropriate to assess the accuracy of a 1-ha mapping unit with 5-ha reference data. The reference data must be collected with the pixel size and/or the minimum mapping unit of the map in mind. Additionally, the same exact classification scheme using the same exact rules must be used to label the reference data and to generate the map. Otherwise, errors will be introduced by classification scheme (definitional) differences and the error matrix created will not be indicative of the true accuracy of the map. Using well-designed field forms that step the collector through the process can be very helpful in ensuring that the reference data are collected at the proper scale and with the same or appropriate classification scheme to accurately assess the map.

1.3.4 Beyond the Error Matrix: Fuzzy Assessment

As remote sensing projects have grown in complexity, so have the associated classification schemes. The classification scheme then becomes a very important factor influencing the accuracy of the entire project. Recently, papers have appeared in the literature that point out some of the limitations of using only an error matrix with a complex classification scheme. A paper by Congalton and Green (1993) recommends the error matrix as a jumping-off point for identifying sources of confusion and not just error in the remotely sensed classification. For example, the variation in human interpretation can have a significant impact on what is considered correct and what is not. As previously mentioned, if photographic interpretation is used as the reference data in an accuracy assessment and the interpretation is not completely correct, then the results of the accuracy assessment will be very misleading. The same statements are true if ground observations, as opposed to ground measurements, are used as the reference data set. As classification schemes get more complex, more variation in human interpretation is introduced. Other factors beyond just variation in interpretation are important also.

In order to deal with ambiguity/variation in remotely sensed maps, Gopal and Woodcock (1994) proposed the use of fuzzy sets to "allow for explicit recognition of the possibility that ambiguity might exist regarding the appropriate map label." In such an approach, it is recognized that instead of a simple system of correct (agreement) and incorrect (disagreement) there can be a variety of responses, such as absolutely right, good answer, acceptable, understandable but wrong, and absolutely wrong. This approach deals well with the ambiguity issue. However, the results are not presented in a standard error matrix format. Therefore, Congalton and Green (1999) and Green and Congalton (2003) presented a fuzzy assessment methodology that not only deals with variation/ambiguity, but also allows for the results of the assessment to be presented in an error matrix.

1.3.5   Error Budget Analysis

Over the last 25 years many papers have been written about the quantification of error associated with remotely sensed and other spatial data (Congalton and Green, 1999). As documented in this chapter, our ability to quantify the total error in a spatial data set has developed substantially. However, little has been done to partition this error into its component parts and construct an error budget. Without this division into parts, it is not possible to evaluate or analyze the impact a specific error has on the entire mapping project. Therefore, it is not possible to determine which components contribute the most errors or which are most easily corrected. Some early work in this area was demonstrated in a paper by Lunetta et al. (1991) and resulted in an often-cited diagram that lists the sources of error accumulating throughout a remote sensing project.

It should be noted that each of the major error sources adds to the total error budget separately, and/or through a mixing process. It is no longer sufficient to always just evaluate the total error. For many applications there is a definite need to identify and understand (1) error sources and (2) the appropriate mechanisms for controlling, reducing, and reporting errors. Perhaps the simplest way to begin to look at an error budget is to create a special error budget analysis table (Congalton and Green, 1999). This table is generated, column-by-column, beginning with a listing of the possible sources of error for the project. Once the various components that comprise the total error are listed, then each component can be assessed to determine its contribution to the overall error. Next, our ability to deal with this error is evaluated. It should be noted that some errors may be very large but are easy to correct while others may be rather small. Finally, an error index can be created directly by multiplying the error contribution potential by the error control difficulty. Combining these two factors allows one to establish priorities for best dealing with individual errors within a mapping project. A template to be used to conduct just such an error budget analysis is presented in Table 1.2.

1.3.6   Change Detection Accuracy Assessment

Much has recently been written in the literature about change detection (Lunetta and Elvidge, 1998; Khorram et al., 1999). This technique is an extremely popular and powerful use of remotely sensed data. Assessing the accuracy of a change detection analysis has all the issues, complications, and difficulties of a single date assessment plus many additional, unique problems. For example, how does one obtain information on reference data for images/maps from the past? Likewise, how can one sample enough areas that will change in the future to generate a statistically valid assessment? Most of the studies on change detection do not present any quantitative results. Without the desired accuracy assessment, it is difficult to determine which change detection methods are best and should be applied to future projects.

Error Source	Error Contribution Potential	Error Control Difficulty	Error Index	Error Priority

Error Contribution Potential: Relative potential for this source as contributing factor to the total error (**1 = low**, **2 = medium**, and **3 = high**).

Error Control Difficulty: Given the current knowledge about this source, how difficult is controlling the error contribution (**1 = not very difficult** to **5 = very difficult**).

Error Index: An index that represents the combination of error potential and error difficulty.

Error Priority: Order in which methods should be implemented to understand, control, reduce, and/or report the error due to this source based on the error index.

Table 1.2 Template for conducting an error budget analysis.

Congalton et al. (1993) provided the first example comparing a single date and change detection error matrix. It should be noted that if a single date error matrix has n map categories, then a change detection error matrix would contain n^2 map categories. This is because we are no longer dealing with a single classification, but rather with a change between two different classifications generated at different times. In a single date error matrix there is one row and column for each map category. However, in a change detection error matrix the question of interest is, What category was this area at T_1 vs. T_2? This comparison uses the exact same logic used for the single classification error matrix; it is just complicated by the two time periods (i.e., the change). As always, the major diagonal indicates correct classification while the off-diagonal elements indicate the errors or confusion.

1.4 SUMMARY

Validation or accuracy assessment is an integral component of most mapping projects incorporating remotely sensed data. In fact, this topic has become so important as to spawn regular conferences and symposia. This emphasis on data quality was not always the case. In the 1970s only a few enlightened scientists and researchers dared ask the question, How good is this map derived from Landsat MSS imagery? In the 1980s, the error matrix became a common tool for representing the accuracy of individual map categories. By the 1990s, most maps derived from remotely sensed imagery were required to meet some minimum accuracy standard. Now, it is important that with all the statistics and spatial analysis available to us that we do not lose track of the primary goal of why we perform an accuracy assessment in the first place.

This chapter presented a review of techniques and considerations necessary to assess or validate maps derived from remotely sensed and other spatial data. Although it is important to perform a visual examination of the map, it is not sufficient. Other techniques, such as nonsite-specific analysis and difference images, can help. Error budgeting is a very useful exercise in helping to realize error and consider ways to minimize it. Quantitative accuracy assessment provides a very powerful mechanism for both descriptive and analytical evaluation of the spatial data. However, given all these techniques and considerations, it is most important that we remember why we are performing the accuracy assessment in the first place.

Both as makers and users of our maps, our goal is to make the best map possible for a given objective. To achieve this goal we must not get lost in all the statistics and analyses but must apply the correct analysis techniques and use the proper sampling approaches. However, all these things will do us no good if we forget about the map we are trying to assess. We must "put the map back in the map assessment process." We must do everything we can to ensure that the assessment is valid for the map and not simply a statistical exercise. It is key that the reference data match the map data, not only in classification scheme but also in sampling unit (i.e., minimum mapping unit) as well. It is also important that we make every effort to collect accurate and timely reference data. Finally, there is still much to do. Many maps generated from remotely sensed data still have no validation or accuracy assessment. There are numerous steps that can be taken to evaluate how good a map is. Now we must move past the age of "it looks good" and move toward the more quantitative assessments outlined in this chapter.

REFERENCES

Bishop, Y., S. Fienberg, and P. Holland, *Discrete Multivariate Analysis: Theory and Practice*, MIT Press, Cambridge, MA, 1975.

Campbell, J., *Introduction to Remote Sensing*, Guilford Press, New York, 1987.

Campbell, J., Spatial autocorrelation effects upon the accuracy of supervised classification of land cover, *Photogram. Eng. Remote Sens.*, 47, 355–363, 1981.

Cliff, A.D. and J.K. Ord, *Spatial Autocorrelation*, Pion Limited, London, 1973.

Cohen, J., A coefficient of agreement for nominal scales, *Educ. Psychol. Meas.*, 20, 37–46, 1960.

Congalton, R.G., A comparison of sampling schemes used in generating error matrices for assessing the accuracy of maps generated from remotely sensed data, *Photogram. Eng. Remote Sens.*, 54, 593–600, 1988a.

Congalton, R.G., Using spatial autocorrelation analysis to explore errors in maps generated from remotely sensed data, *Photogram. Eng. Remote Sens.*, 54, 587–592, 1988b.

Congalton R. and K. Green, A practical look at the sources of confusion in error matrix generation, *Photogram. Eng. Remote Sens.*, 59, 641–644, 1993.

Congalton, R. and K. Green, *Assessing the Accuracy of Remotely Sensed Data: Principles and Practices*, Lewis, Boca Raton, FL, 1999.

Congalton, R., R. Macleod, and F. Short, Developing Accuracy Assessment Procedures for Change Detection Analysis, final report submitted to NOAA CoastWatch Change Analysis Program, Beaufort, NC, 1993.

Congalton, R.G. and R.A. Mead, A quantitative method to test for consistency and correctness in photo-interpretation, *Photogram. Eng. Remote Sens.*, 49, 69–74, 1983.

Congalton, R.G., R.G. Oderwald, and R.A. Mead, Assessing Landsat classification accuracy using discrete multivariate statistical techniques, *Photogram. Eng. Remote Sens.*, 49, 1671–1678, 1983.

Gopal, S. and C. Woodcock, Theory and methods for accuracy assessment of thematic maps using fuzzy sets, *Photogram. Eng. Remote Sens.*, 60, 181–188, 1994.

Green, K. and R. Congalton, An error matrix approach to fuzzy accuracy assessment: The NIMA Geocover project example, in Geospatial Data Accuracy Assessment, Lunetta, R. and J. Lyon, Eds., U.S. Environmental Protection Agency, Report No. EPA/600/R03/064, 2003.

Hay, A.M., Sampling designs to test land-use map accuracy, *Photogram. Eng. Remote Sens.*, 45, 529–533, 1979.

Hord, R.M. and W. Brooner, Land use map accuracy criteria, *Photogram. Eng. Remote Sens.*, 42, 671–677, 1976.

Hudson, W., and C. Ramm, Correct formulation of the kappa coefficient of agreement, *Photogram. Eng. Remote Sens.*, 53, 421–422, 1987.

Khorram, S., G. Biging, N. Chrisman, D. Colby, R. Congalton, J. Dobson, R. Ferguson, M. Goodchild, J. Jensen, and T. Mace, *Accuracy Assessment of Remote Sensing-Derived Change Detection*, American Society for Photogrammetry and Remote Sensing, Bethesda, MD, 1999.

Landis, J. and G. Koch, The measurement of observer agreement for categorical data, *Biometrics*, 33, 159–174, 1977.

Lillesand, T. and R. Kiefer, *Remote Sensing and Image Interpretation*, 3rd ed., John Wiley & Sons, New York, 1994.

Lunetta, R., R. Congalton, L. Fenstermaker, J. Jensen, K. McGwire, and L. Tinney, Remote sensing and geographic information system data integration: error sources and research issues, *Photogram. Eng. Remote Sens.,* 57, 677–687, 1991.

Lunetta, R.L. and C.D. Elvidge, Eds., *Remote Sensing Change Detection: Environmental Monitoring Methods and Applications,* Taylor and Francis, London, 1998.

Meyer, M., J. Brass, B. Gerbig, and F. Batson, ERTS Data Applications to Surface Resource Surveys of Potential Coal Production Lands in Southeast Montana, IARSL Final Research Report 75-1, University of Minnesota, 1975.

Pugh, S. and R. Congalton, Applying spatial autocorrelation analysis to evaluate error in New England forest cover type maps derived from Landsat Thematic Mapper Data, *Photogram. Eng. Remote Sens.,* 67, 613–620, 2001.

Rosenfield, G. and K. Fitzpatrick-Lins, A coefficient of agreement as a measure of thematic classification accuracy, *Photogram. Eng. Remote Sens.,* 52, 223–227, 1986.

Rosenfield, G.H., K. Fitzpatrick-Lins, and H. Ling, Sampling for thematic map accuracy testing, *Photogram. Eng. Remote Sens.,* 48, 131–137, 1982.

Story, M. and R. Congalton, Accuracy assessment: a user's perspective. *Photogram, Eng. Remote Sens.,* 52, 397–399, 1986.

Tortora, R. A note on sample size estimation for multinomial populations, *Am. Stat.,* 32, 100–102, 1978.

van Genderen, J.L. and B.F. Lock, Testing land use map accuracy, *Photogram. Eng. Remote Sens.,* 43, 1135–1137, 1977.

CHAPTER 2

Sampling Design for Accuracy Assessment of Large-Area, Land-Cover Maps: Challenges and Future Directions

Stephen V. Stehman

CONTENTS

2.1 INTRODUCTION

This chapter focuses on the application of accuracy assessment as a final stage in the evaluation of the thematic quality of a land-cover (LC) map covering a large region such as a state or province, country, or continent. The map is assumed to be classified according to a crisp or hard classification scheme, as opposed to a fuzzy classification scheme (Foody, 1999). The standard protocol for accuracy assessment is to compare the map LC label to the reference label at sample locations, where the reference label is assumed to be correct. The source of reference data may be aerial photography, ground visit, or videography. Discussion will be limited to the case in which the assessment unit for comparing the map and reference label is a pixel. Similar issues apply to sampling both pixels and polygons, but a greater assortment of design options has been developed for pixel-based assessments. Most of the chapter will focus on site-specific accuracy, which is accuracy determined on a pixel-by-pixel basis. In contrast, nonsite-specific accuracy provides a comparison aggregated over some spatial extent. For example, in a nonsite-specific assessment, the area of forest mapped for a county would be compared to the true area of forest in that county. Errors of omission for a particular class may be compensated for by errors of commission from other classes such that nonsite-specific accuracy may be high even if site-specific accuracy is poor. Site-specific accuracy may be viewed as spatially explicit, whereas nonsite-specific accuracy addresses map quality in a spatially aggregated framework.

A sampling design is a set of rules for selecting which pixels will be visited to obtain the reference data. Congalton (1991), Janssen and van der Wel (1994), Congalton and Green (1999), and Stehman (1999) provide overviews of the basic sampling designs available for accuracy assessment. Although these articles describe designs that may serve well for small-area, limited-objective assessments, they do not convey the broad diversity of design options that must be drawn upon to meet the demands of large-area mapping efforts with multiple accuracy objectives. An objective here is to expand the discussion of sampling design to encompass alternatives available for more demanding, complex accuracy assessment problems.

The diversity of accuracy assessment objectives makes it important to specify which objectives a particular assessment is designed to address. Objectives may be categorized into three general classes: (1) description of the accuracy of a completed map, (2) comparison of different classifiers, and (3) assessment of sources of classification error. This chapter focuses on the descriptive objective. Recent examples illustrating descriptive accuracy assessments of large-area LC maps include Edwards et al. (1998), Muller et al. (1998), Scepan (1999), Zhu et al. (2000), Yang et al. (2001), and Laba et al. (2002). The foundation of a descriptive accuracy assessment is the error matrix and the variety of summary measures computed from the error matrix, such as overall, user's and producer's accuracies, commission and omission error probabilities, measures of chance-corrected agreement, and measures of map value or utility.

Additional descriptive objectives are often pursued. Because classification schemes are often hierarchical (Anderson et al., 1976), descriptive summaries may be required for each level of the hierarchy. For large-area LC maps, there is frequently interest in accuracy of various subregions, for example, a state or province within a national map, or a county or watershed within a state or regional map. Each identified subregion could be characterized by an error matrix and accompanying summary measures. Describing spatial patterns of classification error is yet another objective. Reporting accuracy for various subsets of the data, for example, homogeneous 3 × 3 pixel blocks, edge pixels, or interior pixels may address this objective. Another potential objective would be to describe accuracy for various aggregations of the data. For example, if a map constructed with a 30-m pixel resolution is converted to a 90-m pixel resolution, what is the accuracy of the 90-m product? Lastly, nonsite-specific accuracy may be of interest. For example, if a primary application of the map were to provide LC proportions for a 5- × 5-km spatial unit (e.g., Jones et al., 2001), nonsite-specific accuracy would be of interest. Nonsite-specific accuracy has typically been thought of as applying to the entire map (Congalton and Green, 1999). However, when viewed in the wider context of how maps are used, nonsite-specific accuracy at various spatial extents becomes relevant.

The basic elements of a statistically rigorous sampling strategy are encapsulated in the specification of a probability sampling design, accompanied by consistent estimation following principles of Horvitz-Thompson estimation. These fundamental characteristics of statistical rigor are detailed in Stehman (2001). Choosing a sampling design for accuracy assessment may be guided by the following additional design criteria: (1) adequate precision for key estimates, (2) cost-effectiveness, and (3) appropriate simplicity to implement and analyze (Stehman, 1999). These criteria hold whether the reference data are crisp or fuzzy and will be prioritized differently for different assessments. Because these criteria often lead to conflicting design choices, the ability to compromise among criteria is a crucial element of the art of sampling design.

2.2 MEETING THE CHALLENGE OF COST-EFFECTIVE SAMPLING DESIGN

Effective sampling practice requires constructing a design that affords good precision while keeping costs low. Strata and clusters are two basic sampling structures available in this regard, and often both are desirable in accuracy assessment problems. Unfortunately, implementing a design incorporating both features may be challenging. This topic will be addressed in the next subsection. A second approach to enhance cost-effectiveness is to use existing data or data collected for purposes other than accuracy assessment (e.g., for environmental monitoring). This topic is addressed in the second subsection.

2.2.1 Strata vs. Clusters: The Cost vs. Precision Paradox

The objective of precise estimation of class-specific accuracy is a prime motivation for stratified sampling. In the typical implementation of stratification in accuracy assessment, the mapped LC classes define the strata, and the design is tailored to enhance precision of estimated user's accuracy or commission error. Stratified sampling requires all pixels in the population to be identified with a stratum. If the map is finished, stratifying by mapped LC class is readily accomplished. Geographic stratification is also commonly used in accuracy assessment. It is motivated by an objective specifying accuracy estimates for key geographic regions (e.g., an administrative unit such as a state or an ecological unit such as an ecoregion), or by an objective specifying a spatially well-distributed sample. It is possible, though rare, to stratify by the cross-classification of land-cover class by geographic region. The drawback of this two-way stratification is that resources are generally not sufficient to obtain an adequate sample size to estimate accuracy precisely in each stratum (e.g., Edwards et al., 1998).

The rationale for cluster sampling is to obtain cost-effectiveness by sampling pixels in groups defined by their spatial proximity. The decrease in the per-unit cost of each sample pixel achieved by cluster sampling may result in more precise accuracy estimates depending on the spatial pattern of classification error. Cluster sampling is a means by which to obtain spatial control (distribution) over the sample. This spatial control can occur at two scales, termed regional and local. Regional spatial control refers to limiting the macro-scale spatial distribution of the sample, whereas local spatial control reflects the logical consequence that sampling several spatially proximate pixels requires little additional effort beyond that needed to sample a single pixel. Examples of clusters achieving regional control over the spatial distribution of the sample include a county, quarter-quad, or 6- × 6-km area. Examples of design structures used to implement local control include blocks of pixels (e.g., 3 × 3 or 5 × 5 pixel blocks), polygons of homogeneous LC, or linear clusters of pixels. Both regional and local controls are designed to reduce costs, and for either option the assessment unit is still an individual pixel.

Regional spatial control is designed to control travel costs or reference data material costs. For example, if the reference data consist of interpreted aerial photography, restricting the sample to a relatively small number of photos will reduce cost. If the reference data are collected by ground visit, regional control can limit travel to within a much smaller total area (e.g., within a sample of counties or 6- × 6-km blocks, rather than among all counties or 6- × 6-km blocks). When used alone, local spatial control may not achieve these cost advantages. For example, a simple random or systematic sample of 3 × 3 pixel blocks providing local spatial control may be widely dispersed across the landscape, therefore requiring many photos or extensive travel to reach the sample clusters.

In practice, both regional and local control may be employed in the same design. The most likely combination in such a multistage design would be to exercise regional control via two-stage cluster sampling and local control via one-stage cluster sampling, as follows. Define the primary sampling unit as the cluster constructed to obtain regional spatial control (e.g., a 6- × 6-km area). The secondary sampling unit would be chosen to provide the desired local spatial control (e.g., 3 × 3 block of pixels). The first-stage sample consists of primary sampling units (PSUs), but not every 3 × 3 block in each sampled PSU is observed. Rather, a second-stage sample of 3 × 3 blocks would be selected from those available in the first-stage sample. The 3 × 3 blocks would not be further subsampled; instead, reference data would be obtained for all nine pixels of the 3 × 3 cluster.

Stratifying by LC class can directly conflict with clustering. The essence of the problem is illustrated by a simple example. Suppose the clusters are 3 × 3 blocks of pixels that, when taken together, partition the mapped region. The majority of these clusters will not consist of nine pixels all belonging to the same LC class. Stratified sampling directs us to select individual pixels from each LC class, in opposition to cluster sampling in which the selection protocol is based on a group of pixels. Because cluster sampling selects groups of pixels, we forfeit the control over the sample allocation that is sought by stratified sampling. It is possible to sample clusters via a stratified design, but it is the cluster, not the individual pixel, that must determine stratum membership.

A variety of approaches to circumvent this conflict between stratified and cluster sampling can be posed. One that should not be considered is to restrict the sample to only homogeneous 3 × 3 clusters. This approach clearly results in a sample that cannot be considered representative of the population, and it is well known that sampling only homogeneous areas of the map tends to inflate accuracy (Hammond and Verbyla, 1996). A second approach, and one that maintains the desired statistical rigor of the sampling protocol, is to employ two-stage cluster sampling in conjunction with stratification by LC class. A third approach in which the clusters are redefined to permit stratified selection will also be described.

The sampling design implemented in the accuracy assessment of the National Land Cover Data (NLCD) map illustrates how cluster sampling and stratification can be combined to achieve cost-effectiveness and precise class-specific estimates (Zhu et al., 2000; Yang et al., 2001; Stehman et al., 2003). The NLCD design was implemented across the U.S. using 10 regional assessments based on the U.S. Environmental Protection Agency's (EPA) federal administrative regions. Within a single region, the NLCD assessment was designed to provide regional spatial control and stratification by LC class. For several regions, the PSU was constructed from nonoverlapping, equal-sized areas of National Aerial Photography Program (NAPP) photo-frames, and in other regions, the PSU was a 6- × 6-km spatial unit. Both PSU constructions were designed to reduce the number of photos that would need to be purchased for reference data collection. A first-stage sample of PSUs was selected at a sampling rate of approximately 2.0%. Stratification by LC class was implemented at the second stage of the design. Mapped LC classes were used to stratify all pixels found within the first-stage sample PSUs. A simple random sample of pixels from each stratum was then selected, typically with 100 pixels per class. This design proved effective for ensuring that all LC classes, including the rare classes, were represented adequately so that estimates of user's accuracies were reasonably precise. The clustering feature implemented to achieve regional control succeeded at reducing costs considerably.

2.2.2 Flexibility of the NLCD Design

The flexibility of the NLCD design permits other options for selecting a second-stage sample. An alternative second-stage design could improve precision of the NLCD estimates (Stehman et al., 2000b), but such improvements are not guaranteed and would be gained at some cost. Precision for the rare LC classes is the primary consideration. Often the rare-class pixels cluster within a relatively small number of PSUs. The simple random selection within each class implemented in the second stage of the NLCD design will result in a sample with representation proportional to the number of pixels of each class within each PSU. That is, if many of the pixels of a rare class are found in only a few first-stage PSUs, many of the 100 second-stage sample pixels would fall within these same few PSUs. This clustering could result in poor precision for the estimated accuracy of this class. Ameliorating this concern is the fact that the NLCD clustering is at the regional level of control. The PSUs were large (e.g., 6 × 6 km), so pixels sampled within the same PSU will not necessarily exhibit strong intracluster correlation. In the case of weak intracluster correlation of classification error, cluster sampling will not result in precision significantly different from a simple random sample of the same size (Cochran, 1977).

Two alternatives may counter the clustering effect for rare-class pixels. One is to select a single pixel at random from 100 first-stage PSUs containing at least one pixel of the rare class. If the class is present in more than 100 PSUs, the first-stage PSUs could be subsampled to reduce the eligible set to 100. If fewer than 100 PSUs contain the rare class, the more likely scenario, the situation is slightly more complicated. A fixed number of pixels may be sampled from each first-stage PSU containing the rare class so that the total sample size for the rare class is maintained at 100. The complication is choosing the sample size for each PSU. This will depend on the number of eligible first-stage PSUs, and also on the number of pixels of the class in the PSU. This design option counters the potential clustering effect of rare-class pixels by forcing the second-stage sample to be widely dispersed among the eligible first-stage PSUs. In contrast to the outcome of the NLCD, PSUs containing a large proportion of the rare class will not receive the majority of the second-stage sample.

The second option to counter clustering of the sample into a few PSUs is to construct a "self-weighting" design (i.e., an equal probability sampling design in which all pixels have the same probability of being included in the sample). The term *self-weighting* arises from the fact that the analysis requires no weighting to account for different inclusion probabilities. At the first stage, 100 sample PSUs would be selected with inclusion probability proportional to the number of pixels of the specified rare class in the PSU. A wide variety of probability proportional to size designs exists, but simplicity would be the primary consideration when selecting the design for an accuracy assessment application. At the second stage, one pixel would be selected per PSU. A consequence of this two-stage protocol is that within each LC stratum, each pixel has an equal probability of being included in the sample (Sarndal et al., 1992), so no individual pixel weighting is needed for the user accuracy estimates. The design goal of distributing the sample pixels among 100 PSUs is also achieved.

2.2.3 Comparison of the Three Options

Three criteria will be used to compare the NLCD design alternatives: (1) ease of implementation, (2) simplicity of analysis, and (3) precision. The actual NLCD design will be designated as "Option 1," sampling one pixel from each of 100 PSUs will be "Option 2," and the self-weighting design will be referred to as "Option 3." Options 1 and 2 are the easiest to implement, and Option 3 is the most complicated because of the potentially complex, unequal probability first-stage protocol. Not only would such a first-stage design be more complex than what is typically done in accuracy assessment, Option 3 requires much more effort because we need the number of pixels of each LC class within each PSU in the region.

Options 1 and 3 share the characteristic of being self-weighting within LC strata. Self-weighting designs are simpler to analyze, although survey sampling computational software would mitigate this analysis advantage. Option 2 is not self-weighting, as demonstrated by the following example. Suppose a first-stage PSU has 1,000 pixels of the rare class and another PSU has 20 pixels of this class. At the first stage under Option 2, both PSUs have an equal chance of being selected. At the second stage, a pixel in the first PSU has a probability of 1/1000 of being chosen, whereas a pixel in the second PSU has a 1/20 chance of being sampled. Clearly, the probability of a pixel's being included in the sample is dependent upon how many other pixels of that class are found within the PSU. The appropriate estimation weights can be derived for this unequal probability design, but the analysis is complicated.

In addition to evaluating options based on simplicity, we would like to compare precision of the different options. Unfortunately, such an evaluation would be difficult, requiring either complicated theoretical analysis or extensive simulation studies based on acquiring reasonably good approximations to spatial patterns of classification error. A key point of this discussion of design alternatives for two-stage cluster sampling is that while the problem can be simply stated and the objectives for what needs to be achieved are clear, determining an optimal solution is elusive. Simple changes in sampling protocol may lead to complications in the analysis, whereas maintaining a simple analysis may require a complex sampling protocol.

2.2.4   Stratification and Local Spatial Control

Clustering to achieve local spatial control also conflicts with the effort to stratify by cover types. Several design alternatives may be considered to remedy this problem. An easily implemented approach is the following. A stratified random sample of pixels is obtained using the mapped LC classes as strata. To incorporate local spatial control and increase the sample size, the eight pixels touching each sampled pixel are also included in the sample. That is, a cluster consisting of a 3×3 block of pixels is created, but the selection protocol is based on the center pixel of the cluster. Two potential drawbacks exist for this protocol. First, the sample size control feature of stratified random sampling is diminished because the eight pixels surrounding an originally selected sample pixel could be any LC type, not necessarily the same type as the center pixel of the block. Sample size planning becomes trickier because we do not know which LC classes will be represented by the surrounding eight pixels or how many pixels will be obtained for each LC class present. This will not be a problem if we have abundant resources because we could specify the desired minimum sample size for each LC class based on the identity of the center pixels. However, having an overabundance of accuracy assessment resources is unlikely, so the loss of control over sample allocation is a legitimate concern.

Second, and more importantly, this protocol creates a complex inclusion probability structure because a pixel may be selected into the sample via two conditions: it is an originally selected center pixel of the 3×3 cluster or it is one of the eight pixels surrounding the initially sampled center pixel. To use the data within a rigorous probability-sampling framework, the inclusion probability determined for each pixel must account for this joint possibility of selection. We require the probability of being selected as a center pixel, the probability of being selected as an accompanying pixel in the 3×3 block, and the probability of being selected by both avenues in the same sample (i.e., the intersection event). The first probability is readily available because it is the inclusion probability of a stratified random sample, n_h/N_h, where n_h and N_h are the sample and population numbers of pixels for stratum h. The other two probabilities are much more complicated. The probability of a pixel's being selected because it is adjacent to a pixel selected in the initial sample depends on the map LC labels of the eight pixels surrounding the pixel in question, and this probability differs among different LC types. Although it is conceptually possible to enumerate the necessary information to obtain these probabilities, it is practically difficult. Finding the intersection probability would be equally complex. Rather than derive the actual inclusion probabilities, we could use the stratified random sampling inclusion probabilities as an easily implemented, but crude, approximation. This would violate the principle of consistent estimation and raise the question of how well such an approximation worked.

A second general alternative is to change the way the stratification is implemented. The problem arises because the strata are defined at the pixel level while the selection procedure is applied to the cluster level. Stratifying at the cluster level, for example a 3×3 block of pixels, resolves this problem but creates another. The nonhomogeneous character of the clusters creates a challenge when deciding to which stratum a block should be assigned if it consists of two or more cover types. Rules to determine the assignment must be specified. For example, assigning the block to the most common class found in the 3×3 block is one possibility, with a tie-breaking provision defined for equally common classes. A drawback of this approach is that few 3 × 3 blocks may be assigned to strata representing rare classes if the rare-class pixels are often found in small patches of two to four pixels. An alternative is to construct a rule that forces greater numbers of blocks into rare-class strata. For example, the presence of a single pixel of a rare class may trigger assignment of that pixel's block to the rare-class stratum. An obvious difficulty of this assignment protocol is what to do if two or more rare classes are represented within the same cluster. Because stratification requires that each block be assigned to exactly one stratum, and all blocks in the region must be assigned to strata, an elaborate set of rules may be needed to encompass all cases. A two-stage protocol such as implemented in the NLCD would reduce the workload of assigning blocks to strata because this assignment would be necessary only for the first-stage sample PSUs, not the entire area mapped. Estimation of accuracy parameters would be straightforward in this approach because each pixel in the 3 × 3 cluster has the same inclusion probability. This is an advantage of this option compared to the first option in which the pixels within a 3 × 3 block may have different inclusion probabilities. As is true for most complex designs, constructing a variance estimator and implementing it via existing software may be difficult.

This discussion of how to resolve design conflicts created by the desire to incorporate both cover type stratification and local spatial control via clustering illustrates that the solutions to practical problems may not be simple. We know how to implement cluster sampling and stratified sampling as separate entities, but we do not necessarily have simple, effective ways to construct a design that simultaneously accommodates both structures. Simple implementation procedures may lead to complex analysis protocols (e.g., difficulty in specifying the inclusion probabilities), and procedures permitting simpler analyses may require complex implementation protocols (e.g., defining strata at the 3 × 3 block level). The situation is even more complex than the treatment in this section indicates. It is likely that these methods focusing on local spatial control will need to be embedded in a design also incorporating regional spatial control. The 3 × 3 pixel clusters would represent subsamples from a larger primary sampling unit such as a 6- × 6-km area. Integrating regional and local spatial control with stratification raises still additional challenges to the design.

The NLCD case study may also be used as the context for addressing concerns related to pixel-based assessments. Positional error creates difficulties with any accuracy assessment because of potential problems in achieving exact spatial correspondence between the reference location and the map location. Typically, the problem is more strongly associated with pixel-based assessments relative to polygon-based assessments, but it is not clear that this association is entirely justified. The effects of positional error are most strongly manifested along the edges of map polygons. Whether the assessment is based on a pixel, polygon, or other spatial unit does not change the amount of edge present in the map. What may be changed by choice of assessment unit is how edges are treated in the collection and use of reference data. For example, suppose a polygon assessment employs an agreement protocol in which the entire map polygon is judged to be either in complete agreement or complete disagreement with the reference data. In this approach, the effect of positional error is greatly diminished because the error associated with a polygon edge may be obscured when blended with the more homogeneous, polygon interior. The positional error problem has not disappeared; it has to some extent been swept under the rug. This particular version of a polygon-based assessment is valid for certain map applications, but not all. For example, if the assessment objective is site-specific accuracy, the assessment must account for possible classification error along polygon boundaries. Defining agreement as a binary outcome based on the entire polygon will not achieve that purpose.

In a pixel-based assessment, provisions should be included to accommodate the reality of positional error when assessing edge or boundary pixels. No option is perfect, because we are dealing with a problem that has no practical, ideal solution. However, the option chosen should address the problem directly. One approach is to construct the reference data protocol so that the potential influence of positional error can be assessed. The protocol may include a rating of location confidence (i.e., how confident is the observer that the reference and map locations correspond exactly?), followed by reporting results for the full reference data as well as subsets of the data defined by the location confidence rating. Readers may then judge the potential effect of positional error by comparing accuracy at various levels of location confidence. A related approach would be to report accuracy results separately for edge and interior pixels. An alternative approach is to define agreement based on more information than comparing a single map pixel to a single reference pixel. In the NLCD assessment, one definition of agreement used was to compare the reference label of the sample pixel with a mode class determined from the map labels of the 3×3 block of pixels centered on the nominal sample pixel (Yang et al., 2001). This definition recognizes the possibility that the actual location used to determine the reference label could be offset by one pixel from the location identified on the map.

Another important feature of a pixel-based assessment is to account for the minimum mapping unit (MMU) of the map. When assigning the reference label, the observer should choose the LC class keeping in mind the MMU established. That is, the observer should not apply tunnel vision restricted only to the area covered by the pixel being assessed, but rather should evaluate the pixel taking into account the surrounding spatial context. In the 1990 NLCD, the MMU was a single pixel. It is expected that NLCD users may choose to define a different MMU depending on their particular application, but the NLCD accuracy assessment was pixel-based because the base product made available was not aggregated to a larger MMU.

The problems associated with positional error are largely specific to the response or measurement component of the accuracy assessment (Stehman and Czaplewski, 1998). However, a few points related to sampling design should be recognized. Although the MMU is a relevant feature of a map to consider when determining the response design protocol, it is important to recognize that a MMU does not define a sampling unit. A pixel, a polygon, or a 3×3 block of pixels, for example, are all legitimate sampling units, but a "1.0-ha MMU" lacks the necessary specificity to define a sampling unit. The MMU does not create the unambiguous definition required of a sampling unit because it permits various shapes of the unit, it does not include specification of how the unit is accounted for when the polygon is larger than the MMU, and it does not lead directly to a partitioning of the region into sampling units. While it may be possible to construct the necessary sampling unit partition based on a MMU, this approach has never been explicitly articulated. When sampling polygons, the basic methods available are simple random, systematic, and stratified (by LC class) random sampling from a list frame of polygons. Less obvious is how to incorporate clustering and spatial sampling methods for polygon assessment units. Polygons may vary greatly in size, so a decision is required whether to stratify by size so as not to have the sample dominated by numerous small polygons. A design protocol of locating sample points systematically or completely at random and including those polygons touched by these sample point locations creates a design in which the probability of including a polygon is proportional to its area. This structure must be accounted for in the analysis and is a characteristic of polygon sampling that has yet to be discussed explicitly by proponents of such designs. Most of the comparative studies of accuracy assessment sampling designs are pixel-based assessments (Fitzpatrick-Lins, 1981; Congalton, 1988a; Stehman, 1992, 1997), and analyses of potential factors influencing design choice (e.g., spatial correlation of error) are also pixel-based investigations (Congalton, 1988b; Pugh and Congalton, 2001).

Problems associated with positional error in accuracy assessment merit further investigation and discussion. Although it is easy to dismiss pixel-based assessments with a "you-can't-find-a-pixel" proclamation, a less superficial treatment of the issue is called for. Edges are a real characteristic of all LC maps, and the accuracy reported for a map should account for this reality. Whether the assessment is based on a pixel or a larger spatial unit, the accuracy assessment should confront the edge feature directly. Although there is no perfect solution to the problem, options exist to specify the analysis or response design protocol in such a way that the effect of positional error on accuracy is addressed. Sampling in a manner that permits evaluating the effect of positional error seems preferable to sampling in a way that obscures the problem (e.g., limiting the sample to homogeneous LC regions).

2.3 EXISTING DATA

It is natural to consider whether existing data or data collected for other purposes could be used as reference data to reduce the cost of accuracy assessment. Such data must first be evaluated to ascertain spatial, temporal, and classification scheme compatibility with the LC map that is the subject of the assessment. Once compatibility has been established, the issue of sampling design becomes relevant. Existing data may originate from either a probability or nonprobability sampling protocol. If the data were not obtained from a probability sampling design, the inability to generalize via rigorous, defensible inference from these data to the full population is a severe limitation. The difficulties associated with nonprobability sampling are detailed in a separate subsection.

The greatest potential for using existing data occurs when the data have a probability-sampling origin. Ongoing environmental monitoring programs are prime candidates for accuracy assessment reference data. The National Resources Inventory (NRI) (Nusser and Goebel, 1997) and Forest Inventory and Analysis (FIA) (USFS, 1992) are the most likely contributors among the monitoring programs active in the U.S. Both programs include LC description in their objectives, so the data naturally fit potential accuracy assessment purposes. Gill et al. (2000) implemented a successful accuracy assessment using FIA data, and Stehman et al. (2000a) discuss use of FIA and NRI data within a general strategy of integrating environmental monitoring with accuracy assessment.

At first glance, using existing data for accuracy assessment appears to be a great opportunity to control cost. However, further inspection suggests that deeper issues are involved. Even when the data are from a legitimate probability sampling design, these data will not be tailored exactly to satisfy all objectives of a full-scale accuracy assessment. For example, the sampling design for a monitoring program may be targeted to specific areas or resources, so coverage would be very good for some LC classes and subregions but possibly inadequate for others. For example, NRI covers nonfederal land and targets agriculture-related questions, whereas the FIA's focus is, obviously, on forested land. To complete a thorough accuracy assessment, it may be necessary to piece together a patchwork of various sources of existing data plus a supplemental, directed sampling effort to fill in the gaps of the existing data coverage. The effort required to cobble together a seamless, consistent assessment may be significant and the statistical analysis of the data complex.

Data from monitoring programs may carry provisions for confidentiality. This is certainly true of NRI and FIA. Confidentiality agreements permitting access to the data will need to be negotiated and strictly followed. Because of limited access to the data, progress may be slow if human interaction with the reference data materials is required to complete the accuracy assessment. For example, additional photographic interpretation for reference data using NRI or FIA materials may be problematic because only one or two qualified interpreters may have the necessary clearance to handle the materials. Confidentiality requirements will also preclude making the reference data generally available for public use. This creates problems for users wishing to conduct subregional assessments or error analyses, to construct models of classification error, or to evaluate different spatial aggregations of the data. It is difficult to assign costs to these features. Existing data obviously save on data collection costs, but there are accompanying hidden costs related to complexity and completeness of the analysis, timeliness to report results, and public access to the data.

2.3.1 Added-Value Uses of Accuracy Assessment Data

In the previous section, accuracy assessment is considered an add-on to objectives of an ongoing environmental monitoring program. However, if accuracy data are collected via a probability sampling design, these data may have value for more general purposes. For example, a common objective of LC studies is to estimate the proportional representation of various cover types and how they change over time. We can use complete coverage maps such as the NLCD to provide such estimates, but these estimates are biased because of the classification errors present. Although the maps represent a complete census, they contain measurement error. The reference data collected for accuracy assessment supposedly represent higher-quality data (i.e., less measurement error), so these data may serve as a stand-alone basis for estimates of LC proportions and areas. Methods for estimating area and proportion of area covered by the various LC classes have been developed (Czaplewski and Catts, 1992; Walsh and Burk, 1993; Van Deusen, 1996). Recognizing this potentially important use of reference data provides further rationale for implementing statistically defensible probability sampling designs. This area estimation application extends to situations in which LC proportions for small areas such as a watershed or county are of interest. A probability sampling design provides a good foundation for implementing small-area estimation methods to obtain the area proportions.

2.4 NONPROBABILITY SAMPLING

Because nonprobability sampling is often more convenient and less expensive, it is useful to review some manifestations of this departure from a statistically rigorous approach. Restricting the probability sample to areas near roads for convenient access or to homogeneous 3×3 pixel clusters to reduce confounding of spatial and thematic error are two typical examples of nonprobability sampling. A positive feature of both examples is that generalization to some population is statistically justified (e.g., the population of all locations conveniently accessible by road or all areas of the map consisting of 3×3 homogeneous pixel blocks). Extrapolation to the full map is problematic. In the NLCD assessment, restricting the sample to 3×3 homogeneous blocks would have represented roughly 33% of the map, and the overall accuracy for this homogeneous subset was about 10% higher than for the full map. Class-specific accuracies could increase by 10 to 20% for the homogeneous areas relative to the full map.

Another prototypical nonprobability sampling design results when the inclusion probabilities needed to meet the consistent estimation criterion of statistical rigor are unknown. Expert or judgment samples, convenience samples (e.g., near roads, but not selected by a probability sampling protocol), and complex, *ad hoc* protocols are common examples. "Citizen participation" data collection programs are another example in which data are usually not collected via a probability sampling protocol, but rather are purposefully chosen because of proximity and ease of access to the participants. This version of nonprobability sampling creates adverse conditions for statistically defensible inference to any population. Peterson et al. (1999) demonstrate inference problems in the particular case of a citizen-based, lake water-quality monitoring program. To support inference from nonprobability samples, the options are to resort to a statistical model, or to simply claim "the sample looks good." In the former case, rarely are the model assumptions explicitly stated or evaluated in accuracy assessment. The latter option is generally regarded as unacceptable, just as it is unacceptable to reduce accuracy assessment to an "it looks good" judgment.

Another use of nonprobability sampling is to select a relatively small number of sample sites that are, based on expert judgment, representative of the population. In environmental monitoring, these locations are referred to as "sentinel" sites, and they serve as an analogy to hand-picked confidence sites in accuracy assessment. In both environmental monitoring and accuracy assessment, judgment samples can play an invaluable role in understanding processes, and their role in accuracy assessment for developing better classification techniques should be recognized. Although nonprobability samples may serve as a useful initial check on gross quality of the data because poorly classified areas may be identified quickly, caution must be exercised when a broad-based, population-level description is desired (i.e., when the objective is to generalize from the sample). Edwards (1998) emphasizes that the use of sentinel sites for population inference in environmental monitoring is suspect. This concern is applicable to accuracy assessment as well.

More statistically formal approaches to nonprobability sampling have been proposed. In the method of balanced sampling, selection of sample units is purposefully balanced on one or more auxiliary variables known for the population (Royall and Eberhardt, 1975). For example, the sample might be chosen so that the mean elevation of the sample pixels matches the mean elevation of all pixels mapped as that LC class (i.e., the population mean). The method is designed to produce a sample robust to violations in the model used to support inference. Most nonprobability sampling designs implemented in accuracy assessment lack the underlying model-based rationale of balanced sampling and instead are the result of convenience, judgment, or poor design. Schreuder and Gregoire (2001) discuss other potential uses of nonprobability sampling data.

2.4.1 Policy Aspects of Probability vs. Nonprobability Sampling

Considering implementation of a nonprobability sampling protocol has policy implications in addition to the scientific issues discussed in the previous section. The policy issues arise because both scientists and managers using the LC map have a vested interest in the map's accuracy. Federal sponsorship to create these maps adds an element of governmental responsibility to ensure, or at least document, their quality. The stakes are consequently high and the accuracy assessment design will need to be statistically defensible. Most government sampling programs responsible for providing national and broad regional estimates are conducted using probability sampling protocols. The Current Population Survey (CPS) (McGuiness, 1994) and National Health and Nutrition Examination Survey (NHANES) (McDowell et al., 1981) are two such programs designed as probability samples. Similarly, national environmental sampling programs are typically based on probability sampling protocols (Olsen et al., 1999).

The expense of LC maps covering large geographic regions combined with the multitude of applications these maps serve elevates the importance of accuracy assessment to a level commensurate with these other national sampling programs. Accordingly, the protocols employed to evaluate the quality of the LC data must achieve standards of sampling design and statistical credibility established by other national sampling programs. These standards of accuracy assessment protocol will exceed those acceptable for more local use, lower-profile maps. The exposure, or perhaps notoriety, accruing to maps such as the NLCD will elicit intense scrutiny of their quality. Concerns related to litigation may become more prevalent as use of LC maps affecting government decisions increases. Map quality may be challenged not only scientifically, but also legally. Because the sampling design is such a fundamental part of the scientific basis of an accuracy assessment, the credibility of this component of accuracy assessment must be ensured. To provide this assurance, the use of scientifically defensible probability sampling protocols should be a matter of policy.

2.5 STATISTICAL COMPUTING

The requirements for statistically rigorous design and analysis will tax the capability of traditional computing practice in accuracy assessment. Stehman and Czaplewski (1998) noted the absence of readily accessible, easy-to-use statistical software that could perform the analyses associated with the more complex sampling designs that will be needed for large-area map assessments. Recent upgrades in computing software have improved this situation. For example, the Statistical Analysis Software (SAS) analysis software now includes survey sampling estimation procedures that can be adapted for accuracy assessment applications. Nusser and Klaas (2003) implemented these procedures to obtain the typical suite of accuracy estimates and accompanying standard errors for complex sampling designs. The SAS procedure accomplishing these tasks is PROC SURVEYMEANS.

Survey sampling software will be invaluable if data from ongoing monitoring programs are to be used for accuracy assessment. For example, suppose NRI data serve as the source of reference data. Two characteristics of the NRI data, confidentiality and the unequal probability design used, may be resolved by the capabilities available in SAS. To adhere to the estimation criterion of consistency, the accuracy estimates must incorporate weights for the sample pixels derived from the unequal inclusion probabilities. The SAS estimation procedures are designed to accommodate these weights. Confidentiality of sample locations can be maintained because the necessary estimation weights need not refer to any location information. The possibility exists that with the location information stripped away, the data could be made available for limited general use for applications requiring only the sample weights, and the map and reference labels. Users would need to conduct their analyses via SAS or another software package that implements design-based estimation procedures incorporating the sampling weights. Analyses ignoring this feature may produce badly misleading results.

The use of SAS for accuracy assessment estimation provides two other advantages. SAS includes estimation of standard errors as standard output. Standard error formulas are complex for the sampling designs combining the advantages of both strata and clusters. Having available software to compute these standard errors is highly beneficial relative to the alternative of writing one's own variance estimation code and having to confirm its validity. Second, SAS readily accommodates the fact that many accuracy estimates, for example producer's accuracy, are ratio estimators (i.e., ratios of two estimates). For ratio estimators, the SAS standard error estimation procedures employ the common practice of using a Taylor Series approximation. The more complex design structures that arise from more cost-effective assessments or use of existing data obtained from an ongoing monitoring program will likely require more sophisticated analysis software than is available in standard GIS and classification software. SAS does not provide everything that is needed, but its capabilities represent a major step forward in computing for accuracy assessment analyses.

2.6 PRACTICAL REALITIES OF SAMPLING DESIGN

In comments directed toward sampling design for environmental monitoring, Fuller (1999) captured the essence of many of the issues facing sampling design for accuracy assessment. These principles are restated, and in some cases paraphrased, to adapt them to accuracy assessment sampling design: (1) every new approach sounds easier than it is to implement and analyze, (2) more will be required of the data at the analysis stage than had been anticipated at the planning stage, (3) objectives and priorities change over time, and (4) the budget will be insufficient.

2.6.1   Principle 1

Every new approach sounds easier than it is. Incorporating existing data for accuracy assessment is a good case in point. While the data may be "free," the analysis and research required to evaluate the compatibility of the spatial units and classification scheme are not without costs. Confidentiality agreements may need to be negotiated and strictly followed, spatial and temporal coverage of the existing data may be incomplete and/or inadequate, and the response time for interaction with the agency supplying the data may be slow because this use of their data may not be a top priority among their responsibilities. Existing data that do not originate from a probability sampling protocol are even more difficult to incorporate into a rigorous protocol and may be useful only as a qualitative check of accuracy and to provide limited anecdotal, case-study information.

2.6.2   Principle 2

More will be required of the data at the analysis stage than had been anticipated at the planning stage. This principle applies to estimating accuracy of subregions and other subsets of the data. That is, a program designed for regional accuracy assessments will be asked to provide state-level estimates and possibly even county-level estimates. Not only will overall accuracy be requested for these small subregions, but also class-specific accuracy within the subregion will be seen as desirable information. Accuracy estimates for other subsets of the data will become appealing. For example, are the classification errors associated with transitions between cover types? How accurate are the classifications within relatively large homogeneous areas of the map? Deriving a spatial representation of classification error is another relevant, but supplemental, objective that places additional requirements on the accuracy assessment analysis that may not have been planned for at the design stage.

2.6.3   Principle 3

Over time, objectives and/or priorities of objectives may change. This may not represent a major problem in accuracy assessment projects, but one example is changing the classification scheme if it is recognized that certain LC classes cannot be mapped well. Another example illustrating this principle occurs when the map is revised (updated) while the accuracy assessment is in progress. Some of the additional analyses described for Principle 2 represent a change in objectives also.

2.6.4   Principle 4

Insufficient budget is a common affliction of accuracy assessments (Scepan, 1999). Resource allocation is dominated by the mapping activity, with scant resources available for accuracy assessment. Adequate resources may exist to obtain reasonably precise, class-specific estimates of accuracy over broad spatial regions. For example, the NLCD accuracy assessment provides relatively low standard errors for class-specific accuracy for each of 10 large regions of the U.S. However, once Principle 2 manifests itself, data that serve well for regional estimates may look woefully inadequate for subregional accuracy objectives. Edwards et al. (1998) and Scepan (1999) recognized these phenomena for state-level and global mapping. In the former case, resources were inadequate to estimate class-specific accuracy with acceptable precision for all three ecoregions found in the state of Utah. In the global application, the data were too sparse to provide precise class-specific estimates for each continent.

Timeliness of accuracy assessment reporting is hampered by the need for the map to be completed prior to drawing an appropriately targeted sample, and any accuracy assessment activity concurrent with map production detracts from timely completion of the map. Managing and quality-checking data is a time-consuming, tedious task for the large datasets of accuracy assessment, and the statistical analysis is not trivial when the design is complex and standard errors are required. Lastly, neither the time nor the financial resources are usually available to support research that would allow tailoring the sampling design to specifically target objectives and characteristics of each individual mapping project. Comparing different sampling designs using data directly relevant to the specific mapping project requires both time and money. Instead of this focused research approach, often design choices must be based on judgment and experience, but without hard data to support the decision.

2.7 DISCUSSION

Sampling design is one of the core challenges facing accuracy assessment, and future developments in this area will contribute to more successful assessments. The goal is to implement a statistically defensible sampling design that is cost-effective and addresses the multitude of objectives that multiple users and applications of the map generate. The future direction of sampling design in accuracy assessment must go beyond the basic designs featured in textbooks (Campbell, 1987; Congalton and Green, 1999) and repeated in several reviews of the field (Congalton, 1991; Janssen and van der Wel, 1994; Stehman, 1999; McGwire and Fisher, 2001; Foody, 2002). While these designs are fundamentally sound and introduce most of the basic structures required of good design (e.g., stratification, clusters, randomization), they are inadequate for assessing large-area maps given the reality of budgetary and practical constraints.

For both policy and scientific reasons, probability sampling is a necessary characteristic of the sampling design. Within the class of probability sampling designs, we must seek to develop or identify methods that resolve the conflicts of a design combining stratifying by LC class and clustering. Protocols incorporating the advantages of two or more of the basic sampling designs need to be implemented when combining data from different ongoing monitoring programs to take advantage of existing data, or when augmenting a general sampling design to increase the sample size for rare classes or small subregions. Sampling methods need to be explored for assessing accuracy for different spatial aggregations of the data and for nonsite-specific accuracy assessments. As is often the case for any developing field of application, sampling design for accuracy assessment may not require developing entirely new methods, but rather learning better how to use existing methods.

Implementing a scientifically rigorous sampling design provides a secure foundation to any accuracy assessment. Accuracy assessment data have little or no value to inform us about the map's utility if the data are not collected via a credible sampling design. Sampling design in accuracy assessment is still evolving according to a progression common in other fields of application. Early innovators identified the need for sound sampling practice (Fitzpatrick-Lins, 1981; Card, 1982; Congalton, 1991). As more familiarity was gained with traditional survey sampling methods, more complex sampling designs could be introduced and integrated into practice. The challenges confronting sampling design for descriptive objectives of accuracy assessment were recognized as daunting, but by no means insurmountable. The platitude that we must choose a sampling design that "balances statistical validity and practical utility" was raised (Congalton, 1991), and specificity was added to this generic recommendation by stating explicit criteria of both validity and utility (Stehman, 2001).

The future direction of accuracy assessment sampling design demands new developments. Practical challenges are a reality. For most, if not all, of these problems, statistical solutions already exist, or the fundamental concepts and techniques with which to derive the solutions can be found in the survey sampling literature. The key to implementing better, more cost-effective sampling procedures in accuracy assessment is to move beyond the parochial, insular traditions characterizing the early stage of accuracy assessment sampling and to recognize more clearly the broad expanse of opportunities offered by sampling theory and practice. The book on sampling design for accuracy assessment is by no means closed. Sampling design in accuracy assessment may have progressed to an advanced stage of adolescence, but it has yet to reach a level of consistency in good practice and sound conceptual fundamentals necessary to be considered a scientifically mature endeavor. More statistically sophisticated sampling designs not only contribute to the value of map accuracy assessments, they are the result of our current needs for more information related to map utility. If our needs were simple and few, the basic sampling designs receiving the bulk of attention in the 1980s and early 1990s would suffice. It is the increasingly demanding questions related to utility of these maps that compel us to seek better, more cost-effective sampling designs. Identifying these designs and implementing them in practice is the future of sampling practice in accuracy assessment.

2.8 SUMMARY

As maps delineating LC play an increasingly important role in natural resource science and policy applications, implementing high-quality, statistically rigorous accuracy assessments becomes essential. Typically, the primary objective of accuracy assessment is to provide precise estimates of overall accuracy and class-specific accuracies (e.g., user's or producer's accuracies). An extended set of objectives exists for most large-area mapping projects because multiple users interested in different applications will employ the map. Constructing a cost-effective accuracy assessment is a challenging problem given the multiple objectives the assessment must satisfy. To meet this challenge, a more integrated sampling approach combining several design elements such as stratification, clustering, and use of existing data must be considered. These design elements are typically found individually in current accuracy assessment practice, but greater efficiency may be gained by more innovatively combining their strengths. To ensure scientific credibility, sampling designs for accuracy assessment should satisfy the criteria defining a probability sample. This requirement places additional burden on how various design elements are integrated. When exploring alternative design options, the apparently simple answers may not be as straightforward as they first appear. Combining basic design structures such as strata and clusters to enhance efficiency has some significant complicating factors, and use of existing data for accuracy assessment has associated hidden costs even if the data are free.

REFERENCES

Anderson, J.R., E.E. Hardy, J.T. Roach, and R.E. Witmer, A Land Use and Land Cover Classification System for Use with Remote Sensor Data, U.S. Geological Survey Prof. Paper 964, U.S. Geological Survey, Washington, DC, 1976.

Campbell, J.B., *Introduction to Remote Sensing*, Guilford Press, New York, 1987.

Card, D.H., Using known map category marginal frequencies to improve estimates of thematic map accuracy, *Photogram. Eng. Remote Sens.*, 48, 431–439, 1982.

Cochran, W.G., *Sampling Techniques*, Wiley, New York, 1977.

Congalton, R.G., A comparison of sampling schemes used in generating error matrices for assessing the accuracy of maps generated from remotely sensed data, *Photogram. Eng. Remote Sens.*, 54, 593–600, 1988a.

Congalton, R.G., A review of assessing the accuracy of classifications of remotely sensed data, *Remote Sens. Environ.*, 37, 35–46, 1991.

Congalton, R.G., Using spatial autocorrelation analysis to explore the errors in maps generated from remotely sensed data, *Photogram. Eng. Remote Sens.*, 54, 587–592, 1988b.

Congalton, R.G. and K. Green, *Assessing the Accuracy of Remotely Sensed Data: Principles and Practices*, CRC Press, Boca Raton, FL, 1999.

Czaplewski, R.L. and G.P. Catts, Calibration of remotely sensed proportion or area estimates for misclassification error, *Remote Sens. Environ.*, 39, 29–43, 1992.

Edwards, D., Issues and themes for natural resources trend and change detection, *Ecol. Appl.*, 8, 323–325, 1998.

Edwards, T.C., Jr., G.G. Moisen, and D.R. Cutler, Assessing map accuracy in an ecoregion-scale cover-map, *Remote Sens. Environ.*, 63, 73–83, 1998.

Fitzpatrick-Lins, K., Comparison of sampling procedures and data analysis for a land-use and land-cover map, *Photogram, Eng, Remote Sens.*, 47, 343–351, 1981.

Foody, G.M., Status of land cover classification accuracy assessment, *Remote Sens. Environ.*, 80, 185–201, 2002.

Foody, G.M., The continuum of classification fuzziness in thematic mapping, *Photogram. Eng. Remote Sensing*, 65, 443–451, 1999.

Fuller, W.A., Environmental surveys over time, *J. Agric. Biol. Environ. Stat.*, 4, 331–345, 1999.

Gill, S., J.J. Milliken, D. Beardsley, and R. Warbington, Using a mensuration approach with FIA vegetation plot data to assess the accuracy of tree size and crown closure classes in a vegetation map of northeastern California, *Remote Sens. Environ.*, 73, 298–306, 2000.

Hammond, T.O. and D.L. Verbyla, Optimistic bias in classification accuracy assessment, *Int. J. Remote Sens.*, 17, 1261–1266, 1996.

Janssen, L.L.F. and F.J.M. van der Wel, Accuracy assessment of satellite derived land-cover data: a review, *Photogram. Eng. Remote Sens.*, 60, 419–426, 1994.

Jones, K.B., A.C. Neale, M.S. Nash, R.D. Van Remotel, J.D. Wickham, K.H. Riitters, and R.V. O'Neill, Predicting nutrient and sediment loadings to streams from landscape metrics: a multiple watershed study from the United States Mid-Atlantic region, *Landscape Ecol.*, 16, 301–312, 2001.

Laba, M., S.K. Gregory, J. Braden, D. Ogurcak, E. Hill, E. Fegraus, J. Fiore, and S.D. DeGloria, Conventional and fuzzy accuracy assessment of the New York Gap Analysis Project land cover maps, *Remote Sens. Environ.*, 81, 443–455, 2002.

McDowell, A., A. Engel, J.T. Massey, and K. Maurer, Plan and Operation of the Second National Health and Nutrition Examination Survey, 1976–1980, Vital and Health Stat. Rep., Series 1(15), National Center for Health Statistics, 1981.

McGuiness, R.A., Redesign of the sample for the Current Population Survey, Employment Earnings, 41, 7–10, 1994.

McGwire, K.C., and P. Fisher, Spatially variable thematic accuracy: Beyond the confusion matrix, in Spatial Uncertainty in Ecology: Implications for Remote Sensing and GIS Applications, Hunsaker, C.T., M.F. Goodchild, M.A. Friedl, and T.J. Case, Eds., Springer, New York, 2001.

Muller, S.V., D.A. Walker, F.E. Nelson, N.A. Auerbach, J.G. Bockheim, S. Guyer, and D. Sherba, Accuracy assessment of a land-cover map of the Kuparuk River Basin, Alaska: considerations for remote regions, Photogram. Eng. Remote Sensing, 64, 619–628, 1998.

Nusser, S.M. and J.J Goebel, The National Resources Inventory: a long-term multi-resource monitoring programme, Environ. Ecol. Stat., 4, 181–204, 1997.

Nusser, S.M. and E.E. Klaas, Survey methods for assessing land cover map accuracy, Environ. Ecol. Stat., 2003, 10, 309–331.

Olsen, A.R., J. Sedransk, D. Edwards, C.A. Gotway, W. Liggett, S. Rathbun, K.H. Reckhow, and L.J. Young, Statistical issues for monitoring ecological and natural resources in the United States, Environ. Monit. Assess., 54, 1–45, 1999.

Peterson, S.A., N.S. Urquhart, and E.B. Welch, Sample representativeness: a must for reliable regional lake condition estimates, Environ. Sci. Technol., 33, 1559–1565, 1999.

Pugh, S.A. and R.G. Congalton, Applying spatial autocorrelation analysis to evaluate error in New England forest-cover-type maps derived from Landsat Thematic Mapper data, Photogram. Eng. Remote Sens., 67, 613–620, 2001.

Royall, R.M. and K.R. Eberhardt, Variance estimates for the ratio estimator, Sankhya C (37), 43–52, 1975.

Sarndal, C.E., B. Swensson, and J. Wretman, Model-Assisted Survey Sampling, Springer-Verlag, New York, 1992.

Scepan, J., Thematic validation of high-resolution global land-cover data sets, Photogram. Eng. Remote Sens., 65, 1051–1060, 1999.

Schreuder, H.T. and T.G. Gregoire, For what applications can probability and non-probability sampling be used? Environ. Monit. Assess., 66, 281–291, 2001.

Stehman, S.V., Basic probability sampling designs for thematic map accuracy assessment, Int. J. Remote Sens., 20, 2423–2441, 1999.

Stehman, S.V., Comparison of systematic and random sampling for estimating the accuracy of maps generated from remotely sensed data, Photogram. Eng. Remote Sens., 58, 1343–1350, 1992.

Stehman, S.V., Estimating standard errors of accuracy assessment statistics under cluster sampling, Remote Sens. Environ., 60, 258–269, 1997.

Stehman, S.V., Statistical rigor and practical utility in thematic map accuracy assessment, Photogram. Eng. Remote Sens., 67, 727–734, 2001.

Stehman, S.V. and R.L. Czaplewski, Design and analysis for thematic map accuracy assessment: fundamental principles, Remote Sens. Environ., 64, 331–344, 1998.

Stehman, S.V., R.L. Czaplewski, S.M. Nusser, L.Yang, and Z. Zhu, Combining accuracy assessment of land-cover maps with environmental monitoring programs, Environ. Monit. Assess., 64, 115–126, 2000a.

Stehman, S.V., J.D. Wickham, L. Yang, and J.H. Smith, Accuracy of the national land-cover dataset (NLCD) for the eastern United States: statistical methodology and regional results, Remote Sens. Environ., 86, 500–516, 2003.

Stehman, S.V., J.D. Wickham, L. Yang, and J.H. Smith, Assessing the accuracy of large-area land cover maps: Experiences from the Multi-resolution Land-Cover Characteristics (MRLC) project, in Accuracy 2000: Proceedings of the 4th International Symposium on Spatial Accuracy Assessment in Natural Resources and Environmental Sciences, Heuvelink, G.B.M. and M.J.P.M. Lemmens, Eds., Delft University Press, The Netherlands, 2000b, pp. 601–608.

USFS (U.S. Forest Service), Forest Service Resource Inventories: An Overview, USGPO 1992-341-350/60861, U.S. Department of Agriculture, Forest Service, Forest Inventory, Economics, and Recreation Research, Washington, DC, 1992.

Van Deusen, P.C., Unbiased estimates of class proportions from thematic maps, Photogram Eng. Remote Sens., 62, 409–412, 1996.

Walsh, T.A. and T.E. Burk, Calibration of satellite classifications of land area, *Remote Sens. Environ.*, 46, 281–290, 1993.

Yang, L., S.V. Stehman, J.H. Smith, and J.D. Wickham, Thematic accuracy of MRLC land cover for the eastern United States, *Remote Sens. Environ.*, 76, 418–422, 2001.

Zhu, Z., L. Yang, S.V. Stehman, and R.L. Czaplewski, Accuracy assessment for the U. S. Geological Survey regional land-cover mapping program: New York and New Jersey region, *Photogram. Eng. Remote Sens.*, 66, 1425–1435, 2000.

CHAPTER 3

Validation of Global Land-Cover Products by the Committee on Earth Observing Satellites

Jeffrey T. Morisette, Jeffrey L. Privette, Alan Strahler, Philippe Mayaux, and Christopher O. Justice

CONTENTS

3.1 INTRODUCTION

3.1.1 Committee on Earth Observing Satellites

The Committee on Earth Observation Satellites (CEOS) is an international organization charged with coordinating international civil space-borne missions designed to observe and study planet Earth. Current membership is composed of 41 space agencies and other national and international organizations. It was created (1984) in response to a recommendation from the Economic Summit of Industrialized Nations Working Group on Growth, Technology, and Employment's Panel of Experts on Satellite Remote Sensing, which recognized the multidisciplinary nature of satellite Earth observation and the value of coordination across all proposed missions. The main goals of CEOS are to ensure that: (1) critical scientific questions relating to Earth observation and global change are covered and (2) satellite missions do not unnecessarily overlap (http://www.ceos.org). The first goal can be achieved by providing timely and accurate information from satellite-derived products. Proper use of these products, in turn, relies on our ability to ascertain their uncertainty. The second goal is achieved through coordination among CEOS members.

As validation efforts are an integral part of "satellite missions," part of the CEOS mission is to reduce the likelihood of unnecessary overlap in validation efforts. The particular CEOS work related to validation falls within the Working Group on Calibration and Validation (WGCV), which is one of two standing working groups of CEOS (the other is the Working Group on Information Systems and Services, WGISS). The ultimate goal of the WGCV is to ensure long-term confidence in the accuracy and quality of Earth observation data and products through (1) sensor-specific calibration and validation and (2) geophysical parameter and derived-product validation.

To ensure long-term confidence in the accuracy and quality of Earth observation data and products, the WGCV provides a forum for calibration and validation information exchange, coordination, and cooperative activities. The WGCV promotes the international exchange of technical information and documentation; joint experiments; and the sharing of facilities, expertise, and resources (http://wgcv.ceos.org). There are currently six established subgroups within WGCV: (1) atmospheric chemistry, (2) infrared and visible optical sensors (IVOS), (3) land product validation (LPV), (4) terrain mapping (TM), (5) synthetic aperture radar (SAR), and (6) microwave sensors subgroup (MSSG).

Each subgroup has a specific mission. For example, the relevant subgroup for global land product validation is LPV. The mission of LPV is to increase the quality and economy of global satellite product validation by developing and promoting international standards and protocols for field sampling, scaling, error budgeting, and data exchange and product evaluation and to advocate mission-long validation programs for current and future earth-observing satellites (Justice et al., 2000). In this chapter, by considering the lessons learned from previous and current programs, we describe a strategy to utilize LPV for current and future global land-cover (LC) validation efforts.

3.1.2   Approaches to Land-Cover Validation

Approaches to LC validation may be divided into two primary types: statistical approaches and confidence-building measures. Confidence-building measures include studies or comparisons made without a firm statistical basis that provide confidence in the map. When presented with a LC map product, users typically first carry out "reconnaissance measures" by examining the map to see how well it conforms to regional landscape attributes, such as mountain chains, valleys, or agricultural regions. Spatial structure is inspected to ensure that the map has sensible patterns of LC that are without excessive "salt-and-pepper" noise or excessive smoothness and generalization. Land–water boundaries are checked for continuity to reveal the quality of multidate registration. The map is carefully examined for gross errors, such as cities in the Sahara or water on high mountain slopes. If the map seems reasonable based on these and similar criteria, validation can proceed to more time-consuming confidence measures. These include ancillary comparisons, in which specific maps or datasets are compared to the map. However, such comparisons are not always straightforward, since ancillary materials are typically prepared from input data acquired at a different time. Also, map scales and LC units used in the ancillary materials may not be directly comparable to the map of interest.

The Global Land Cover 2000 program has established a systematic approach for qualitative confidence building in which a global map is divided into small cells, each of which is examined carefully for discrepancies. This procedure is described more fully in section 2.1.

Statistical approaches may be further broken down into two types: model-based inference and design-based inference (Stehman, 2000, 2001). Model-based inference is focused on the classification process, not on the map *per se*. A map is viewed as one realization of a classification process that is subject to error, and the map's accuracy is characterized by estimates of errors in the classification process that produced it. For example, the Moderate Resolution Imaging Spectroradiometer (MODIS) LC product provides a confidence value for each pixel that measures of how well the pixel fits the training examples presented to the classifier. Design-based inference uses statistical principles in which samples are acquired to infer characteristics of a finite population, such as the pixels in a LC map. The key to this approach is probability-based sampling, in which the units to be sampled are drawn with known probabilities. Examples include random sampling, in which all possible sample units have equal probability of being drawn, or stratified random sampling, in which all possible sample units within a particular stratum have equal probability of being drawn.

Probability-based samples are used to derive consistent estimates of population parameters that equal the population parameters when the entire population is included in the sample. Consistent estimators commonly used in LC mapping from remotely sensed data include the proportion of pixels correctly classified (global accuracy); "user's accuracy," which is the probability that a pixel is truly of a particular cover to which it was classified; and "producer's accuracy," which is the probability that a pixel was mapped as a member of a class of which it is truly a member. These estimators are typically derived from a confusion matrix, which tabulates true class labels with those assigned on the map according to the sample design.

While design-based inference allows proper calculation of these very useful consistent estimators, it is not without its difficulties. Foremost is the difficulty of verifying the accuracy of the label assigned to a sampled pixel. In the case of a global map, it is not possible to go to a randomly assigned location on the Earth's surface. Thus, the accuracy of a label is typically assessed using finer-resolution remotely sensed data. In this case, accuracy is assessed by photointerpretation, which is subject to its own error. Registration errors also occur and commonly restrict or negate a pixel-based assessment strategy.

Another practical problem may lie in the classification scheme itself. Sometimes the LC types are not mutually exclusive or are difficult to resolve. For example, in the International Geosphere/Biosphere Project (IGBP) legend, permanent wetland may also be forest (Loveland et al., 1999). Or, the pixel may fall on a golf course. Is it grassland, savanna, agriculture, urban, or built-up land? A related problem is that of mixed pixels. Where fine-resolution data show a selected pixel to contain more than one cover class, how is a correct label to be assigned?

Additionally, the classification error structure as assessed by the consistent estimators above may not be the most useful measure of classification accuracy. Some errors are clearly more problematic than others. For example, confusing forest with water is probably a more serious error than confusing open and closed shrubland for many applications. This problem leads to the development of "fuzzy" accuracies that better meet users' needs (Gopal and Woodcock, 1994).

A final concern is that a design-based sample designed to validate a specific map cannot necessarily be used to validate another. A proper design-based validation procedure normally calls for stratified sampling so that accuracies may be established for each class with equal certainty. With stratified sampling, the probability of selection of all pixels within the same class is equal. If a stratified sample is overlain on another map, the selected pixels do not retain this property, thus introducing bias. Whereas an unstratified (random or regular) sample does not suffer from this problem, very large sample sizes are typically required to gain sufficient samples from small classes to establish their accuracies with needed precision.

While the foregoing discussion described the major elements for validating LC maps, particularly at the global scale, it is clear that a proper validation plan requires all three. Confidence-building measures are used at early stages both to refine a map that is under construction and to characterize the general nature of errors of a specific map product. Model-based inference, implemented during the classification process, can provide users with a quantitative assessment of each classification decision. Design-based inference, although costly, provides unbiased map accuracy statements using consistent estimators.

3.1.3   Lessons Learned from IGPB DISCover

The IGBP DISCover LC dataset, produced from 1.1-km spatial resolution AVHRR data by Loveland et al. (2000), remains a milestone in global LC classification using satellite data. The validation process used incorporated a global random sample stratified by cover type. Selected pixels were examined at high spatial resolution using Landsat and SPOT data in a design that featured multiple photographic interpreters classifying each pixel. Although not without difficulties, the validation process was very successful, yielding the first global validation of a global thematic map.

Recent research by Estes et al. (1999) summarized the lessons learned in the IGBP DISCover validation effort that apply to current and future global LC validation efforts. A primary conclusion was that the information of coarse-resolution satellite datasets is limited by such factors as multidate registration, atmospheric correction, and directional viewing effects. These limits in turn impose limits on the accuracies achievable in any global classification scenario. It should be noted that coarse-resolution satellite imaging instruments continue to produce data of improved quality. For example, data from MODIS that are used to develop LC products include nadir-looking surface reflectances that are obtained at multiple spatial resolutions (250, 500, and 1000 m).

Second, LC products developed using the spectral and temporal information available from coarse-resolution satellite imagers will always be an imperfect process, given the high intrinsic variance found in the global range (variability) of cover types. While the natural variation within many cover types is large, new instruments may yield new data streams that increase the certainty of identifying them uniquely. Among these are measures of vegetation structure derived from multi-angular observations, measures of spatial variance obtained from finer-resolution channels, and ancillary datasets such as land surface temperature.

A third lesson concerns the quality and availability of fine-resolution imagery for use in validation. Not only were Landsat and SPOT images costly, they were also very scarce for some large and ecologically important regions, such as Siberian conifer forest. However, the present Landsat 7 acquisition policy, which includes acquiring at least four relatively cloud-free scenes per year for every path and row, coupled with major price decreases, has eased this problem significantly for future validation efforts. However, the recent degradation of Enhanced Thematic Mapper Plus (ETM+) capabilities may significantly reduce future data acquisition capabilities.

A fourth lesson documented that interpreter skill and the quality of ancillary data are major factors that significantly affect assessment results. Best results were obtained using local interpreters who were familiar with the region of interest. The most important observation was that proper validation was an essential component of the mapping process and required a significant amount of the total effort. Roughly one third of the mapping resources were expended equally to each of the following: (1) data assembly, (2) data classification, and (3) quality and accuracy assessment of the result. Supporting agencies need to understand that a map classification is not completed until it is properly validated.

3.2 VALIDATION OF THE EUROPEAN COMMISSION'S GLOBAL LAND-COVER 2000

The general objective of the European Commission's Global Land Cover (GLC) 2000 was to provide a harmonized global LC database. The year 2000 was considered a reference year for environmental assessment in relation to various activities, and in particular the United Nation's Ecosystem-related International Conventions. To achieve this objective GLC 2000 made use of the **VEGA 2000** dataset: a dataset of 14 months of preprocessed daily global data acquired by the VEGETATION instrument aboard SPOT 4. These data were made available through a sponsorship from members of the VEGETATION program (http://www.gvm.sai.jrc.it/glc2000/defaultGLC2000.htm).

The validation of the GLC 2000 products incorporated confidence building based on a comparison with ancillary data and quantitative accuracy assessment using a stratified random sampling design and high-resolution sites. First, the draft products were reviewed by experts and compared with reference data (thematic maps, satellite images, etc.). These quality controls met two important objectives: (1) the elimination of macroscopic errors and (2) the improvement of the global acceptance by the customers associated in the process. Each validation cell (200 × 200 km) was systematically compared with reference material and documented in a database containing intrinsic properties of the GLC 2000 map (thematic composition and spatial pattern) and identified errors (wrong labels or limits).

This design-based inference had the objective of providing a statistical assessment of the accuracy by class and was based on a comparison with high-resolution data interpretations. It was characterized by: (1) random stratification by cover class, (2) a broad network of experts with local knowledge, (3) a decentralized approach, (4) visual interpretation of the higher-resolution imagery, and (5) interpretations based on the hierarchal classification scheme (Di Gregorio, 2000). Both the confidence building and design-based components occurred sequentially. Confidence building started with problematic areas (as expected by the map producer). This allowed for the correction of macro-errors found during the check. Then, a systematic review of the product using the same procedure was conducted before implementing the final quantitative accuracy assessment.

3.3 VALIDATION OF THE MODIS GLOBAL LAND-COVER PRODUCT

A team of researchers at Boston University currently produces a global LC product at 1-km spatial resolution using data from the MODIS instrument (Friedl et al., 2002). The primary product is a map of global LC using the IGBP classification scheme, which includes 17 classes that are largely differentiated by the life-form of the dominant vegetation layer. Included with the product is a confidence measure for each pixel as well as the second-most-likely class label. Input data are MODIS surface reflectance obtained in seven spectral bands coupled with an enhanced vegetation index product also derived from MODIS. These are obtained at 16-d intervals for each 1-km pixel. The classification is carried out using a decision tree classifier operating on more than 1300 global training sites identified from high-resolution data sources, primarily Landsat Thematic Mapper and Enhanced Thematic Mapper Plus (ETM+). The product is produced at 3- to 6-mo intervals using data from the prior 12-mo period (http://geography.bu.edu/landcover/userguidelc/intro.html).

The validation plan for the MODIS-derived LC product incorporates all approaches identified in section 3.1.2. Confidence-building exercises are used to provide a document accompanying the product that describes its strengths and weaknesses in qualitative terms for specific regions. A Web site also accumulates comments from users, providing feedback on specific regions. Confidence-building exercises also include comparisons with other datasets, including the Landsat Pathfinder for the humid tropics, United Nation's Food and Agricultural Organization (FAO) forest resource assessment, the European Union's Co-ordination of Information on the Environment (CORINE) database of LC for Europe, and the U.S. interagency-sponsored Multi-Resolution Land Characteristics (MRLC) database.

Model-based inference of classification accuracy is represented by the layer of per-pixel confidence values, which quantifies the posterior probability of classification for each pixel. This probability is first estimated by the classifier, which uses information on class signatures and separability obtained during the building of the decision tree using boosting (Friedl et al., 2002) to calculate the classification probability. This probability is then adjusted by three weighted prior probabilities associated with (1) the global frequency of all classes taken from the prior product, (2) the frequency of class types within the training set, and (3) the frequency of classes within a 200- × 200-pixel moving window. The result is a posterior probability that merges present and prior information and is used to assign the most likely class label to each pixel. The posterior probabilities are then summarized by cover type and region to convey information to users about the quality of the classification.

A form of design-based inference is used in the preparation of a confusion matrix taken from the classification of training sites. In this process, all training sites are divided into five equal sets. The classifier is trained using four of the five sets and then classifies the unseen sites in the fifth set. This procedure is repeated for each set as the unseen set, yielding a pair of labels — "true" and "as classified" — for every training site. Cross-tabulation of the two labels for the training site collection yields a confusion matrix that provides estimates of global, user's, and producer's accuracies. This matrix is provided in the documentation of the LC data product.

Note that these estimated accuracies will be biased because the training sites are not chosen randomly, and thus they may not properly reflect the variance encountered across the full extent of the true LC class. However, in selection of training sites, every effort is made to identify sites that do reflect the full range of variance of each class. Accordingly, the accuracies obtained are thought to be reasonable characterizations of the true accuracies, even though they cannot be shown to be proper unbiased estimators. In a final application of design-based inference, the MODIS team plans to conduct a random stratified sample of its LC product at regular intervals. The methodology will be similar to that of the IGBP DISCover validation effort (see section 3.1.3). However, funds have not yet been secured to support this costly endeavor.

3.4 CEOS LAND PRODUCT VALIDATION SUBGROUP

The lessons learned from previous and ongoing projects point to several areas where LPV can help with validation efforts. Perhaps most fundamental is that CEOS/WGCV/LPV provides a forum to discuss these issues and develop and maintain a standardized protocol. Indeed, the authors are all involved with LPV, and it was through this association that this chapter was developed. There is also the opportunity to communicate on LC classification systems; although each project will have its own system, coordination between the two projects results in synergy between the two systems (Thomlinson et al., 1999; Di Gregorio and Jansen, 2000). Here we present methods by which LPV can help address the specific lessons learned from IGBP in the context of the two current projects. Table 3.1 lists the various subgroups and their corresponding URLs.

Table 3.1   CEOS Land-Cover Validation Participants and Contributions

Entity	Role in Global Land-Cover Validation

CEOS
Working Group on Calibration and Validation Land Product
  Validation subgroup
  http://www.wgcvceos.org/
	Coordinates validation activities of CEOS
	  members

Global Observation of Forest Cover
Global Observation of Land Dynamics
  http://www.fao.org/gtos/gofc-gold/index.html
	Coordinates regional networks to provide
	  "local" expertice

European Commission's Global Land Cover 2000 http://www.gvm.sai.jrc.it/glc2000/defaultGLC2000.htm	Produces data
NASA's Global Land Cover product http://edcdaac.usgs.gov/modis/mod12q1.html	Produces data

EOS Land Validation Core Sites
  http://modis.gsfc.nasa.gov/MODIS/LAND/VAL/CEOS_WGCV/
  lai_intercomp.html
	Sites under consideration for CEOS Land
	  Product Validation Core Sites

VALERI (VAlidation of Land European Remote sensing
  Instruments)
  http://www.avignon.inra.fr/valeri/
	Sites under consideration for CEOS Land
	  Product Validation Core Sites

CEOS "LAI-intercomparison"
  http://landval.gsfc.nasa.gov/LPVS/BIO/lai_intercomp.html
	Sites under consideration for CEOS Land
	  Product Validation Core Sites

As satellite sensors and related algorithms continue to improve, many of the technical obstacles addressed above may be overcome. However, it is essential that groups producing global LC products have a thorough awareness of technology improvements across the range of satellite sensors, including optical, microwave LIDAR, and SAR. Such awareness can be supported through LPV's interaction with the other WGCV subgroups, including discussions at the semiannual WGCV plenary meetings and utilizing the projects and publications available through the other subgroups of WGCV. Further, coordination of various LC products can help determine the most suitable approach to using multiple products. For example, the MODIS product has been operationally produced since 2001. Careful examination of this product as well as the GLC 2000 product could lend insight into the best way to use both in a complementary fashion.

3.4.1   Fine-Resolution Image Quality and Availability

Data sharing of high-resolution imagery may be one of the most immediate and concrete ways in with LPV can support global land-product validation. Using the NASA Earth Observing System Land Validation Core Sites (Morisette et al., 2002) as an example, the LPV and WGISS subgroups are establishing an infrastructure for a set of "CEOS Land Product Validation Core Sites." The initial sites being considered for this project are shown in Plate 3.1, which represents an agglomeration of three entities: the EOS Land Validation Core Sites, the VAlidation of Land European Remote sensing Instruments (VALERI) project, and the CEOS "LAI Inter-comparison" activity (Table 3.1). The concept is to establish a set of sites where high-resolution data will be archived and proved free or at minimal cost over locations where field and/or tower measurements are continuously or periodically collected (Plate 3.1). These core sites are intended to serve as validation sites for multiple satellite products. Specific products appropriate for validation depend on the individual field tower measurement parameters (Morisette et al., 2002). Practically, the limited number of sites (approximately 50), which are not based on a random sample, cannot be used for statistical inferences on a global product. However, in terms of LC validation, the high-resolution data from these sites would allow a set of common "confidence-building sites" that could be shared by GLC 2000 and MODIS as well as future global LC mapping efforts. LC product comparisons with high-resolution data and cross-comparison with other global LC products over the core sites would provide substantive information for initial quality control. Additionally, within a given site a random sample could be collected and design-based inference carried out for that particular "subpopulation." So, while the core site concept has limitations with respect to statistical inference, the opportunities for data sharing and initial cross-comparison at a set of core sites seems worthwhile.

3.4.2   Local Knowledge Requirements

The LPV was strategically designed to complement the objectives of the Global Observation of Forest Cover/Land Dynamics (GOFC/GOLD) program (http://www.fao.org/gtos/gofc-gold/). This partnership provides a context for validation activities (through LPV) within the specific user group (GOFC/GOLD). GOFC/GOLD is broken down into three implementation teams that include: (1) LC characteristics and change, (2) fire-related products, and (3) biophysical processes. Initial activities of LPV have also focused on these three areas through topical workshops and initial projects.

A major component of GOFC/GOLD is to build on "regional networks." These networks involve local and regional partners who are interested in using the global products and serve to provide feedback to the data producers. This regional network concept has proven to be a significant resource to support validation efforts. The IGBP experience indicates that the knowledge gained through regional collaborators is critical. LPV can use the regional networks as an infrastructure to gain local expertise for product validation. This infrastructure can provide assistance with the difficult and labor-intensive task of design-based inference planned for both MODIS and GLC 2000.

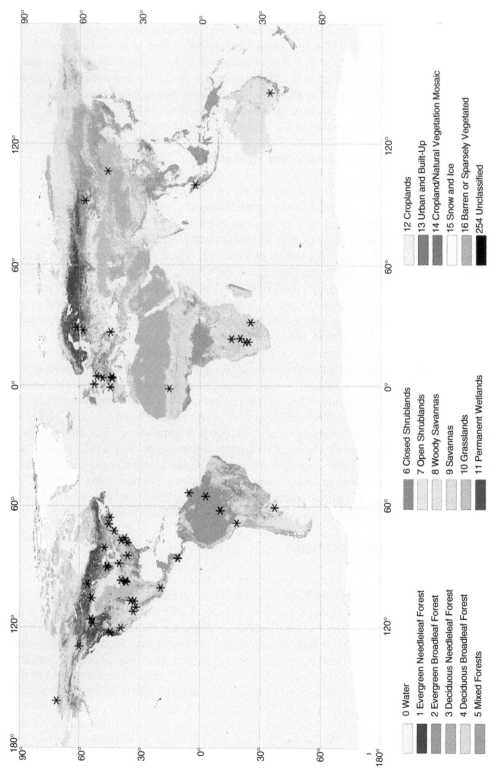

Plate 3.1   (See color insert following page 114.) CEOS land-cover product evaluation core site locations.

3.4.3 Resource Requirements

The LPV has been established with the realization that proper validation requires a significant scientific effort. Indeed, the subgroup has been established to conduct global validation activities as efficiently as possible. The validation approaches described here have all been conceived to minimize the resource requirements for global LC validation. To this end, the LPV has capitalized on the most current sensor technologies (high-resolution) and exploited data-sharing opportunities with both the CEOS core sites and the use of GOFC/GOLD regional networks to reduce the cost and effort of global validation efforts. The LPV subgroup is collaborating with the MODIS LC and GLC 2000 programs to help realize and develop these suggestions. This, in turn, can be applied to future global LC products.

3.5 SUMMARY

This chapter presents the approach for the use of the CEOS to coordinate the validation efforts of global land products. This premise is based on experience from previous global validation through the IGBP, which depended on the goodwill, support, cooperation, and collaboration of interested organizations and institutions. Two global LC efforts are now underway: (1) NASA's MODIS Global LC product and (2) the European Commission's GLC 2000. These validation efforts will likewise require coordination and collaboration — much of which has been or is being established. In this chapter we discussed issues pertaining to validation of global LC products, presented a brief overview of the validation strategy for the two current efforts, then described a mutually beneficial strategy for both to realize some efficiencies by using CEOS to further coordinate their validation efforts. This strategy should be applicable to other global LC mapping efforts, such as those being developed for the GOFC/GOLD and beyond.

ACKNOWLEDGMENTS

Thanks are extended to Yves-Louis Desnos, as chair of the CEOS Working Group on Calibration and Validation, for continued attention to the CEOS Core Site concept. The authors would like to acknowledge John Hodges, Boston University, for providing Plate 3.1. Also, reviews from Ross Lunetta and anonymous reviewers were helpful and appreciated.

REFERENCES

Di Gregorio, A. and L.J.M. Jansen, *Land Cover Classification System (LCCS): Classification Concepts and User Manual.* Food and Agriculture Organization of the United Nations, Rome (http://www.fao.org/DOCREP/003/X0596E/X0596E00.HTM), 2000.

Estes, J., A. Belward, T. Loveland, J. Scepan, A. Strahler, J. Townshend, and C. Justice, The way forward, *Photogram. Eng. Remote Sens.*, 65, 1089–1093, 1999.

Friedl, M.A., D.K. McIver, J.C.F. Hodges, X.Y. Zhang, D. Muchoney, A.H. Strahler, C.E. Woodcock, S. Gopal, A. Schneider, A. Cooper, A. Baccini, F. Gao, and C. Schaaf, Global land cover mapping from MODIS: algorithms and early results, *Remote Sens. Environ.*, 83, 287–302, 2002.

Gopal, S. and C. Woodcock, Theory and methods for accuracy assessment of thematic maps using fuzzy sets, *Photogram. Eng. Remote Sens.*, 60, 181–188, 1994.

Justice, C., A. Belward, J. Morisette, P. Lewis, J. Privette, and F. Baret, Developments in the "validation" of satellite sensor products for the study of land surface, *Int. J. Remote Sens.*, 21, 3383–3390, 2000.

Loveland, T. R., B.C. Reed, J.F. Brown, D.O. Ohlen, Z. Zhu, L. Yang, and J.W. Merchant, Development of a global land cover characteristics database and IGBP DISCover from 1-km AVHRR data, *Int. J. Remote Sens.*, 21, 1303–1330, 2000.

Loveland, T.R., Z. Zhu, D.O. Ohlen, J.F. Brown, B.C. Reed, and Y. Limin, An analysis of the IGBP global land-cover characterization process, *Photogram. Eng. Remote Sens.*, 65, 1021–1032, 1999.

Morisette J.T., J.L. Privette, and C.O. Justice, A framework for the validation of MODIS land products, *Remote Sens. Environ.*, 83, 77–96, 2002.

Stehman, S.V., Practical implications of design-based sampling inference for thematic map accuracy assessment, *Remote Sens. Environ.*, 72, 34–45, 2000.

Stehman, S.V., Statistical rigor and practical utility in thematic map accuracy assessment, *Photogram. Eng. Remote Sens.*, 67, 727–734, 2001.

Thomlinson J.R., P.V. Bolstad, and W.B. Cohen, Coordinating methodologies for scaling landcover classifications from site-specific to global: steps toward validation global map products, *Remote Sens. Environ.*, 70, 16–28, 1999.

CHAPTER **4**

In Situ Estimates of Forest LAI for MODIS Data Validation

John S. Iiames, Jr., Andrew N. Pilant, and Timothy E. Lewis

CONTENTS

4.1 INTRODUCTION

Satellite remote sensor data are commonly used to assess ecosystem conditions through synoptic monitoring of terrestrial vegetation extent, biomass, and seasonal dynamics. Two commonly used vegetation indices that can be derived from various remote sensor systems include the Normalized Difference Vegetation Index (NDVI) and Leaf Area Index (LAI). Detailed knowledge of vegetation

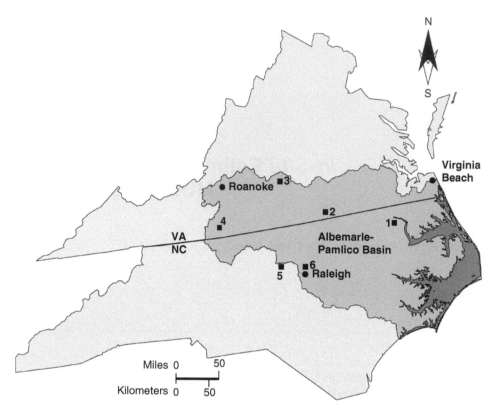

Figure 4.1   LAI field validation site locations within the Albemarle-Pamlico Basin in southern Virginia and northern North Carolina. (1) Hertford; (2) South Hill; (3) Appomattox; (4) Fairystone; (5) Duke FACE; (6) Umstead.

index performance is required to characterize both the natural variability across forest stands and the intraannual variability (phenology) associated with individual stands. To assess performance accuracy, *in situ* validation procedures can be applied to evaluate the accuracy of remote sensor-derived indices. A collaborative effort was established with researchers from the U.S. Environmental Protection Agency (EPA), National Aeronautics and Space Administration (NASA), academia, and state and municipal governmental organizations, and private forest industry to evaluate the Moderate Resolution Imaging Spectroradiometer (MODIS) NDVI and LAI products across six validation sites in the Albemarle-Pamlico Basin (APB), in North Carolina and Virginia (Figure 4.1).

The significance of LAI and NDVI as source data for process-based ecological models has been well documented. LAI has been identified as the variable of greatest importance for quantifying energy and mass exchange by plant canopies (Running et al., 1986) and has been shown to explain 80 to 90% of the variation in the above-ground forest net primary production (NPP) (Gholz, 1982; Gower et al., 1992; Fassnacht and Gower, 1997). LAI is an important biophysical state parameter linked to biological productivity and carbon sequestration potential and is defined here as one half the total green leaf area per unit of ground surface area (Chen and Black, 1992). NPP is the rate at which carbon is accumulated by autotrophs and is expressed as the difference between gross photosynthesis and autotrophic respiration (Jenkins et al., 1999).

NDVI has been used to provide LAI estimates for the prediction of stand and foliar biomass (Burton et al., 1991) and as a surrogate to estimate stand biomass for denitrification potential in forest filter zones for agricultural nonpoint source nitrogenous pollution along riparian waterways (Verchot et al., 1998). Interest in tracking LAI and NDVI changes includes the role forests play in the sequestration of carbon from carbon emissions (Johnsen et al., 2001) and the formation of tropospheric ozone from biogenic emissions of volatile organic compounds naturally released into the atmosphere (Geron et al., 1994). The NDVI has commonly been used as an indicator of biomass (Eidenshink and Haas, 1992) and vegetation vigor (Carlson and Ripley, 1997). NDVI has been applied in monitoring seasonal and interannual vegetation growth cycles, land-cover (LC) mapping, and change detection. Indirectly, it has been used as a precursor to calculate LAI, biomass, the fraction of absorbed photosynthetically active radiation (fAPAR), and the areal extent of green vegetation cover (Chen, 1996).

Direct estimates of LAI can be made using destructive sampling and leaf litter collection methods (Neumann et al., 1989). Direct destructive sampling is regarded as the most accurate approach, yielding the closest approximation of "true" LAI. However, destructive sampling is time-consuming and labor-intensive, motivating development of more rapid, indirect field optical methods. A subset of field optical techniques include hemispherical photography, LiCOR Plant Canopy Analyzer (PCA) (Deblonde et al., 1994), and the Tracing Radiation and Architecture of Canopies (TRAC) sunfleck profiling instrument (Leblanc et al., 2002). *In situ* forest measurements serve as both reference data for satellite product validation and as baseline measurements of seasonal vegetation dynamics, particularly the seasonal expansion and contraction of leaf biomass.

The development of appropriate ground-based sampling strategies is critical to the accurate specification of uncertainties in LAI products (Tian et al., 2002). Other methods that have been implemented to assess the MODIS LAI product have included a spatial cluster design and a patch-based design (Burrows et al., 2002). Privette et al. (2002) used multiple parallel 750-m TRAC sampling transects to assess LAI and other canopy properties at scales approaching that of a single MODIS pixel. Also, a stratified random sampling (SRS) design element provided sample intensification for less frequently occurring LC types (Lunetta et al., 2001).

4.1.1 Study Area

The study area is the Albemarle-Pamlico Basin (APB) of North Carolina and Virginia (Figure 4.1). The APB has a drainage area of 738,735 km^2 and includes three physiographic provinces: mountain, piedmont, and coastal plain, ranging in elevation from 1280 m to sea level. The APB subbasins include the Albemarle-Chowan, Roanoke, Pamlico, and Neuse River basins. The Albemarle-Pamlico Sounds compose the second-largest estuarine system within the continental U.S. The 1992 LC in the APB consisted primarily of forests (50%), agriculture (27%), and wetlands (17%). The forest component is distributed as follows: deciduous (48%), conifer (33%), and mixed (19%) (Vogelmann et al., 1998).

4.2 BACKGROUND

4.2.1 TRAC Measurements

The TRAC sunfleck profiling instrument consists of three quantum PAR sensors (LI-COR, Lincoln, NE, Model LI-190SB) mounted on a wand with a built-in data logger (Leblanc et al., 2002) (Figure 4.2). The instrument is hand-carried along a linear transect at a constant speed, measuring the downwelling solar photosynthetic photon flux density (PPFD) in units of micromoles per square meter per second. The data record light–dark transitions as the direct solar beam is alternately transmitted and eclipsed by canopy elements (Figure 4.3). This record of sunflecks and shadows is processed to yield a canopy gap size distribution and other canopy architectural parameters, including LAI and a foliage element clumping index.

From the downwelling solar flux recorded along a transect, the TRACWin software (Leblanc et al., 2002) computes the following derived parameters describing forest canopy architecture: (1)

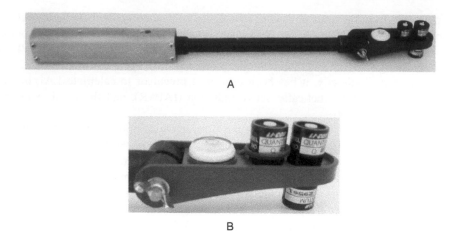

A

B

Figure 4.2   Photograph of (A) TRAC Instrument (length ~ 80 cm) and (B) PAR detectors (close-up).

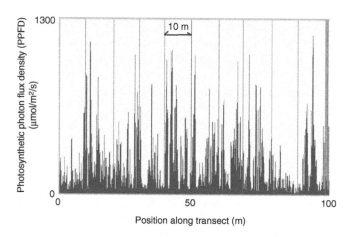

Figure 4.3   TRAC transect in loblolly pine plantation (site: Hertford). Peaks (black spikes) are canopy gaps. Computed parameters for this transect were gap fraction = 9%; clumping index (Ω_e) = 0.94; PAI = 3.07; L_e = 4.4 (assuming γ = 1.5, α = 0.1, and mean element width = 50 mm).

canopy gap size (physical dimension of a canopy gap), (2) canopy gap fraction (percentage of canopy gaps), (3) foliage element clumping index, $\Omega_e(\theta)$, (4) plant area index (LAI, which includes both foliage and woody material), and (5) LAI with clumping index (Ω_e) incorporated. Note that in each case the parameters are for the particular solar zenith angle θ at the time of data acquisition, defining an inclined plane slicing the canopy between the moving instrument and the sun.

Parameters entered into the TRACWin software to invert measured PPFD to the derived output parameters include the mean element width (the mean size of shadows cast by the canopy), the needle-to-shoot area ratio (γ) (within-shoot clumping index), woody-to-total area ratio (α), latitude/longitude, and time. Potential uncertainties were inherent in the first three parameters and will be assessed in future computational error analyses.

Solar zenith and azimuth influence data quality. Optimal results are achieved with a solar zenith angle θ between 30 and 60 degrees. As θ approaches the horizon ($\theta > 60°$), the relationship between LAI and light extinction becomes increasingly nonlinear. Similarly, best results are attained when TRAC sampling is conducted with a solar azimuth perpendicular to the transect azimuth. Sky condition is a significant factor for TRAC measurements. Clear, blue sky with unobstructed sun is optimal. Overcast conditions are unsuitable; the methodology requires distinct sunflecks and shadows.

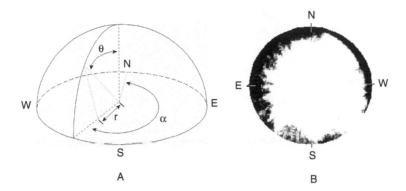

Figure 4.4 Illustration of (A) a hemispherical coordinate system. Such a system is used to convert a hemispherical photograph into a two-dimensional circular image (B), where the zenith () is in the center, the horizon at the periphery, east is to the left, and west is to the right. In a equiangular hemispherical projection, distance along a radius (r) is proportional to zenith angle (Rich, 1990).

The TRAC manual (Leblanc et al., 2002) lists the following as studies validating the TRAC instrument and approach: Chen and Cihlar (1995), Chen (1996), Chen et al. (1997), Kucharik et al. (1997), and Leblanc (2002). TRAC results were compared with direct destructive sampling, which is generally regarded as the most accurate sampling technique.

4.2.2 Hemispherical Photography Measurements

Hemispherical photography is an indirect optical method that has been used in studies of forest light transmission and canopy structure. Photographs taken upward from the forest floor with a 180° hemispherical (fish-eye) lens produce circular images that record the size, shape, and location of gaps in the forest overstory. Photographs can be taken using 35-mm film cameras or digital cameras. A properly classified fish-eye photograph provides a detailed map of sky visibility and obstructions (sky map) relative to the location where the photograph was taken. Various software programs, such as Gap Light Analyzer (GLA), were available to process film or digital fish-eye camera images into a myriad of metrics that reveal information about the light regimes beneath the canopy and the productivity of the plant canopy. These programs rely on an accurate projection of a three-dimensional hemispherical coordinate system onto a two-dimensional surface (Figure 4.4). Accurate projection requires calibration information for the fish-eye lens that is used and any spherical distortions associated with the lens. GLA used in this analysis was available for download at http://www.ecostudies.org/gla/ (Frazer et al., 1999).

The calculation of canopy metrics depends on accurate measures of gap fraction as a function of zenith angle and azimuth. The digital image can be divided into zenith and azimuth "sky addresses" or sectors (Figure 4.5). Each sector can be described by a combined zenith angle and azimuth value. Within a given sector, gap fraction is calculated with values between zero (totally "obscured" sky) and one (totally "open" sky) and is defined as the proportion of unobscured sky as seen from a position beneath the plant canopy (Delta-T Devices, 1998).

4.2.3 Combining TRAC and Hemispherical Photography

LAI calculated using hemispherical photography or other indirect optical methods does not account for the nonrandomness of canopy foliage elements. Hence, the term *effective leaf area index* (L_e) is used to refer to the leaf area index estimated from optical measurements including hemispherical photography. L_e typically underestimates "true" LAI (Chen et al., 1991). This underestimation is due in part to nonrandomness in the canopy (i.e., foliage "clumping" at the scales of tree

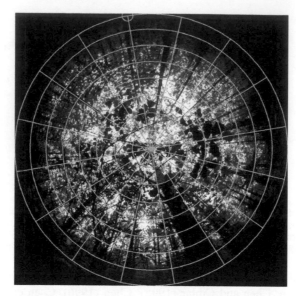

Figure 4.5   Sky-sector mapping using GLA image analysis software. Eight zenith by 18 azimuth sectors are shown.

crown), whorls, branches, and shoots. The TRAC instrument was developed at the Canada Centre for Remote Sensing (CCRS) to address canopy nonrandomness (Chen and Cihlar, 1995). In the APB study, hemispherical photography (L_e) and TRAC measurements (foliage clumping index) were combined to provide a better estimate of LAI following the method of Leblanc et al. (2002).

4.2.4   Satellite Data

MODIS was launched in 1999 aboard the NASA Terra platform (EOS-AM) and in 2002 aboard the Aqua platform (EOS-PM) and provides daily coverage of most of the earth (Justice et al., 1998; Masuoka et al., 1998). MODIS sensor characteristics include a spectral range of 0.42 to 14.35 µm in 36 spectral bands, variable pixel sizes (250, 500, and 1000 m), and a revisit interval of 1 to 2 days. Landsat Enhanced Thematic Mapper Plus (ETM+) images were acquired at various dates throughout the year and were used for site characterization and in subsequent analysis for linking field measurements of LAI with MODIS LAI. ETM+ data characteristics include a spectral range of 0.45 to 12.5 µm; pixel sizes of 30 m (multispectral), 15 m (panchromatic), and 60 m (thermal); and a revisit interval of 16 d. They also play a vital role in linking meter-scale *in situ* LAI measurements with kilometer-scale MODIS LAI imagery. IKONOS is a high-spatial-resolution commercial sensor that was launched in 1999 that provides 4.0-m multispectral (four bands, 0.45 to 0.88 µm) and 1-m panchromatic data (0.45 to 0.90 µm) with a potential revisit interval of 1 to 3 d.

4.2.5   MODIS LAI and NDVI Products

Numerous land, water, and atmospheric geophysical products are derived from MODIS radiance measurements. Two MODIS land products established the primary time-series data for this research: NDVI (MOD13Q1) (Huete et al., 1996) and LAI/FPAR (MOD15A2) (Knyazikhin et al., 1999). The NDVI product was a 16-d composite at a nominal pixel size of 250 m. The LAI product was an 8-d composite product with a pixel size of 1000 m. Both products were adjusted for atmospheric effects and viewing geometry (bidirectional reflectance distribution function, BRDF). The NDVI product used in this study was produced using the standard MODIS-NDVI algorithm (Huete et al., 1996).

The MODIS LAI product algorithms were considerably more complex. The primary approach for calculating LAI involved the inversion of surface reflectance in two to seven spectral bands and comparison of the output to biome-specific look-up tables derived from three-dimensional canopy radiative transfer modeling. All terrestrial LC was assigned to six global biomes, each with distinct canopy architectural properties that drove photon transport equations. The six biomes included grasses and cereal crops, shrubs, broadleaf crops, savannas, broadleaf forests, and needle forests. The secondary technique was invoked when insufficient high-quality data were available for a given compositing period (e.g., cloud cover, sensor system malfunction) and calculated LAI based on empirical relationships with vegetation indices. However, a deficiency inherent with the second approach was that NDVI saturates at high leaf biomass (LAI values between 5 and 6). The computational approach used for each pixel was included with the metadata distributed with each data set.

4.3 METHODS

Here we describe a field sampling design and data acquisition protocol implemented in 2002 for measuring *in situ* forest canopy properties for the analysis of correspondence to MODIS satellite NDVI and LAI products. The study objective was to acquire field measurement data to evaluate LAI and NDVI products using *in situ* measurement data and indirectly using higher-spatial-resolution imagery sensors including Landsat Enhanced Thematic Mapper Plus (ETM⁺) and IKONOS.

4.3.1 Sampling Frame Design

Six long-term forested research sites were established in the APB (Table 4.1). The objective was to collect ground-reference data using optical techniques to validate seasonal MODIS NDVI and LAI products. Baseline forest biometrics were also measured for each site. Five sites were located in the Piedmont physiographic region and one site (Hertford) in the coastal plain. The Hertford and South Hill sites were composed of homogeneous conifer forest (loblolly pine), Fairystone mixed deciduous forest (oak/hickory), and Umstead mixed conifer and mixed forest, and both Duke and Appomattox sites contained homogeneous stands of conifer and deciduous forest managed under varying silvicultural treatments (e.g., thinning). At Duke and South Hill, university collaborators monitored LAI using direct means (destructive harvest and leaf litter); their data were employed to validate the field optical techniques used in this study.

The fundamental field sampling units are referred to as quadrants and subplots (Figure 4.6). A quadrant was a 100- × 100-m grid with five 100-m east–west TRAC sampling transects and five interspersed transects for hemispherical photography (lines A–E). The TRAC transects were spaced at 20-m intervals (north–south), as were the interleaved hemispherical photography sampling transects. A subplot consisted of two 50-m transects intersecting at the 25-m center point. The two

Table 4.1 Location Summary for Six Validation Sites in the Albemarle-Pamlico Basin

Site	State	Location (lat., long.)	Elevation (m)	Physiographic Region	Ownership	Area
Appomattox	VA	37.219, −78.879	165–215	Piedmont	Private	1200 m² (144 ha)
Duke FACE	NC	35.975, −79.094	165–180	Piedmont	Private	1200 m² (144 ha)
Fairystone	VA	36.772, −80.093	395–490	Upper Piedmont	State	1200 m² (144 ha)
Hertford	NC	36.383, −77.001	8–10	Coastal Plain	Private	1200 m² (144 ha)
South Hill	VA	36.681, −77.994	90	Piedmont	Private	1200 m² (144 ha)
Umstead	NC	35.854, −78.755	100–125	Piedmont	State	1200 m² (144 ha)

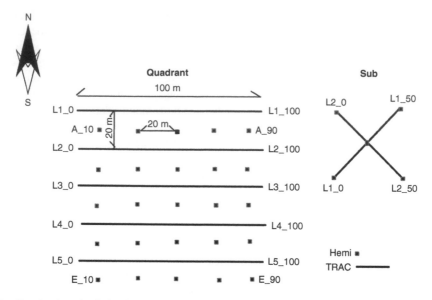

Figure 4.6 Quadrant and subplot designs used in the Albemarle-Pamlico Basin study area.

transects were oriented at 45° and 135° to provide flexibility in capturing TRAC measurements during favorable morning and afternoon solar zenith angles.

Quadrants were designed to approximate an ETM+ 3 × 3 pixel window. Subplots were designed to increase sample site density and were selected on the basis of ETM+ NDVI values to sample over the entire range of local variability. Quadrants and subplots were geographically located on each LAI validation site using real-time (satellite) differentially corrected GPS to a horizontal accuracy of ± 1 m. TRAC transects were marked every 10 m with a labeled, 46-cm wooden stake. The stakes were used in TRAC measurements as walking-pace and distance markers. Hemispherical photography transects were staked and marked at the 10-, 30-, 50-, 70-, and 90-m locations. Hemispherical photographs were taken at these sampling points.

The APB quadrant design was similar to a measurement design used in a Siberian LAI study in the coniferous forest of Krasnoyarsk, Russia (Leblanc et al., 2002). Here, each validation site had a minimum of one quadrant. Multiple quadrants at Fairystone were established across a 1200- × 1200-m oak–hickory forest delineated on a georeferenced ETM+ image to approximate a MODIS pixel (1 km²), with a 100-m perimeter buffer to partially address spatial misregistration of a MODIS pixel (Figure 4.7). The stand was quartered into 600- × 600-m units. The northwest corner of a LAI sampling quadrant was assigned within each quarter block using a random number generator.

A SRS design was used to select ground reference data spanning the entire range of LAI–NDVI values. Fairystone sites were stratified based on a NDVI surface map calculated from July 2001 ETM+ imagery. Analysis of the resulting histogram allowed for the identification of pixels beyond ± 1 standard deviation. From these high/low NDVI regions, eight locations (four high, four low) were randomly selected from each of the four 600- × 600-m units. Subplots were established at these points to sample high or low and midrange NDVI regions within each of the four quadrants.

4.3.2 Biometric Mensuration

The measurement of crown closure was included in quadrant sampling to establish the relationship between LAI and NDVI. Wulder et al. (1998) found that the inclusion of this textural information strengthened the LAI:NDVI relationship, thus increasing the accuracy of modeled LAI estimates. Crown closure was estimated directly using two field-based techniques: the vertical tube

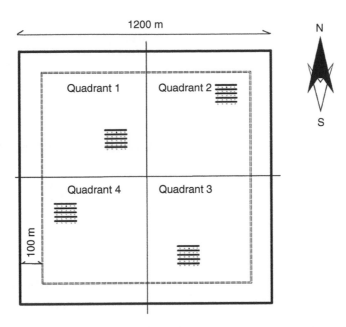

Figure 4.7 Multiple quadrant design used at the Fairystone and Umstead sites. The 1200- × 1200-m region approximates a MODIS LAI pixel, with a 100-m buffer on each edge. Quadrants are randomly located within each 600- × 600-m quarter.

(Figure 4.8) and the spherical densiometer (Figure 4.9) (Becker et al., 2002). Measurement estimates were also performed using the TRAC instrument and hemispherical photography.

Measurements of forest structural attributes (forest stand volume, basal area, and density) were made at each quadrant and subplot using a point sampling method based on a 10-basal-area-factor prism. Point sampling by prism is a plotless technique (point-centered) in which trees are tallied on the basis of their size rather than on frequency of occurrence on a plot (Avery and Burkhart, 1983). Large trees at a distance had a higher probability of being tallied than small trees at that same distance. Forest structural attributes measured on trees that fell within the prism angle of view included (1) diameter at breast height (dbh) at 1.4 m, (2) tree height, (3) tree species, and (4) crown position in the canopy (dominant, codominant, intermediate, or suppressed).

At each quadrant, forest structural attributes were sampled at the 10-, 50-, and 90-m stations along the A, C, and E hemispherical photography transects (Table 4.2). Point sampling was performed at the subplot 25-m transect intersection. Physical site descriptions were made at each

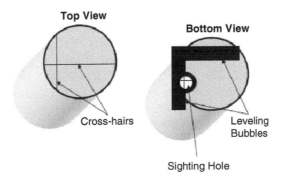

Figure 4.8 Schematic of vertical tube used for crown closure estimation.

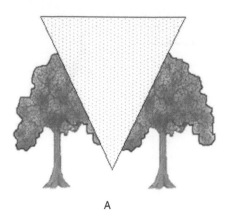

A   B

Figure 4.9   Illustration of (A) a spherical densiometer 60° field of view and (B) convex spherical densiometer (courtesy of Ben Meadows).

Table 4.2   Vegetation Summary for Six Validation Sites in the Albemarle-Pamlico Basin

Site	Type	%	Over TPH	Under TPH	Avg. Ht (m)	Avg. dbh (cm)	CC% Dom	CC% Sup	BA/H (m²/ha)
Appomattox	Pine	25	1250	3790	15.9	21.6	71	34	36.7
	Hardwood	25	1255	—	21.3	24.3	—	—	22.9
	Pine-Thinned	50	313	—	16.9	23.2	—	—	11.5
Duke	Hardwood	30	—	—	—	—	—	—	—
Fairystone	Hardwood	100	725–1190	—	15.5–19.5	8.5–11.5	—	—	12.6–13.1
Hertford	Pine	100	1740	2830	14.3	18.5	71	29	37.3
South Hill	Pine	100	—	—	—	—	—	—	—
Umstead	Pine	30	—	—	—	—	—	—	—
	Hardwood	70	—	—	—	—	—	—	—

Note: Over TPH = trees per hectare for trees greater than 5.08 cm dbh; Under TPH = trees per hectare less than 5.08 cm in dbh; Avg. Ht = average height; Avg dbh = average diameter at breast height; CC% Dom = crown closure for dominant crown class determined by vertical tube method; CC% Sup = crown closure for suppressed crown class determined by fixed radius plot method; BA/H = basal area per hectare.

quadrant and subplot by recording slope, aspect, elevation, and soil type. Digital images were recorded at the zero-meter station of each TRAC transect during each site visit for visual documentation. Images were collected at 0°, 45°, and 90° from horizontal facing east along the transect line.

4.3.3 TRAC Measurements

The TRAC instrument was hand-carried at waist height (~ 1 to 1.5 m) along each transect at a constant speed of 0.3 m/sec. The operator traversed 10 m between survey stakes in 30 sec, monitoring speed by wristwatch. The spatial sampling interval at 32 Hz at a cruising speed of 0.3 m/sec was approximately 10 mm (i.e., 100 samples/m). To the degree possible, transects were sampled during the time of day at which the solar azimuth was most perpendicular to the transect azimuth. Normally, quadrants were traversed in an east–west direction, but if the solar azimuth at the time of TRAC sampling was near 90° or 270° (early morning or late afternoon in summer), quadrants were traversed on a north–south alignment.

PPFD measurements were made in an open area before and after the undercanopy data acquisition for data normalization to the maximum solar input. Generally, large canopy gaps provided an approximation of the above-canopy PPFD, used to define the above canopy solar flux at times when access to open areas was limited. Under uniform sky conditions, above-canopy solar flux was interpolated between measured values. Under partially cloudy conditions, the operator stopped recording photon flux during cloud eclipse of the solar beam.

Operators performed a check on the data in the field immediately after download to a portable computer. Typically, this involved plotting the PPFD in graphical form and comparing the number of segments collected to the number of 10-m intervals traversed. An important quality assurance measure was the use of paper and computer forms for data entry. To ensure that all relevant ancillary data (i.e., weather conditions, transect orientation, operator names, data file names) were captured in the field, operators filled out paper forms on-site for TRAC, hemispherical photography, and biometric measurements. These data forms were then entered into a computer database via prescribed forms, preferably immediately after data collection. This was a simple but valuable step to ensure that critical data acquisition and processing parameters were not inadvertently omitted from field notes. The computer forms provided a user interface to the relational database containing all the metadata for the APB project.

4.3.4 Hemispherical Photography

Two Nikon Coolpix 995 digital cameras with Nikon FC-E8 fish-eye converters were used in conjunction with TRAC at all six APB research sites. Exposures were set to automatic with normal file compression (approximately 1/8) selected at an image size of 1600×1200 pixels. Hemispherical images were not collected while the sun was above the horizon, unless the sky was uniformly overcast. Images were primarily captured at dawn or dusk to avoid the issue of nonuniform brightness, resulting in the foliage being "washed out" in the black-and-white binary image.

The camera was mounted on a tripod and leveled over each wooden stake along each A through E photo transect. The height of the camera was adjusted to approximately breast height (1.4 m) and leveled to ensure that the "true" horizon occurred at a 90° zenith angle in the digital photographic image. The combination of two bubble levelers, one mounted on the tripod and the other on the lens cap, ensured the capture of the "true" horizon in each photograph. Using a hand-held compass, the camera was oriented to true north so that the azimuth values in the photograph corresponded to the true orientation of the canopy architecture in the forest stand. Orientation did not affect any of the whole-image canopy metrics (i.e., LAI, canopy openness, or site openness) calculated by GLA. However, comparison of metrics derived by hemispherical photography, TRAC, densiometer, or forest mensuration measurements required accurate image orientation.

After the images were captured in the field they were downloaded from the camera disk, placed in a descriptive file directory structure, and renamed to reflect the site and transect point. A GLA configuration file (image orientation, projection distortion and lens calibration, site location coordinates, growing-season length, sky-region brightness, and atmospheric conditions) was created for each site. Next, images were registered in a procedure that defined an image's circular area and location of north in the image. Image registration entailed entering pixel coordinates (image size- and camera-dependent) for the initial and final X and Y points. The FC-E8 fish-eye lens used in this study had an actual field of view greater than 180° (~185°). The radius of the image was reduced accordingly so that the 90° zenith angle represented the true horizon. Frazer et al. (2001) described the procedure for calibrating a fish-eye lens. Calibration results were entered into the GLA configuration file (Canham et al., 1994).

The analyst-determined threshold setting in GLA adjusted the number of black ("obscured" sky) and white ("unobscured" sky) pixels in the working image. This was perhaps the most subjective setting in the entire measurement process and potentially the largest source of error in the calculation of LAI and other canopy metrics from hemispherical photographs. As a rule of thumb, the threshold value was increased so that black pixels appeared that were not represented by canopy elements in the registered color image. The threshold was then decreased from this point until the black dots or blotches disappeared and the black-and-white working image was a reasonable representation of the registered color image (Frazer et al., 1999).

4.3.5   Hemispherical Photography Quality Assurance

The height at which each hemispherical photograph was taken represented a potential source of positional errors (~ 5 to 10 cm). At relatively level sampling points, the tripod legs and center shaft were fully extended to attain a height that approximated breast height. However, at sites with steep and/or uneven slopes, the camera height may have varied between repetitive measurement dates due to variations in the extension of the tripod legs, possibly resulting in inclusion or exclusion of near-lens vegetation.

Several comparisons of hemispherical photographic estimates of LAI with direct estimates in broad leaf and conifer forest stands have been reported (Neumann et al., 1989; Chason, 1991; Chen and Black, 1991; Deblonde et al., 1994; Fassnacht et al., 1994; Runyon et al., 1994). These comparisons all showed that there was a high correlation between the indirect and direct methods, but the indirect methods were biased low. This was because the clumping factor was not accounted for using a random foliar distribution model (Chen et al., 1991).

To assess analyst repeatability, a set of 31 hemispherical photographic images collected in eastern Oregon were analyzed and threshold values charted using SAS QC software (SAS, 1987). Two analysts in the APB study repeatedly analyzed the 31 images to develop an ongoing quality control assessment of precision compared to the Oregon assessment.

4.4 DISCUSSION

4.4.1   LAI Accuracy Assessment

Chen (1996) provided an estimate of errors in optical measurements of forest LAI using combined TRAC and LiCOR 2000 PCA instruments. We assumed that the PCA was equivalent to digital hemispherical photography for this discussion. Chen states that, based on error analysis, carefully executed optical measurements can provide LAI accuracies of close to or better than 80% compared to destructive sampling. The approximate errors accumulated as follows: PCA measurements (3 to 5%); estimate of needle-to-shoot area ratio (γ) (5 to 10%); estimate of foliage element clumping index (3 to 10%); estimate of woody-to-total area ratio (5 to 12%). These factors sum to an approximate total error of 15 to 40% in ground-based optical instrument estimates of LAI.

Chen (1996) also reports that the highest accuracy (~ 85%) (relative to destructive sampling) "can be achieved by carefully operating the PCA and TRAC, improving the shoot sampling strategy and the measurement of woody-to-total area ratio." A crucial issue for this analysis was to better understand the robustness of published values of needle-to-shoot area ratio (γ) and woody-to-total area ratio (α), because direct sampling of these quantities was logistically infeasible in this research effort. Published values have been used in this analysis (Leblanc et al., 2002).

4.4.2   Hemispherical Photography

Figure 4.10 presents a chronosequence of hemispherical photographic images taken at the midpoint (50 m) of the C transect at the Hertford site at five different dates in 2002. The images were the registered black-and-white bitmap images produced by GLA. The date and LAI Ring 5 values were displayed to the right of each image. LAI Ring 5 represented a 0° to 75° field of view. In the March 5, 2002, image, near-lens understory foliage was observed in the lower-left portion. However, in subsequent images, the large-leafed obstruction was absent. The reason for the disappearance of this understory image component was unclear. The tripod height may have been adjusted to place the camera above the near-lens foliar obstruction, or perhaps field-crew effects may have resulted in the disappearance of the obstruction. The presence of the near-lens foliage in the March 5 image may account for the somewhat elevated LAI value before leaf-out.

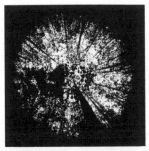

Hertford, VA - TRANSECT C-50

05 March 2002

LAI Ring 5 = 1.6

Hertford, VA - TRANSECT C-50

05 April 2002

LAI Ring 5 = 1.7

Hertford, VA - TRANSECT C-50

05 June 2002

LAI Ring 5 = 2.29

Hertford, VA - TRANSECT C-50

05 July 2002

LAI Ring 5 = 2.13

Hertford, VA - TRANSECT C-50

05 August 2002

LAI Ring 5 = 1.88

Figure 4.10 Chronosequence of hemispherical photographs taken at the Hertford site along transect C and the 50-m midpoint. Dates and LAI Ring 5 values are shown to the right of each image.

The orientation of the camera can be assessed by noting the position of the large tree bole that originates from the five-o'clock position in the image. The April 9, 2002, image places the bole closer to the 4:30 position. However, as mentioned previously, camera orientation does not affect whole-image calculations of LAI or canopy openness. Orientation was important only if it became necessary to match TRAC data with a particular sector of the hemispherical photograph.

L_e values derived by hemispherical photography increased over the course of phenological development at the Hertford site. A decrease in L_e from 2.13 to 1.88 was observed between the July 25 and August 5, 2002, images. The decrease may have partly been a result of the understory removal operation that occurred between July 25 and 30, 2002. However, decreases of this magnitude were observed at other APB sites in mid to late summer, when no understory canopy removal was performed. The Hertford site was primarily coniferous forest. Needle loss due to the extreme drought conditions experienced in the APB study area may partially account for the observed decrease in L_e.

4.4.3   Satellite Remote Sensing Issues

The MODIS LAI product was produced at 1-km^2 spatial resolution. Inherent in this product were a number of spatial factors that may contribute to uncertainty in the final accuracy of this analysis. MODIS pixels were nominally 1 km^2 at nadir but expanded considerably as the scan moved off nadir toward the edges of the 2330-km-wide swath. As a result, off-nadir pixels sampled a larger area on the ground than near-nadir pixels. The compositing scheme partially compensated for this by preferentially selecting pixels closer to nadir. Mixed pixels contained more than one LC type. In the APB study region, the landscape exhibits varying degrees of fragmentation, producing a mosaic of parcels on the ground. Within a 1-km^2 block, agricultural, urbanized, and forested LC types may be mixed to such a degree that assigning a single LAI value is questionable. There were also angular effects to consider. The NDVI and LAI products were adjusted for the bidirectional reflectance distribution function (BRDF; MODIS product MOD43). Still, angular effects produced by variable viewing geometry may have degraded the accuracy or interpretability of the results.

An important issue was that of spatial scaling from *in situ* reference data measurements (m^2) to MODIS products (1 km^2). ETM$^+$ data provided the link between *in situ* measurements and MODIS measurements. Quadrants correspond to a ground region of a 3×3 ETM$^+$ pixel window (90×90 m), and the subplots correspond to a region of approximately 2×2 pixels (60×60 m). The Landsat data were precision-registered to ground coordinates using ground control points, providing an accuracy between 0.5 and 1 pixel. Once the *in situ* data and Landsat image were coregistered, an ETM$^+$ LAI map was generated by establishing a regression relationship between *in situ* LAI and Landsat NDVI. Then, the Landsat LAI map could be generalized from a 30-m resolution to a 1-km^2 resolution. Overall accuracy was influenced by accuracy of coregistered data sets, interpolation methods used to expand *in situ* measurements to ETM$^+$ NDVI maps, and the spatial coarsening approach applied to scale the ETM$^+$ imagery to the MODIS scale of 1-km^2 pixels.

4.5 SUMMARY

Research efforts at the U.S. Environmental Protection Agency's National Exposure Research Laboratory and National Center for Environmental Assessment include development of remote sensing methodologies for detection and identification of landscape change. This chapter describes an approach and techniques for estimating forest LAI for validation of the MODIS LAI product, in the field using ground-based optical instruments. Six permanent field validation sites were established in the Albemarle-Pamlico Basin of North Carolina and Virginia for multitemporal measurements of forest canopy and biometric properties that affect MODIS NDVI and LAI products. LAI field measurements were made using hemispherical photography and TRAC sunfleck profiling in the landscape context of vegetation associations and physiography and in the temporal context of the annual phenological cycle. Results of these field validation efforts will contribute to a greater understanding of phenological dynamics evident in NDVI time series and will provide valuable data for the validation of the MODIS LAI product.

ACKNOWLEDGMENTS

The authors express their sincere appreciation to Mark Murphy, Chris Murray and Maria Maschauer for their assistance in the field. We thank Conghe Song for sharing of field instruments, Malcolm Wilkins for research support, Paul Ringold for providing hemispherical photographic images for use in the quality assurance and quality control aspects of the study, and Ross Lunetta, David Holland, and Joe Knight for assistance in study design. International Paper Corporation, Westvaco Corporation, the states of Virginia and North Carolina, Duke University, and North Carolina State University provided access to sampling sites. We also thank three anonymous reviewers for helpful comments on this manuscript. The U.S. Environmental Protection Agency funded and partially conducted the research described in this chapter. It has been subject to the Agency's programmatic review and has been approved for publication. Mention of any trade names or commercial products does not constitute endorsement or recommendation for use.

REFERENCES

Avery, T.E. and H.E. Burkhart, *Forest Measurements*, 3rd ed., McGraw-Hill, New York, 1983.

Becker, M.L., R.G. Congalton, R. Budd, and A. Fried, A GLOBE collaboration to develop land cover data collection and analysis protocols, *J. Sci. Ed. Tech.*, 7, 85–96, 2002.

Burrows, S.N., S.T. Gower, M.K. Clayton, D.S. MacKay, D.E. Ahl, J.M. Norman, and G. Diak, Application of geostatistics to characterize leaf area index (LAI) from flux tower to landscape scales using a cyclic sampling design, *Ecosystems*, 5, 667–679, 2002.

Burton, A.J., K.S. Pregitzer, and D.D. Reed, Leaf area and foliar biomass relationships in northern hardwood forests located along an 800-km acid deposition gradient, *For. Sci.*, 37, 1041–1059, 1991.

Canham, C.D., A.C. Finzi, S.W. Pascala, and D.H. Burbank, Causes and consequences of resource heterogeneity in forests: interspecific variation in light transmission by canopy trees, *Can. J. For. Res.*, 24, 337–349, 1994.

Carlson, T.N. and D.A. Ripley, On the relation between NDVI, fractional vegetation cover, and leaf area index, *Remote Sens. Environ.*, 62, 241–252, 1997.

Chason, J.W., A comparison of direct and indirect methods for estimating forest canopy leaf area, *Agric. For. Meteorol.*, 57, 107–128, 1991.

Chen, J.M., Optically-based methods for measuring seasonal variation of leaf area index in boreal conifer stands, *Agric. For. Meteorol.*, 80, 135–163, 1996.

Chen, J.M. and T.A. Black, Defining leaf area index for non-flat leaves, *Plant Cell Environ.*, 15, 421–429, 1992.

Chen, J.M. and T.A. Black, Measuring leaf area index of plant canopies with branch architecture, *Agric. For. Meteorol.*, 57, 1–12, 1991.

Chen, J.M., T.A. Black, and R.S. Adams, Evaluation of hemispherical photography for determining plant area index and geometry of a forest stand, *Agric. For. Meteorol.*, 56, 129–143, 1991.

Chen, J.M. and J. Cihlar, Plant canopy gap-size analysis theory for improving optical measurements of leaf-area index, *Appl. Optics*, 34, 6211–6222, 1995.

Chen, J.M., P.M. Rich, S.T. Gower, J.M. Norman, and S. Plummer, Leaf area index of boreal forests: theory, techniques, and measurements, *J. Geophys. Res.*, 102(D24), 29429–29443, 1997.

Deblonde, G., M. Penner, and A. Royer, Measuring leaf area index with the LI-COR LAI-2000 in pine stands, *Ecology*, 75, 1507–1511, 1994.

Delta-T Devices, Ltd., Hemiview User Manual (2.1), Cambridge, U.K., 1998.

Eidenshink, J.C. and R.H. Haas, Analyzing vegetation dynamics of land systems with satellite data, *Geocarto Int.*, 7, 53–61, 1992.

Fassnacht, K.S. and S.T. Gower, Interrelationships among the edaphic and stand characteristics, leaf area index, and above-ground net primary production of upland forest ecosystems in north central Wisconsin, *Can. J. For. Res.*, 27, 1058–1067, 1997.

Fassnacht, K.S., S.T. Gower, J.M. Norman, and R.E. McMurtrie, A comparison of optical and direct methods for estimating foliage surface area index in forests, *Agric. For. Meteorol.*, 71, 183–207, 1994.

Frazer, G.W., C.D. Canham, and K.P. Lertzman, Gap Light Analyzer (GLA): Imaging software to extract canopy structure and gap light transmission indices from true-color fisheye photographs, users manual and program documentation (Version 2), Simon Frazer University, Burnaby, British Columbia, Canada, Simon Frazer University and the Institute of Ecosystem Studies, 1999.

Frazer, G.W., R.A. Fournier, J.A.Trofymow, and R.J. Hall, A comparison of digital and film fisheye photography for analysis of forest canopy structure and gap light transmission, *Agric. For. Meteorol.*, 109, 249–263, 2001.

Geron, C.D., A.B. Guenther, and T.E. Pierce, An improved model for estimating emissions of volatile organic compounds from forests in the eastern United States, *J. Geophys. Res.*, 99, 12773–12791, 1994.

Gholz, H.L., Environmental limits on aboveground net primary production, leaf area and biomass in vegetation zones of the Pacific Northwest, *Ecology*, 53, 469–481, 1982.

Gower, S.T., K.A. Vogt, and C.C. Grier, Carbon dynamics of Rocky Mountain Douglas-fir: influence of water and nutrient availability, *Ecological Monogr.*, 62, 43–65, 1992.

Huete, A., Justice, C.O., and W. van Leeuwen, MODIS Vegetation Index (MOD13) Algorithm Theoretical Basis Document (Version 2), http://eospso.gsfc.nasa.gov/atbd/modistables.html, 1–142, 1996.

Jenkins, J.C., S.V. Kicklighter, J.D. Ollinger, J.D. Aber, and J.M. Melillo, Sources of variability in net primary production predictions at a regional scale: a comparison using PnET-II and TEM 4.0 in Northeastern US forests, *Ecosystems*, 2, 555–570, 1999.

Johnsen, K.H., D. Wear, R. Oren, R.O. Teskey, F. Sanchez, R. Will, J. Butnor, D. Markewitz, D. Richter, T. Rials, H.L. Allen, J. Seiler, D. Ellsworth, C. Maier, G. Katul, and P.M. Dougherty, Meeting global policy commitments: carbon sequestration and southern pine forests, *J. For.*, 99, 14–21, 2001.

Justice, C.O., E. Vermote, J.R.G. Townshend, R. Defries, D.P. Roy, D.K. Hall, V.V. Salomonson, J.L. Privette, G. Riggs, A. Strahler, W. Lucht, R.B. Myneni, Y. Knyazikhin, S.W. Running, R.R. Nemani, Z. Wan, A.R. Huete, W. van Leeuwen, R.E. Wolfe, L. Giglio, J.P. Muller, P. Lewis, and M.J. Barnsely, The Moderate Resolution Imaging Spectroradiometer (MODIS): land remote sensing for global change research, *IEEE Trans. Geosci. Rem. Sens.*, 36, 1228–1249, 1998.

Knyazikhin, Y., J. Glassy, J.L. Privette, Y. Tian, A. Lotsch, Y. Zhang, Y. Wang, J.T. Morisette, P. Votava, R.B. Myneni, R.R. Nemani, and S.W. Running, MODIS Leaf Area Index (LAI) and Fraction of Photosynthetically Active Radiation Absorbed by Vegetation (FPAR) Product (MOD15) Algorithm Theoretical Basis Document, http://eospso.gsfc.nasa, 1999.

Kucharik, C.J., J. M. Norman, and L.M. Murdock, Characterizing canopy nonrandomness with a multiband vegetation imager (MVI), *J. Geophys. Res.*, 102, 29455–29473, 1997.

Leblanc, S.G., Correction to the plant canopy gap size analysis theory used by the Tracing Radiation and Architecture of Canopies (TRAC) instrument, *Appl. Optics*, 41 (36), 7667–7670, 2002.

Leblanc, S.G., J.M. Chen, and M. Kwong, Tracing Radiation and Architecture of Canopies: TRAC Manual (Version 2.1.3), 3rd Wave Engineering, Nepean, Ontario, Canada, 2002.

Lunetta, R.S., J.S. Iiames, J. Knight, R.G. Congalton, and T.H. Mace, An assessment of reference data variability using a "Virtual Field Reference Database," *Photogram. Eng. Remote Sens.*, 67, 707–715, 2001.

Masuoka, E., A. Fleig, R.E. Wolfe, and F. Patt, Key characteristics of MODIS data products, *IEEE Trans. Geosci. Remote Sens.*, 36, 1313–1323, 1998.

Neumann, H.H., G. Den Hartog, and R.H. Shaw, Leaf area measurements based on hemispheric photographs and leaf-litter collection in a deciduous forest during autumn leaf-fall, *Agric. For. Meteorol.*, 45, 325–345, 1989.

Privette, J.L., R.B. Myneni, Y. Knyazikhin, M. Mukelabai, G. Rogerts, Y. Tian, Y. Wang, and S.G. Leblanc, Early spatial and temporal validation of MODIS LAI product in the Southern Africa Kalahari, *Remote Sens. Environ.*, 83, 232–243, 2002.

Rich, P.M., Characterizing plant canopies with hemispherical photographs, *Rem. Sens. Rev.*, 5, 13–29. 1990.

Running, S.W., D.L. Peterson, M.A. Spanner, and K.A. Tewber, Remote sensing of coniferous forests leaf-area, *Ecology*, 67, 273–276, 1986.

Runyon, J., R.H. Waring, S.N. Goward, and J.M. Welles, Environmental limits on net primary production and light use efficiency across the Oregon Transect, *Ecol. Appl.*, 4, 226–237, 1994.

SAS (Statistical Analysis Software), *SAS/Graph Guide for Personal Computers*, Version 6, SAS Institute, Inc., Cary, NC, 1987.

Tian, Y., C.E. Woodcock, Y. Wang, J.L. Privette, N.L. Shabanov, Y. Zhang, W. Buermann, B. Veikkanen, T. Hame, Y. Knyazikhin, and R.B. Myneni, Multiscale analysis and MODIS LAI product. II. sampling strategy, *Remote Sens. Environ.*, 83, 431–441, 2002.

Verchot, L.V., E.C. Frankling, and J.W. Gilliam, Effects of agricultural runoff dispersion on nitrate reduction in forested filter zone soils, *Soil Sci. Soc. Am. J.*, 62, 1719–1724, 1998.

Vogelmann, J. E., T. Sohl, S.M. Howard, and D.M. Shaw, Regional land cover characterization using Landsat Thematic Mapper data and ancillary data sources, *Environ. Monitor. Assess.*, 51, 415–428, 1998.

Wulder, M.A., F.L. Ellsworth, S.E. Franklin, and M.B. Lavigne, Aerial image texture information in the estimation of northern deciduous and mixed wood forest Leaf Area Index (LAI), *Remote Sens. Environ.*, 64, 64–76, 1998.

CHAPTER 5

Light Attenuation Profiling as an Indicator of Structural Changes in Coastal Marshes

Elijah Ramsey III, Gene Nelson, Frank Baarnes, and Ruth Spell

CONTENTS

5.1 INTRODUCTION

To best respond to natural and human-induced stresses, resource managers and researchers require remote sensing techniques that can map the biophysical characteristics of natural resources on regional and local scales. The implementation of advanced measurement techniques would provide significant improvements in the quantity, quality, and timeliness of biophysical data useful in understanding the sensitivity of vegetation communities to external influences. In turn, this biophysical data would provide resource planners with a rational decision-making system for resource allocation and response action development planning.

Remotely sensed imagery can be analyzed to provide an accurate, instantaneous, synoptic view of the spatial characteristics of vegetation environments (Ustin et al., 1991; Wickland, 1991). By simultaneously recording reflectances in the visible to short-wave region of the electromagnetic spectrum, the canopy reflectance associated with these spatial characteristics may be used to provide information on the biophysical characteristics of vegetation (Goudriaan, 1977). To predict the vegetation response to external stresses, it is essential to identify biophysical characteristics observable by remote sensing techniques that have well-defined connections to vegetative community type and condition.

In complex vegetation communities, canopy structure and leaf spectral properties are biophysical characteristics that can vary in response to changes in vegetation type, environmental conditions, and vegetation health. These changes can modify the spectrum of light reflected from the canopy and thus directly influence the remotely sensed signal. Transformed into reflectance, variations in the image are directly related to changes in the canopy properties broadly defined by the leaf composition, canopy structure, and background reflectance. Direct links, however, cannot be inferred unless vegetation type covaries directly and uniquely with these canopy parameters, or when one canopy property dominates the canopy reflectance (e.g., leaf reflectance). Historically, limited ground-based observations circumvented the need for directly incorporating variation in canopy properties into the remote sensing classification by defining reflectance ranges (e.g., class ranges) that incorporate within-type canopy variability and acceptable between-vegetation-type classification errors. Currently, the trend is to transform the temporal patterns revealed in the remote sensing data into quantitative rate determinations to support qualitative judgments of external effects on these resources (Lulla and Mausel, 1983). As we strive to extract more detailed and accurate information about vegetation class variability, a greater understanding is needed of how each canopy property (e.g., canopy structure) influences the canopy reflectance portion of the remotely sensed signal.

Leaf spectral properties have been directly related to vegetation type and stress and are general indicators of the leaf chlorophyll, water content, and leaf biomass. Numerous studies have related the canopy structure variable of leaf area index (LAI) to vegetation type, health, and phenology (Goudriaan, 1977). In essence, to map vegetation type, and especially to monitor status, it is necessary to relate, both individually and in aggregate, changes in leaf spectral properties and structural and background parameters to changes in the canopy reflectance. In the pursuit of extracting more detailed and accurate information about vegetation type and status from remote sensing data, our goal is to provide an accurate assessment of canopy structure that will not covary with leaf spectral and background properties with respect to location or time. As part of this goal, the canopy structure indicator must be ultimately linkable to the remote sensing signal in complex wetland and adjacent upland forest environments. Our challenge is to provide this information based upon routine measurements that are cost-effective and easily implemented into operational resource management and verified and calibrated with current operational ground-based measurements (Teuber, 1990; Nielsen and Werle, 1993).

This chapter will examine light attenuation profiling as an indicator of changes in marsh canopy structures. Reported here are techniques that were tested and implemented to gain a useful measure of canopy light attenuation over space and time. Within the constraints of the data collected, the consistency, reliability, and comparability of the collected light attenuation data are related to the (1) area sampling frequency (horizontal spacing between profile samples), (2) canopy profile (vertical) sampling frequencies, (3) exclusion of atypical canopy structures, and (4) collections at different sun elevations. In addition, we present some relationships observed between and within coastal wetland types and changes in the canopy structural properties. These relationships are presented to indicate the spatial and temporal stability of these biophysical indicators as related to mapping and monitoring with remote sensing imaging.

5.1.1 Marsh Canopy Descriptions

Measurements of canopy light attenuation and canopy reflectance spectra were collected at 20 marsh sites (30 × 30 m) in coastal Louisiana and at 15 marsh sites in the Big Bend area of coastal

Florida (Ramsey et al., 1992a,b, 1993). To provide a description of marsh characteristics, a few data subsets were selected based on the presence of marsh grasses that dominate three of the gulf coast wetland zones (Chabreck, 1970): *Juncus roemerianus* (Juncus R.) and *Spartina alterniflora* (Spartina A.) for saline marsh, *Spartina patens* (Spartina P.) for intermediate (brackish) marsh, and *Panicum hemitomon* (Panicum H.) for fresh marsh (Chabreck, 1970). Juncus R. dominates the landscape and makes up the majority of biomass in marshes of the northeast gulf coast and Spartina A. dominates the north-central gulf coast marshes (Stout, 1984). In these marshes, except for sites recovering from recent burns, canopies usually contain a high proportion of dead canopy material (Hopkinson et al., 1978; Ramsey et al., 1999).

After reaching maturity, turnover rates of both live and dead biomass can remain nearly constant, showing no clear seasonal pattern. Although mostly vertical, Juncus R. and Spartina A. (relatively less vertical and more leafy) canopy structures vary depending on local conditions (e.g., flushing strength), and dominant leaf orientation can change from top to bottom. Spartina P. and Panicum H. marshes dominate the interior marshes of Louisiana. Generally, Spartina P. canopies are hummocky with vertical shoots rising above a layer of thick and logged dead material. As in Juncus R. and Spartina A. canopies, Spartina P. canopies seem to have a low turnover with little seasonal pattern in live and dead composition. Panicum H. canopies exhibit yearly turnover. Beginning with nearly vertical shoots in the late winter to early spring, the canopy gains height and increasingly adds mixed orientations until maturity in the late spring to summer, then senesces in winter.

5.2 METHODS

Light measurements were collected along transects centered at flag markers, as were all measurements describing the canopy characteristics. A 30- × 30-m area was used to encompass the spatial resolution of Landsat Thematic Mapper (TM) and similar Earth resource sensors. Additional recordings and observations collected at each site included upwelling radiance from a helicopter platform, canopy species type, percentage of cover, and height; photography; and estimates of live and dead biomass percentages.

5.2.1 Field Collection Methods

Canopy light attenuation measurements were acquired using a Decagon Sunfleck Ceptometer (Decagon Devices, 1991). The ceptometer measures both photosynthetically active radiation (PAR) (400 to 700 nm) and the canopy gap fraction (sunflecks). Canopy light attenuation curves were derived from PAR measurements. The ceptometer probe has 80 light sensors (calibrated to absolute units) placed at equal intervals along an 80-cm probe covered with a diffuse plate. The narrow probe (approximately 1.3 × 1.3 cm) is constructed with a hard and pointed plastic tip so that it can be inserted horizontally with minimal disruption of the marsh canopy. After inserting the probe into the canopy and obtaining a horizontal probe orientation relative to gravity (bubble level), the 80 sensors are scanned and an average light intensity value for the probe is calculated, displayed, and recorded (Decagon Devices, 1991). At each site, measurements for estimating PAR canopy reflectance and the fraction of direct beam PAR (1 – skylight/direct sun irradiance) were collected. A correction for PAR canopy reflectance was not included in the calculations. Disregarding this correction, in general, results in less than a 5% error in the intercepted radiation (Decagon Devices, 1991). The direct beam fraction was used to estimate the leaf area and angle distributions. Normally, measurements were collected when clouds did not obstruct or influence (intensified by cloud reflection) the downwelling sunlight; sky conditions were documented.

The depiction in Figure 5.1 presents our standard method of depicting light falloff with depth in the canopy. Each point on the graph reflects the mean of all light measurements collected throughout the site at the associated height above ground level. Error bars showing plus and minus

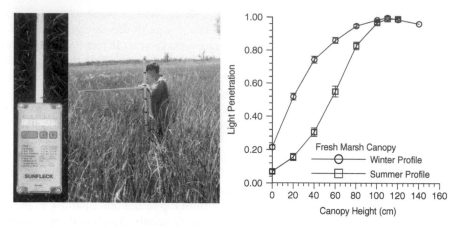

Figure 5.1   Left: A Ceptometer showing the sunflect (22.5) and PAR (299) readings. Middle: Ruth Spell is shown measuring the above PAR intensity after collecting readings with canopy depth at one point along one of the two transect directions (Figure 5.2). The above canopy PAR intensity (shown) is used to normalize PAR measurements at each canopy depth. Right: The resulting summer and winter profiles of a fresh marsh canopy site showing the light penetration (PAR at each depth/above-canopy PAR) with depth averaged over the 22 measurement points along the two transect directions. The standard errors of the 22 measurements at each canopy depth also are shown as horizontal bars attached to each symbol.

one standard error (65% confidence interval) depict the variance about the mean at each canopy height. The light attenuation curve represents the percentage of above-canopy PAR sunlight (abscissa) reaching varying depths in the canopy (ordinate) throughout the site area. The curve typifies light attenuation in an undisturbed and fully formed Panicum H. marsh.

5.2.1.1   Area Frequency Sampling

The three considerations combined to set the distance between profile collections included: (1) the estimated canopy spatial variability, (2) the decision to sample 30-m transects in two cardinal directions, and (3) the necessity to restrict the site occupation time. Early analyses of the collected light attenuation data suggested profiles collected every 3 m and averaged over the 30-m transects provided the best compromise to all sampling considerations (Figure 5.2).

5.2.1.2   Vertical Frequency Sampling

Two vertical sampling distances were compared to assess the accurate and reliable portrayal of canopy light attenuation. The earliest measurements were collected relative to the canopy height, not at a constant height above the ground level. In a few of these early site occupations the relative top, middle, and bottom measurements collected every 3 m along the 30-m transects were supplemented with light attenuation profiles (every 20 cm) collected at three to four transect locations.

5.2.1.3   Atypical Canopy Structures

At each profile location, sky condition and canopy structure were recorded. Indicator flags were inserted into the database to indicate whether (1) sunny or (2) cloudy sky existed and whether the profile location was (1) undisturbed, (2) a partial gap or hole, or (3) completely lodged. These flags could be used during the data processing to exclude or include any combination of sky and canopy conditions. In almost all cases only sunny sky conditions were processed. Similarly, undisturbed canopy was most often solely processed for generation of PAR light attenuation profiles typifying each site. In relation to remotely sensed data, however, all canopy conditions will be incorporated

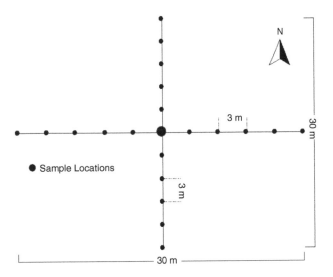

Figure 5.2 Light attenuation profile locations every 3 m along the 30-m east to west and south to north transects.

in the reflectance returned to the sensor. Ability to include or exclude atypical structures is expected to enhance remote sensing reflectance and canopy structure comparisons.

5.2.1.4 Changing Sun Zenith

Light attenuation measurements collected at different times correspond to different sun elevations or zeniths. In order to relate PAR recordings at different sun zeniths, we used the following relationships equating the beam transmittance coefficient to the product of leaf area and canopy extinction coefficient (Decagon Devices, 1991):

$$\ln\tau_{\Theta1}/\ln\tau_{\Theta2} = K_{\Theta1}/K_{\Theta2} = \rho \tag{5.1}$$

where $\ln\tau_{\Theta1}$ = the log of the canopy transmission coefficient (τ) at sun zenith 1 (Θ_1), $\ln\tau_{\Theta2}$ = the log of τ at Θ_2, $K_{\Theta1}$ = the extinction coefficient at Θ_1, $K_{\Theta2}$ = at Θ_2, and ρ = the Θ correction factor. Although this relationship is based on the penetration probability without interference and is directly relevant to sunfleck measurements, we tested the application of the sun zenith normalization to PAR measurements. The canopy extinction coefficient expression (K) taken from Decagon Devices (1991) was presented by Campbell (1986) as:

$$K = (x^2 + \tan^2\Theta)^{1/2}/[x + 1.774(x + 1.182)^{-0.733}] \tag{5.2}$$

where x equates to the ratio of area projected by an average canopy element on a horizontal to vertical plane. An x of 1000 defines a horizontal, an x of 1 a spherical, and an x of 0 a vertical leaf distribution, and the Θ represents the sun zenith angle.

Accounting for no change in leaf area and x between light attenuation measurements and after simplification, a correction factor for off-nadir sun angles is constructed as follows:

$$\rho = (x^2 + \tan^2\Theta_1)^{1/2}/(x^2 + \tan^2\Theta_2)^{1/2} = K_{\Theta1}/K_{\Theta2}. \tag{5.3}$$

where the sun zenith at the time of measurement is Θ_2. Assuming x is 1.0 (spherical) and choosing a standard zenith angle of $\Theta_1 = 0$ (sun directly overhead), then:

$$\rho = (1.0/(1.0 + \tan^2\Theta_2)^{1/2} \qquad (5.4)$$

where normalization of PAR measurements to a sun nadir zenith (= 0.0) was estimated to be:

$$PAR(\Theta_1) = PAR(\Theta_2)^\rho \qquad (5.5)$$

5.3 RESULTS

5.3.1 Vertical Frequency Sampling

Two examples were selected to illustrate noticeable differences between samples taken only at the top, bottom, and middle canopy positions and at every 20 cm above the bottom (Figure 5.3). These early measurements were collected during July and August when the canopies reached full growth. In both the Spartina P. (more hummocky and logged) and Panicum H. (more vertical and leafy) marsh sites, similar differences were revealed between curves associated with the higher and lower (relative) frequency depth sampling. Light attenuation was overpredicted nearer the top of the canopy and underpredicted in the lower canopy. Even though measurement techniques were further refined following these early collections to expressly test field sampling techniques, these and similar results laid the basis for the vertical sampling frequency used throughout field collections in all marsh types. The choice of vertical sampling frequency relied on what was necessary to obtain our primary purpose: to detect and monitor canopy structure differences between and within wetland types that might influence variability in the remote sensing image data. An additional consideration was our goal to use the data we collected to estimate what influence these structural differences have on the canopy reflectance and whether these differences could be detected at some level with remote sensing data. We felt light penetration collections limited to a few positions in the canopy profile would severely jeopardize our ability to fulfill this purpose and to reach our goal.

Following these initial tests, our standard collection technique was to profile light intensities from the ground level to above the canopy in 20-cm increments at 3-m intervals along each transect. To ensure proper measurement height at each profile, a pole marked in 20-cm increments was driven into the ground until the zero mark was at ground-surface level (flood or nonflood). In Spartina P. canopies the pole was placed between grass clumps. Profile measurements were collected perpendicular to the transect direction and toward the hemisphere containing the sun. At each site occupation, either 11 or 22 (most often) PAR recordings (one or two transects) were taken at each profile height. The above-canopy PAR measured at the associated profile location normalized each recorded PAR.

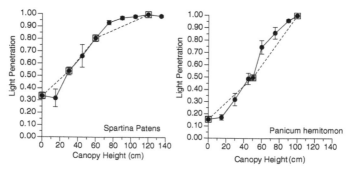

Figure 5.3 Aggregate site profiles (*Spartina patens* and *Panicum hemitomon*) associated with PAR intensity collections at top, middle, and bottom canopy depths (shown as □ with a dashed line) and at every 20 cm (shown as ● with a solid line).

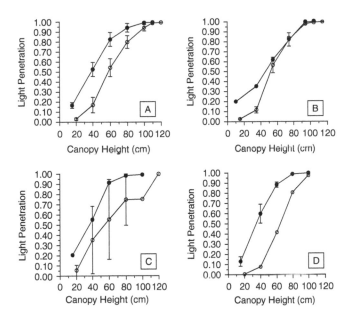

Figure 5.4 Light attenuation profiles ([●] 9 and [○] 10 September) associated with a *Panicum hemitomon* marsh representing the aggregate of (A) all profiles collected every 3 m along the 30-m transects, (B) only profiles associated with undisturbed canopy locations, (C) only profiles associated with partial canopy gaps, and (D) only profiles associated with severely lodged canopy locations.

5.3.2 Atypical Canopy Structures

Examples of disturbed or logged conditions at one or more locations within a site were found in all marsh types. Relative to the number of occurrences, however, Juncus R. marshes contained the fewest of these and Panicum H. and Spartina P. the most. Juncus R. canopies were most often affected by wrack deposits or the subsequent marsh dieback or by fire and the subsequent recovery. Animal herbivory and fire often affected Panicum H. marshes, but water with higher salinity deposited by a storm surge seemed to have a lingering impact evident in the poststorm collections. In the 2 years of data collections, a major storm and a fire affected Spartina A. sites. To a lesser extent, Spartina P. sites were affected by storms and fire. The typical hummocky nature of this marsh limited the usefulness of the logged indicator.

The first example contains light attenuation curves generated from two occupations 1 day apart of a Panicum H. marsh site that was severely affected by animal activity following the occupation (Figures 5.4A and B). Other than the magnitude of variance depicted by the error bars, little evidence was present in the affected curves (Figures 5.4C and D) indicating the widespread abnormal canopy structure. In fact, neglecting that only 1 day elapsed between collections, aggregating all profile locations results in fairly reasonable profiles (Figure 5.4A). Excluding all affected profiles left few observations in the undisturbed sample set; however, the aggregate of these remaining profiles showed a more consistent depiction of canopy structure of little or no change in canopy structure (Figure 5.4B).

A second example shows curves depicting site occupations in a lightly affected Panicum H. marsh chronologically from October (before full senescence) to February (after senescence and removal of most dead material), September (substantially before the initiation of senescence), and December (after full senescence but before total dead material removal). Although differences between the undisturbed (Figure 5.5B) and aggregated (Figure 5.5A) sequences were not dramatic and only two of the occupations contained severely logged locations, inclusion of locations with partial gap (Figures 5C and D) reduces the clarity of the trend consistent with expected seasonal

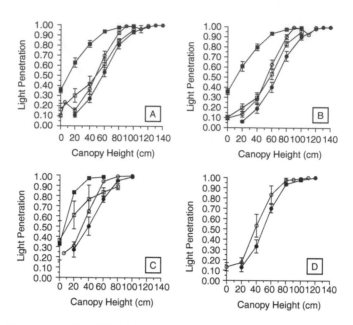

Figure 5.5   Light attenuation profiles ([□] 16 October, [■] 7 February, 9 [●] September, and [○] 11 December) associated with a *Panicum hemitomon* marsh representing the aggregate of (A) all profiles, (B) only undisturbed canopy locations, (C) partial canopy gaps, and (D) only severely lodged canopy locations.

changes. Similar to the second example, curves are shown depicting site occupations in a lightly affected Juncus R. marsh chronologically in April, September, November, and March (Figures 5.6A, B, and C). Although little canopy structural change is expected in these marshes, curves including only undisturbed profiles (Figure 5.6B) indicate a slight tendency for increased canopy light attenuation in the spring vs. late summer and winter seasons. A final example shows the effect

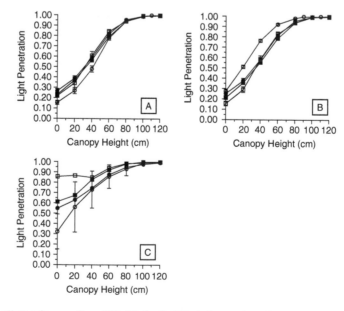

Figure 5.6   Light attenuation profiles ([□] 26 April, [■] 4 September, [●] 3 November, and [○] 9 March) associated with a *Spartina alterniflora* marsh representing the aggregate of (A) all profiles, (B) only undisturbed canopy locations, and (C) partial canopy gaps.

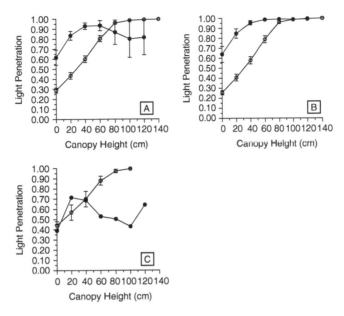

Figure 5.7   Light attenuation profiles ([●] 30 March and [○] 9 July) associated with a *Spartina patens* marsh representing the aggregate of (A) all profiles, (B) only undisturbed canopy locations, and (C) partial canopy gaps.

of herbivory on a Spartina P. marsh closely preceding the March occupation and regrowth toward recovery by July (Figures 5.7A, B, and C). Inclusion of locations with partial canopy (Figure 5.7C) fully distorted the consistency shown in the undisturbed profiles (Figure 5.7B).

5.3.3   Changing Sun Zenith

Normalization of canopy light penetration measurements to a nadir sun zenith was more successful in more vertical canopies such as Juncus R. and less effective in more lodged and horizontal canopies such as Spartina P. Canopy light penetration recordings collected at one Juncus R. site at four different times and sun zeniths show a number of results of applying the sun zenith correction factor (Figures 5.8A and B). First, as sun zenith increased the rate of PAR fall-off with canopy depth increased. After application of the sun zenith correction factor, the PAR fall-off was greatly decreased, and conversely the magnitude of PAR reaching lower within the canopy was vastly increased. Second, after correction, light attenuation profiles associated with the first three

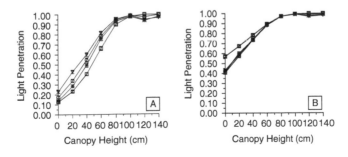

Figure 5.8   Aggregate light attenuation profiles of a *Juncus roemerians* marsh associated with PAR collections at a sun zenith angles of [∇] 54°, [○] 59°, [✳] 66°, and [□] 75° (A) without normalization and (B) with normalization of PAR recordings to a nadir collection.

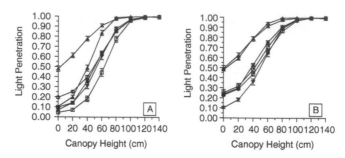

Figure 5.9   Aggregate light attenuation profiles of a *Juncus roemerians* marsh associated with PAR collections at different sun zenith angles the first year on [▲] 25 September, the second year on [△] 29 April and [□] 17 September, the third year on [✳] 13 January and [○] 24 June, and the final year on [▽] 22 September (1994) (A) without normalization and (B) with normalization of PAR recordings to a nadir collection.

occupation times completely aligned. The latest occupation indicated a limit to the correction. Sun zenith angles higher than about 60° overcorrected the rate of PAR fall-off through the canopy. This led to excluding PAR canopy penetration measurements to times when the sun zenith was at least 60° or higher.

A second example shows the application of the correction to PAR profile measurements during nearly 3 years of occupations (Figure 5.9A and Figure 5.9B). The Juncus R. site was recovering from a burn when the first occupation occurred in September. Occupations then followed chronologically in April, September, January, June, and, finally, in September, 2 years after the initial occupation. Inspection of the two plot series shows the consistency and accuracy resulting from the application of the sun zenith correction factor. In contrast to the uncorrected plot series, the corrected canopy PAR attenuation profiles became increasingly steep and reached lower levels of PAR with time-since-burn.

The third example shows a series of PAR attenuation profiles created from measurements collected at a Spartina A. site during 1 year and 9 months of site occupations beginning in February (Figures 5.10A and B). The corrected and uncorrected series are dramatically different. After February, the site occupations chronologically occurred in April, July, September, November, and March, and again in September in the following year. The July occupation took place about 1 month before a hurricane affected this site (August 26), and the first September occupation occurred about 10 days after the hurricane. The canopy light attenuation profile deepened each occupation before the hurricane, but consistent with the effect of the hurricane, the attenuation profile shallowed sharply immediately after the impact. PAR attenuation profiles changed little for a year after the dramatic change in September.

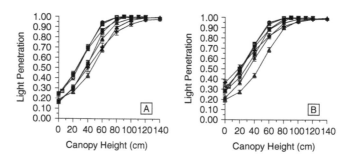

Figure 5.10   Aggregate light attenuation profiles of a *Spartina alterniflora* marsh associated with PAR collections at different sun zenith angles on [✛] 16 February, [▲] 26 April, [△] 10 July, [□] 4 September, and [✳] 3 November and the following year on [○] 9 March and [▽] 2 September (A) without normalization and (B) with normalization of PAR recordings to a nadir collection.

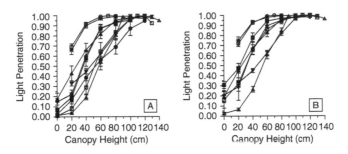

Figure 5.11 Aggregate light attenuation profiles of a *Panicum hemitomon* marsh associated with PAR collections at different sun zenith angles on [■] 7 February, [✦] 13 March, [▲] 21 April, [△] 09 October, and [□] 11 December and the following year on [✳] 22 January, [○] 22 April, and [▽] 6 July (A) without normalization and (B) with normalization of PAR recordings to a nadir collection.

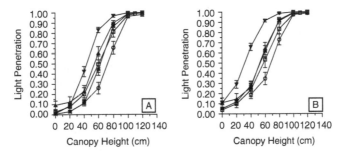

Figure 5.12 Aggregate light attenuation profiles of a *Spartina patens* marsh associated with PAR collections at different sun zenith angles on [△] 5 August, [○] 21 October, and [✳] 30 December and the following year on [▽] 21 July at a sun zenith of 49°, and [□] 21 July at 18° (A) without normalization and (B) with normalization of PAR recordings to a nadir collection.

PAR attenuation profiles over nearly a year and a half at a Panicum H. site highlight the effectiveness of the sun zenith factor correction and depict typical seasonal changes and canopy recovery following a burn (Figures 5.11A and B). Site occupations occurred chronologically beginning in February and subsequently occurred in March, April, September, December, and the following year in January, April, and July. Although differences between the corrected and uncorrected profiles were subtler than in the Juncus R. marshes, the sun zenith-corrected profile sequencing is a more convincing depiction of expected canopy structure changes over the selected time period. From the earliest occupation in February, PAR attenuation increased and the profiles deepened until October, after which the December profile shows the expected decrease in PAR attenuation. After the December occupation, the site was burned, as is confirmed in the February and subsequent April profiles. The final July profile shows the canopy recovering to the late summer and early fall profile.

A final series of profiles associated with a Spartina P. site illustrate the increased consistency in the sun zenith-corrected vs. -uncorrected profiles (Figures 5.12A and B). The final two profiles in these series again show the zenith angle limitation of the correction technique in more vertical canopies. Canopy PAR penetration measurements collected at very low sun zeniths are not comparable to those collected at sun zenith angles at least < 60°. The movement of the PAR attenuation profile from shallower to deeper with decreasing sun zenith angle in the Spartina P. canopy was similar to what occurred in the Juncus R. marsh canopy (Figure 5.9A and Figure 5.9B).

5.4 DISCUSSION

Light penetration field measurements were described and discussed in terms of their completeness, reliability or consistency and accuracy to characterize canopy structure. Marsh types discussed included *Spartina alterniflora* and *Juncus roemerianus* (saline marshes), *Spartina patens* (brackish marsh), and *Panicum hemitomon* (fresh marsh). Our primary purpose was to devise simple operational field collection methods and postanalysis procedures that could detect and monitor canopy structure differences between and within wetland types that might influence variability in the remote sensing image data. Our goal was to improve reliability and accuracy of current classifications based on remote sensor data, and as a consequence to extend the biophysical detail extractable with remotely sensed data.

Our first objective was to test the light penetration measurements for reliability, accuracy and comparability over time and space and within and between marsh types. A ceptometer device that measures PAR along an 80-cm probe was chosen for the light penetration measurements. In addition to the choice of measuring device, four variables related to field data collection and postdata analysis design were examined with respect to fulfilling data reliability and accuracy but also to maximizing the potential for operational use for remote sensing calibration and assessment of classification accuracy. These variables were the (1) horizontal (planar) and (2) vertical (canopy profile) spatial sampling frequencies, (3) description and possible exclusion of atypical canopy structures, and (4) normalization of measurements at different sun elevations.

At each site, we used 30-m transects in the north and south and east and west directions. Initially, sample locations were at the site center, the transect extremes, and midpoints; however, sample variability indicated higher sample frequency was needed. Within the early testing, sampling protocol along transects was changed to collecting light penetration measurements every 3 m. This 30- × 30-m site area and higher 3-m transect sampling frequency helped ensure more accurate depiction of the local variability and more reliable mean and variance measures and matched or encompassed the spatial resolution of most common resource remote sensing sensors. Similarly, after testing a relative canopy profile sampling of top, middle, and bottom, the vertical sampling frequency was standardized to light penetration measurements every 20 cm from the ground surface to above the canopy. At each profile location, the above-canopy light recordings were used to normalize lower canopy light recordings, transforming light absolute magnitudes to percentage penetrations. The relative profile sampling (top, middle, bottom) did not adequately replicate the canopy light attenuation profile compared to the 20-cm sampling frequency, especially in fully formed canopies. The 20-cm sampling interval was selected to ensure our goals to increase the extractable canopy detail and improve the predictability of canopy structure with remote sensing data.

To improve the reliability, accuracy, and detail of the canopy light attenuation data, at each profile location the state of the sky condition (sunny or cloudy) and canopy structure (undisturbed, partial gap, or completely lodged) was recorded. In this analysis, only sunny sky conditions were processed. Observed differences relative to including or excluding disturbed canopy profiles were mostly attributable to the level of canopy disturbance. In most cases, excluding the disturbed profiles at each site from the aggregate site light attenuation profile increased our ability to compare aggregate profiles taken at the same site during multiple occupations. If the aggregate profiles were more comparable within a site after exclusion, comparison between sites and marsh types also would improve with exclusion. The inclusion of all profiles, each designated with a flag as to the sky and canopy condition, allows us to view and analyze selected aspects (include or exclude) of the canopy variability and thus greatly enhances our ability to understand and relate remote sensing reflectance to canopy structure.

Although more noticeable in *Juncus roemerianus* and less in *Spartina patens* canopies, in all marshes as sun zenith increased the rate of light fall-off with canopy depth increased. To ensure comparability of aggregate site light attenuation profiles, canopy light penetration measurements were normalized to a nadir sun zenith. The successfulness of the normalization seemed to be associated with the preferred orientation of the marsh canopy. In the most vertical canopies such as *Juncus roemerianus*, the correction worked well up to a sun zenith around 60°. In highly lodged or horizontal canopies such as *Spartina patens*, the light penetration seemed less effective and restricted to sun zenith angles < 49°. Nontheless, normalized light attenuation profiles were more consistent with the expected changes in canopy structure. Even in this normally highly lodged canopy, the normalization improved the comparability and accuracy of the generated light attenuation profiles. In between these two extremes of vertical to lodged canopy orientations, improvement of *Spartina alterniflora* aggregate attenuation profiles was somewhat similar to that associated with the more vertical *Juncus roemerianus* canopy measurements. *Panicum hemitomon* improvement, however, seemed dependent on the seasonal stage of canopy development. In the early stage of regrowth and spring "green-up" the canopy is more vertical and therefore more conducive to normalization. As the canopy transforms and becomes a dense mixture of leaf orientations, the normalization tends to be less successful.

In all marshes, application of the sun zenith correction factor decreased the perceived light falloff with canopy depth and increased the recorded light intensity reaching lower within the canopy. After normalization the light attenuation profiles taken at different sun zeniths were more closely aligned with each other and with the expected progression of canopy structure. A limitation of the normalization seemed to be about a sun zenith of 60° in vertical canopies and less than 49° in more horizontal canopies; however, these normalizations used a spherical canopy orientation parameter ($x = 1$, $\rho = (x/(x + \tan^2\Theta_2)^{1/2})$. In most cases, preferred orientations deviate highly from spherical in *Spartina patens* and, depending on the season, *Panicum hemitomon* canopies. Inclusion of more appropriate values of x (0 = vertical, 1 = spherical, 1000 = horizontal) could improve the reliability and accuracy of light attenuation profiles in more horizontal canopies such as *Spartina patens*. Future analyses will examine inclusion of more appropriate orientation parameters.

It is difficult to relate canopy structure (as defined by light attenuation) to canopy reflectance without further analysis, but a few observations are possible. In combination, the light attenuation profiles show at least one major difference between marsh grass structures: the amount of vertical to lodged grass. The relative amounts remain: (1) relatively stable throughout the year as shown in the fairly vertical canopies of *Juncus roemerianus* and *Spartina alterniflora*; (2) relatively transitional, as in *Panicum hemitomon,* typifying a more vertical canopy in the winter and early spring, a thicker, lodged canopy in summer, and a transition back to less lodged material through fall; and (3) highly variable, as in *Spartina patens,* which shows typically a highly lodged, hummocky character. Even though local areas may show a fairly consistent trend or pattern in light attenuation profiles, high variability in light attenuation seems more common within *Spartina patens* marshes than within other marsh types.

For remote sensing, structural influences would be the least variable in the *Juncus roemerianus* and *Spartina alterniflora* marshes, but background variability may be relatively higher because of the higher base light levels throughout the year. More variable influence of canopy structure on spectral reflectance may be expected in the *Panicum hemitomon* marsh, with possibly higher influences of background in the winter and early spring. Structure in the *Panicum hemitomon* marsh is closely related to the seasonal occurrences of "green-up" and senescence. Less light penetration in the summer because of increased lodging would decrease spectral information from deeper within the canopy, but in the winter dieback background influences may be higher through this more open canopy. Further, without the ability to separate structure and leaf optical influences on canopy reflectance in these marshes, it would be difficult or impossible to detect what canopy property was changing, as both were dramatically varying during these periods. Higher structural variability within *Spartina patens* marshes would be expected to cause variability in canopy reflectances, with reflectances least affected by structural variation in the summer and fall periods. During winter and spring, however, increased high base light levels in *Spartina patens* marshes could further complicate interpretation of canopy reflectance variability.

5.5 SUMMARY

Light penetration field measurements were tested and described in terms of their completeness, reliability or consistency and accuracy to characterize canopy structure. A ceptometer device measuring photosynthetic active radiation (PAR) along a 1-m probe was chosen for the light penetration measurements. Marsh types included *Spartina alterniflora* and *Juncus roemerianus* (saline marshes), *Spartina patens* (brackish marsh), and *Panicum hemitomon* (fresh marsh). Four variables related to field data collection and postdata analysis design were examined with respect to fulfilling data reliability and accuracy and maximizing the potential for operational use for remote sensing calibration and assessment of classification accuracy. These variables included: (1) the horizontal (planar) and (2) vertical (canopy profile) spatial sampling frequencies, (3) the description and possible exclusion of atypical canopy structures, and (4) the normalization of measurements at different sun elevations.

Early testing showed 30-m transects in the north and south and east and west directions combined with light penetration measurements every 3 m helped ensure more accurate depiction of the local variability and matched or encompassed the spatial resolution of most common resource remote sensing sensors. Similarly, vertical light attenuation profiles derived from sampling the canopy every 20 cm from the ground surface to above the canopy improved reliability, consistency, and completeness of repeated measurements. Accounting for the state of the canopy as undisturbed, partial gap, or completely lodged at each profile location was found to increase the comparability and detail of PAR attenuation profiles taken at the same site during multiple occupations and between sites and marsh types.

In all marshes, as sun zenith increased the rate of light fall-off with canopy depth increased, although this effect was more noticeable in *Juncus roemerianus* and less in *Spartina patens* canopies. To remove the sun zenith influences, a method was tested to normalize canopy PAR penetration measurements to a nadir sun zenith. The success of the removal was linked to the spherical canopy leaf orientation used by the normalization. PAR normalizations seemed more successful when used in more vertical canopies such as *Juncus roemerianus* and *Spartina alterniflora*, least successful in highly lodged canopies such as *Spartina patens*, and more dependent on seasonal canopy development in marshes such as *Panicum hemitomon*. In all marshes, application of the normalization increased alignment of PAR attenuation profile taken at different sun zeniths and alignment with the expected progression of canopy structure over time.

ACKNOWLEDGMENTS

We thank Joe White and James Burnett of the U.S. Fish and Wildlife Service for access to the St. Marks National Wildlife Refuge, Florida, and John Fort and Doug Scott for help in field logistics and data collections. We are grateful to U.S. Geological Survey personnel Allison Craver, Kevin McRae, Dal Chappell, Richard Day, and Steve Laine for the many hours of work on planning field logistics and data collections on this study and also thank Beth Vairin for editing this manuscript. Mention of trade names or commercial products is not an endorsement or recommendation for use by the U.S. Government.

REFERENCES

Campbell, G., Extinction coefficients for radiation in plant canopies calculated using an ellipsoidal inclination angle distribution. *Agric. For. Meteorol.* 36, 317–321, 1986.

Chabreck, R., Marsh Zones and Vegetative Types in the Louisiana Coastal Marshes, Ph.D. dissertation, Louisiana State University, Baton Rouge, 1970.

Decagon Devices, Sunfleck Ceptometer Reference Guide, Decagon Devices, Inc., Pullman, WA, 1991.

Goudriaan, J., Crop Micrometeorology: A Simulation Study, Centre for Agricultural Publishing and Documentation, Wageningen, 1977.

Hopkinson, C., J. Gosselink, and R. Parrondo, Aboveground production of seven marsh plant species in coastal Louisiana, *Ecology*, 59, 760–769, 1978.

Lulla, K. and P. Mausel, Ecological applications of remotely sensed multispectral data, in *Introduction to Remote Sensing of the Environment*, Richasen, B.F., Jr., Ed., Kendall/Hall Publishing, Dubuque, IA, 1983, pp. 354–377.

Nielsen, C. and D. Werle, Do long-term space plans meet the needs of the Mission to Planet Earth? *Space Policy*, February 11–16, 1993.

Ramsey, E. III, G. Nelson, S. Sapkota, S. Laine, J. Verdi, and S. Krasznay, Using multiple-polarization L-band radar to monitor marsh burn recovery, *IEEE Trans. Geosci. Remote Sens.*, 37, 635–639, 1999.

Ramsey, E. III, R. Spell, and R. Day, Light attenuation and canopy reflectance as discriminators of gulf coast wetland types, Proceedings of the International Symposium on Spectral Sensing Research, Maui, Hawaii, November 15–20, 1992, 1992a.

Ramsey, E. III, R. Spell, and R. Day, Measuring and monitoring wetland response to acute stress by using remote sensing techniques, in Proceedings of the 25th International Symposium on Remote Sensing and Global Environmental Change, Graz, Austria, April 4–8, 1993.

Ramsey, E. III, R. Spell, and J. Johnston, Preliminary analysis of spectral data collected for the purpose of wetland discrimination, Technical Papers of the ASPRS/ACSM Conference, Albuquerque, NM, 1, 386–394, 1992b.

Stout, J., The Ecology of Irregularly Flooded Salt Marshes of the Northeastern Gulf of Mexico: A Community Profile, U.S Fish and Wildlife Service Biological Report, 85(7.1), 1984.

Teuber, K.B., Use of AVHRR imagery for large-scale forest inventories, *For. Ecol. Manage.*, 33/34, 621–631, 1990.

Ustin, S., C. Wessman, B. Curtiss, E. Kasischke, J. Way, and B. Vanderbilt, Opportunities for using the EOS imaging spectrometers and synthetic aperture radar in ecological models, *Ecology*, 72, 1934–1945, 1991.

Wickland, D., Mission to planet Earth: the ecological perspective, *Ecology*, 72, 1923–1933, 1991.

CHAPTER 6

Participatory Reference Data Collection Methods for Accuracy Assessment of Land-Cover Change Maps

John Sydenstricker-Neto, Andrea Wright Parmenter, and Stephen D. DeGloria

CONTENTS

6.1 INTRODUCTION

Development strategies aimed at settling the landless poor and integrating Amazonia into the Brazilian national economy have led to the deforestation of between 23 and 50 million ha of primary forest. Over 75% of the deforestation has occurred within 50 km of paved roads (Skole and Tucker, 1993; INPE, 1998; Linden, 2000). Of the cleared areas, the dominant land-use (LU) practice continues to be conversion to low-productivity livestock pasture (Fearnside, 1987; Serrão and Toledo, 1990). Meanwhile, local farmers and new migrants to Amazonia continue to clear primary forest for transitory food, cash crops, and pasture systems and eventually abandon the land as it loses productivity. Though there are disagreements on the benefits and consequences of this practice

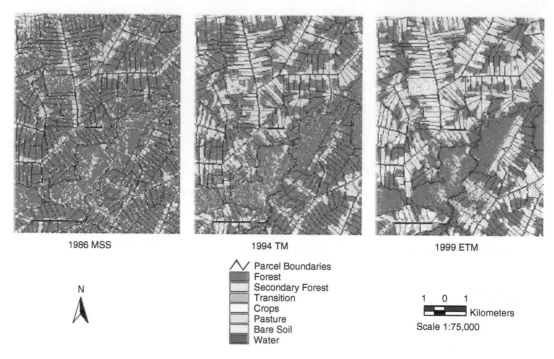

Plate 6.1   (See color insert following page 114.) Land-cover classification for three time periods between 1986 and 1999.

from economic, agronomic, and environmental perspectives, there is a need to link land-cover (LC) change in Amazonia with more global externalities.

Rehabilitating the productivity of abandoned pasture lands has the potential to convert large areas from sources to sinks of carbon (C) while providing for the well-being of people in the region and preserving the world's largest undisturbed area of primary tropical rainforest (Fernandes et al., 1997). Primary forests and actively growing secondary forests sequester more C, cycle nutrients more efficiently, and support more biodiversity than abandoned pastures (Fearnside, 1996; Fearnside and Guimaraes, 1996). Results from research on LU options for agriculture in Amazonia point to agrosilvopastoral LU systems involving rotations of adapted crops, pasture species, and selected trees as being particularly appropriate for settlers of western Amazonia (Sanchez and Benites, 1987; Szott et al., 1991; Fernandes and Matos, 1995). Coupled with policies that encourage the sustainability of these options and target LU intensifications, much of the vast western Amazonia could be preserved in its natural state (Sanchez, 1987; Vosti et al., 2000).

Many studies have focused on characterizing the spatial extent, pattern, and dynamics of deforestation in the region using various forms of remotely sensed data and analytical methods (Boyd et al., 1996; Roberts et al., 1998; Alves et al., 1999; Peralta and Mather, 2000). Given the importance of secondary forests for sequestering C, the focus of more recent investigations in the region has been on developing spectral models and analytical techniques in remote sensing to improve our ability to map these secondary forests and pastures in both space and time, primarily in support of global C modeling (Lucas et al., 1993; Mausel et al., 1993; Foody et al., 1996; Steininger, 1996; Asner et al., 1999; Kimes et al., 1999).

The need to better integrate the human and biophysical dimensions with the remote sensing of LC change in the region has been reported extensively (Moran et al., 1994; Frohn et al., 1996; Rignot et al., 1997; Liverman et al., 1998; Moran and Brondizio, 1998; Rindfuss and Stern, 1998; Wood and Skole, 1998; McCracken et al., 1999; Vosti et al., 2000; http://www.uni-bonn.de/ihdp/lucc/). Most investigations that integrate remote sensing, agroecological, or socioeconomic dimensions focus on the prediction of deforestation rates and the estimation of land-cover/land-use (LCLU) change at a regional scale.

Local stakeholders have seldom been involved in remote sensing research in the area. This is unfortunate because municipal authorities and local organizations represent a window of opportunity to improve frontier governance (Nepstad et al., 2002). These stakeholders have been increasingly called upon to provide new services or fill gaps in services previously provided by federal and state government. Small-scale farmer associations are key local organizations because some of the obstacles to changing current land use patterns and minimizing deforestation cannot be instituted by farmers working individually but are likely to require group effort (Sydenstricker-Neto, 1997; Ostrom, 1999).

6.1.1 Study Objectives

The objectives of our study were to: (1) determine LC change in the recent colonization area (1986–1999) of Machadinho D'Oeste, Rondonia, Brazil; (2) engage community stakeholders in the processes of mapping and assessing the accuracy of LC maps; and (3) evaluate the relevance of LC maps (inventory) for understanding community-based LU dynamics in the study area. The objectives were defined to compare stakeholder estimates and perceptions of LC change in the region to what could be measured through the classification of multispectral, multitemporal, remotely sensed data. We were interested in learning whether there would be increased efficiencies, quality, and ownership of the inventory and evaluation process by constructively engaging stakeholders in local communities and farmer associations. In this chapter, we focus our presentation on characterizing and mapping LC change between 1994 and 1999.

6.1.2 Study Area

Established in 1988, the municipality of Machadinho D'Oeste (8502 km²) is located in the northeast portion of the State of Rondonia, western Brazilian Amazonia (Figure 6.1). The village of Machadinho D'Oeste is 150 km from the nearest paved road (BR 364 and cities of Ariquemes and Jaru) and 400 km from Porto Velho, the state capital. When first settled, the majority of the area was originally composed of untitled public lands. A portion of the area also included old, privately owned rubber estates (*seringais*), which were expropriated (Sydenstricker-Neto, 1992).

The most recent occupation of the region occurred during the mid-1980s with the development of the Machadinho Colonization Project (PA Machadinho) by the National Institute for Colonization and Agrarian Reform (INCRA). In 1984, the first parcels in the south of the municipality were delivered to migrant farmers, and since then the area has experienced recurrent migration inflows. From hundreds of inhabitants in the early 1980s, Machadinho's 1986 population was estimated to be 8,000, and in 1991 it had increased to 16,756 (Sydenstricker-Neto and Torres, 1991; Sydenstricker-Neto, 1992; IBGE, 1994). In 2000, the demographic census counted 22,739 residents. This amounted to an annual population increase during the decade of the 1990s of 3.5%. Although Machadinho is an agricultural area by definition, 48% of its population lives in the urban area (IBGE, 2001).

Despite the importance of colonization in Machadinho, forest reserves comprise 1541 km², or 18.1%, of its area. Most of these reserves became state extractive reserves in 1995, but there are also state forests for sustained use. Almost the entire area of the reserves is covered with primary forest (Olmos et al., 1999).

In biophysical terms, Machadinho's landscape combines areas of altiplano with areas at lower elevation between 100 and 200 m above sea level. The major forest cover types are tropical semideciduous forest and tropical flood plain forest. The weather is hot and humid with an average annual temperature of 24°C and relative humidity between 80 and 85%. A well-defined dry season occurs between June and August and annual precipitation is above 2000 mm. The soils have medium

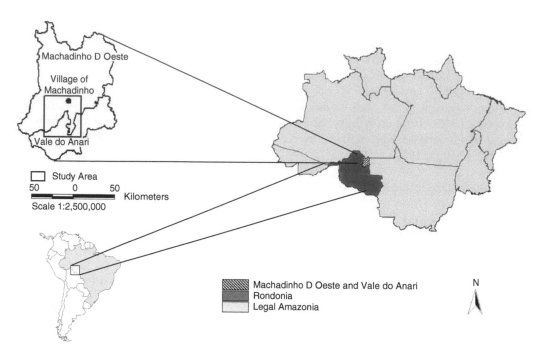

Figure 6.1   Legal Amazonia, Rondonia, and study area, Brazil.

to low fertility and most of them require high inputs for agriculture development (EMBRAPA/SNLCS, 1982; Brasil, MIRAD-INCRA e SEPLAN – Projeto RADAMBRASIL, 1985).

The study area is 215,000 ha and is divided between the municipalities of Machadinho D'Oeste (66%) and the north of Vale do Anari (34%). It includes more recent colonization areas, but its core comprises the first phase (land tracts 1 and 2) of the former Machadinho Settlement settled in 1984 and 1985. These two land tracts have a total area of 119,400 ha. The land tracts have multiple uses: 3,000 ha are designated for urban development, 35,165 ha are in extractive reserves, and 81,235 ha are divided into 1,742 parcels (average size 46 ha) distributed to migrant farmers by INCRA (Sydenstricker-Neto, 1992).

6.2 METHODS

6.2.1 Imagery

Landsat Multi-Spectral Scanner (MSS), Thematic Mapper (TM), and Enhanced Thematic Mapper (ETM+) digital images were acquired for the study area (path 231/row 67) for one date in 1986, 1994, and 1999. The 1994 and 1999 TM images were 30-m resolution and the 1986 MSS image was resampled to 30 m to match the TM images. The images were acquired during the dry season (July or August) of each year to minimize cloud cover. The Landsat images used for LC analysis were the best available archived scenes.

The 1986 MSS image (August 10) and the 1999 ETM+ image (August 6) were obtained from the Tropical Rainforest Information Center (TRFIC) at Michigan State University. The 1994 TM image (July 15) was provided by the Center for Development and Regional Planning (CEDEPLAR) at the Federal University of Minas Gerais (UFMG) in Brazil. Although a TM image for the 1986 date was available, random offset striping made this scene unusable. The MSS image acquired on the same date was used instead, though thin clouds obscured part of the study area.

The geometrically corrected 1999 ETM+ image provided by TRFIC had the highest geometric accuracy as determined using Global Positioning System (GPS) coordinates collected in the field and resulting in a root mean square error (RMSE) of less than one pixel. Therefore, we coregistered the 1986 and 1994 images to the "base" 1999 ETM+ image using recognizable fixed objects (such as road intersections) in ERDAS Imagine 8.4. We used nine "fixed" locations, known as ground control points (GCPs), to register both images. For the 1986 and 1994 MSS images, the RMSEs were 0.54 and 0.47 pixels, respectively.

Additional image processing included the derivation of tasseled-cap indices for each image. Tasseled-cap transformed spectral bands 1, 2, and 3 (indices of brightness, greenness, and wetness, respectively) were calculated for the TM images using Landsat-5 coefficients published by Crist et al. (1986). Although Huang et al. (2002) recommended using a reflectance-based tasseled-cap transformation for Landsat 7 (ETM+) based on at-satellite reflectance, those recommended tasseled-cap coefficients for Landsat 7 were not published at the time of this study. Tasseled-cap bands 1 and 2 (brightness and greenness) were calculated for the MSS image using coefficients published by Kauth and Thomas (1976). These investigators have shown tasseled-cap indices to be useful in differentiating vegetation types on the landscape, and the tasseled-cap indices were therefore included in this analysis of mapping LC. Image stacks of the raw spectral bands and tasseled-cap bands were created in ERDAS Imagine 8.4. This resulted in one 6-band image for 1986 (MSS spectral bands 1, 2, 3, 4, and tasseled-cap bands 1 and 2), a 10-band image for 1994 (TM spectral bands 1–7 and tasseled-cap bands 1, 2, and 3), and an 11-band image for 1999 (ETM+ spectral bands 1–8 and tasseled-cap bands 1–3). The 15-m panchromatic band in the 1999 ETM+ image was not used in this analysis.

6.2.2 Reference Data Collection

As in many remote areas in developing countries, data sources for producing and assessing accuracy of LC maps for our study area were limited. Upon project initiation (2000) no suitable LC reference data were available. Historical aerial photographs were not available for discriminating between LC types for our study area. In this context, satellite imagery was the only spatially referenced data source for producing reliable LC maps for the area.

Because we wanted to document LC change from the early stages of human settlement and development (beginning in 1985), when major forest conversion projects were established, our objective was to compile retrospective data to develop and validate a time series of LC maps. The challenge of compiling retrospective data became an opportunity to engage community stakeholders in the mapping process and "bring farmers into the map." We decided to enlist the help of farmers, who are very knowledgeable about land occupation practices and the major forces of land use dynamics, to be our source for contemporary and retrospective data collection. Also, by engaging the locals early in the process, we could examine the advantages and limitations of this strategy for future resource inventory projects in the region conducted by researchers and local stakeholders.

We utilized a seven-category LC classification scheme as defined in Table 6.1. The level of detail of this classification scheme is similar to that of others used in the region and should permit some level of comparative analysis with collaborators and stakeholders (Rignot et al., 1997; de Moraes et al., 1998).

In August 2000, with the assistance of members of nine small-scale farmer associations in the study area, we collected field data to assist in the development of spectral models of each cover type for image classification and to validate the resulting LC maps. All associations that we contacted participated in the mapping project. Initially, we met with the leadership of each association and presented our research goals and objectives, answered questions, and invited members of each association to participate in the study. After developing mutual trust and actively engaging the association, data collection groups were formed averaging 12 individuals per association (total over 100 individuals). Special effort was made to include individuals in each group who were long-

Table 6.1   Land-Cover Classification Scheme and Definitions

Land Cover	Definition
Primary forest	Mature forest with at least 20 years growth
Secondary forest	Secondary succession at any height and less than 20 years growth
Transition	Area recently cleared, burned, or unburned and not currently in use
Pasture	Area planted with grass, ranging from overgrazed to bushy
Crops	Area with agriculture, including perennial and annual crops
Bare soil	Area with no vegetation or low, sparse vegetation
Water	Waterbody, including major rivers, water streams, and reservoirs

term residents and who were knowledgeable about historical LU practices in the region. Nearly half of the members in each of the nine groups were farmers who settled prior to 1986.

An introductory meeting was conducted with each group to provide a hard copy (false-color composite) of the 1999 ETM+ image with parcel boundaries overlaid and to solicit comments and observations regarding farm locations, significance of color tones on the image, and clarification of LU practices and associated cover types. All participants were then asked to indicate retrospective and current LU for their parcels and for other parcels with which they were familiar. Any questions that could not be answered by individuals were referred to the group for discussion, elaboration, and decision making. For each identified cover type, we annotated and labeled polygons on stable acetate overlaid on the false-color composite image. Each polygon consisted of a homogeneous area labeled as one of seven LC types for each year corresponding to the dates of the Landsat images used in the study.

Notes were taken during the interview process to indicate the date each farmer started using the land, areas of the identified LC types for each of the 3 years considered in the study (1986, 1994, 1999), changes over time, level of uncertainty expressed by participants while providing information for each annotated polygon, and other information farmers considered relevant. After each meeting, the research team traveled the main roads in the area just mapped by the farmer association and compared the identified polygons with what could be observed. The differences between the cover type provided by the farmers and what was observed were minimal. In areas where such meetings could not be organized, the research team traveled the feeder roads and annotated the contemporary LC types that could be confidently identified.

Field data were collected for over 1500 polygons, including all seven LC types of interest. We considered this to be an adequate sample for image classification and validation of our maps. Although an effort was made to ensure all land cover types were well represented in the database, some types such as bare soil were represented by a relatively small sample sizes ($n < 200$ pixels).

6.2.3   Data Processing

More than 1000 polygons identified during the farmer association interviews were screen-digitized and field notes about the polygons were compiled into a table of attributes. Independent random samples of polygons for each of the seven land-cover types were selected for use in image classifier training and land-cover map validation. Although the number of homogenous polygons annotated in the field was large, polygons varied greatly in size from < 5 to >1000 ha and were not evenly distributed among the seven cover types (Table 6.2). For cover types that had a large number of polygons, half of the polygons were used for classifier training and half for map validation. For two cover types, however, the polygon samples were so large in area (and therefore contained so many pixels) that they could not be used effectively because of software limitations. The primary forest and pasture cover type polygons were therefore randomly subdivided so that only one half of the pixels were set aside for both classifier training and for map validation (i.e., one quarter of the total eligible data pixels were used for each part of the analysis). However, this approach did not yield a sufficient number of sample polygons for some of the more rare cover

Table 6.2 Number of Pixels Sampled for Classifier Training and Map Validation for the 1999 Image

Land-Cover Class	Total No. of Polygons	Total No. of Pixels	No. of Pixels/Polygons	
			Mean	Variance
Forest	189	16,755	89	5,349
Secondary forest	108	3,060	28	401
Transition	43	10,054	33	917
Crops	306	2,693	63	1,358
Pasture	261	4,496	17	120
Bare soil	17	140	8	18
Water	106	1,705	16	244
Total	1,030	38,903	38	2,089

types (i.e., < 1% land area). To address this issue, we randomly sampled individual pixels within these polygons of the rare cover types and equally partitioned the pixels into the two groups used for classifier training and map validation.

6.2.4 Image Classification

Spectral signature files were generated to be used in supervised classification using a maximum likelihood algorithm. The spectral signatures included both image and tasseled-cap bands created for each image of each analysis year. LC maps were produced for each of the 3 years containing all seven LC types in each of the resulting maps. Postclassification 3 × 3 pixel majority convolution filter was applied to all three LC maps to eliminate some of the speckled pattern (noise) of individual pixels. The result of this filter was to eliminate pixels that differed in LC type from their neighbors, which tended thereby to eliminate both rare cover types and those that exist in small patches on the landscape (such as crops). However, we concluded that the filtering process introduced an unreasonable amount of homogeneity onto the landscape and obscured valuable information relevant to the spatial pattern of important cover types within our unit of analysis, which was the land parcel. All subsequent analyses were performed on the unfiltered LC maps for all three dates of imagery.

6.2.5 Accuracy Assessment

We assessed the accuracy of the three LC maps at the pixel level using a proportional sampling scheme based on the distribution of validation sample points (pixels) for each of the cover types in the study. This methodology was efficiently applied in this study because the distribution of our field-collected validation sample points was representative of the distribution in area of each cover type in the study area (Table 6.2).

The proportional sample of pixels used for the accuracy assessment for each year was selected by first taking into account the cover type having the smallest area based on the number of validation pixels we had for that cover type. Once the number of pixels in the validation data set was determined for the cover type occupying the smallest area, the total number of validation pixels to be used for each analysis year was calculated by the general formula:

$$S_t = N_s/P_s \tag{6.1}$$

where S_t = the total number of validation pixels to be sampled for use in accuracy assessment, N_s = the number of pixels in the land cover type with the smallest number of validation pixels, and P_s = the proportion of the classified map predicted to be the cover type with the smallest amount of validation pixels.

The total number of validation pixels to be used to assess the accuracy for each cover type was then calculated by the general formula:

$$V_c = S_t \times P_c \tag{6.2}$$

where V_c = the total number of validation pixels to be used for a specific cover type, S_t = the total number of validation pixels to be sampled for use in accuracy assessment, and P_c = the proportion of the classified map predicted to be that cover type.

To illustrate this proportional sampling accuracy method, we describe the forest cover type for the 1999 map. The cover type with the smallest number of validation pixels in 1999 was the bare soil cover type with a total of 79 validation pixels (N_s). Of the total number of pixels in the 1999 classified map (8,970,395), the bare soil cover type was predicted to be 201,267 pixels, or a proportion of 0.0224 of the total classified map (P_s). Using Equation 6.1 above, the resulting sample size of validation pixels to be used for accuracy assessment of the 1999 LC map (S_t) was 3,521 pixels. In the 1999 map, the forest cover type was predicted to cover 68.6% of the classified map (i.e., 6,155,275 pixels out of 8,970,395 total pixels). Using Equation 6.2 above, the sample size of validation pixels to be used for the forest cover type (V_c) was then 2,414 (i.e. 3,521 × 0.686).

Once the validation sample sizes were chosen for each cover type, a standard accuracy assessment was performed whereby the cover type of each of the validation pixels was compared with the corresponding cover type on the classified map. Agreement and disagreement of the validation data set pixels with the pixels on the classified map were calculated in the form of an error matrix wherein the producer's, user's, and overall accuracy were evaluated.

6.3 RESULTS AND DISCUSSION

6.3.1 Classified Imagery and Land-Cover Change

Presentation and discussion of accuracy assessment results will focus only on the 1994 and 1999 LC maps. (The 1986 map was not directly comparable because it was based on coarser resolution and resampled MSS data and because it contained cirrus cloud cover over parts of the study.) A visual comparison of 1986–1999 LC maps shows significant change. Plate 6.1 presents the classified imagery with parcel boundaries overlaid for a portion of the study area near one of the major feeder roads. In 1986, approximately 2 years after migrant settlement, only some initial clearing was observed near roads; however, 13 years later (1999) there were significant open areas and only a small number of parcels that remained mostly covered with primary forest. The extensive deforestation illustrated in Plate 6.1 is confirmed by the numeric data presented in Table 6.3. In 1994, 147,380 ha, or 68.5% of the total study area (215,000 ha), was covered in primary forest.

Table 6.3 Land-Cover Change in Study Area, Rondonia 1994–1999

Class	Area (ha) 1994	Area (ha) 1999	Change in Area 1994–1999 (ha)	Percentage of Area 1994	Percentage of Area 1999	Percentage of Change 1994–1999
Forest	147,380	117,573	−29,806	68.5	54.6	−20.2
Secondary forest	27,759	30,732	2,973	12.9	14.3	10.7
Transition	2,234	5,555	3,321	1.0	2.6	148.6
Crops	12,072	27,833	15,760	5.6	12.9	130.5
Pasture	16,253	22,386	6,133	7.6	10.4	37.7
Bare soil	5,183	6,823	1,640	2.4	3.2	31.6
Water	4,251	4,252	1	2.0	2.0	0.0
Total	215,132	215,154		100.0	100.0	

Table 6.4 Land-Cover Change Matrix and Transitions in Study Area, Rondonia 1994–1999

1994	1999 Forest	Sec. Forest	Transition	Crops	Pasture	Bare Soil	Water	Total percentage	Total area (ha)
Forest	**48.9**	8.3	1.8	5.9	2.3	1.2	0.0	68.5	147,380
Sec. forest	4.8	**3.5**	0.3	2.5	1.3	0.5	0.0	12.9	27,759
Transition	0.1	0.2	**0.0**	0.4	0.2	0.1	0.0	1.0	2,234
Crops	0.3	1.2	0.2	**2.1**	1.4	0.4	0.0	5.6	12,072
Pasture	0.2	0.7	0.1	1.3	**4.5**	0.8	0.0	7.6	16,253
Bare soil	0.3	0.4	0.1	0.8	0.7	**0.2**	0.0	2.4	5,183
Water	0.0	0.0	0.0	0.0	0.0	0.0	**2.0**	2.0	4,251
Total percentage	54.6	14.3	2.6	12.9	10.4	3.2	2.0	100.0	
Total area (ha)	117,553	30,731	5,554	27,833	22,386	6,823	4,252		215,132

Note: No change 1994–1999: 61.1%.

The amount of primary forest decreased in 1999 by 30,000 ha, a negative change of 20.2% in primary forested area. The area of deforestation observed between 1994 and 1999 was more than twice that estimated for the 1986 to 1994 period (not shown). This represented a 4.5 times increase compared to the 1986–1994 deforestation rate. Table 6.3 presents the change in LC for 1994–1999 as both percentage of area and percentage of change.

All the nonforest cover types increased in area between 1994 and 1999. This was largely at the expense of primary forest. Increases in secondary forest had the dominant "gain" in area during this period, with a total increase in area of almost 31,000 ha in 1999, followed by slightly smaller increases in crops and pasture (27,832 ha and 22,386 ha, respectively). The most significant increases on a proportional basis occurred with the crops and pasture cover types; both increased over 200% during this time period.

The increase in pasture area was inflated by a tremendous deforestation event totaling approximately 5000 ha in 1995 in the southeastern portion of the study site. Subsequent to clearing, the area was partially planted with grass and later divided into small scale farm parcels in 1995 to 1996, creating a new settlement called Pedra Redonda. The most important and broadly distributed crop among the small-scale farms was coffee *(Coffea robusta)*, which received special incentives through subsidized federal government loans and the promotional campaign conducted by the State of Rondonia "Plant Coffee" (1995 to 1999).

The LC change matrix provides more detailed change information, including the distribution of deforested areas into different agricultural uses (Table 6.4). For 1994 to 1999 we determined that 61.1% of the area did not undergo LC change. This metric was calculated by summing the percentages along the major diagonal of the matrix. Note that primary forest showed the greatest decrease in area while concurrently exhibiting the largest area unchanged (48.9%), due to the large area occupied by this cover type. For the remaining cover types, the change was significantly greater (as shown throughout the diagonal of the matrix) because of the proportionally smaller area occupied by these cover types.

The 8.3% conversion rate of primary forest to secondary forest indicates that some recently deforested areas remained in relative abandonment, allowing vegetation to partially recover in a relatively short period of time (Table 6.4). An increase in classes such as transition and bare soil also indicates the same trend of new areas incorporated into farming and their partial abandonment as well. Of areas that were primary and secondary forest in 1994, crops were the most dominate change category (> 8%) followed by pasture (< 4%). While the change in LC mapped from the image classification fits with what we expect to see in the region, it is important to differentiate (when possible) real change from misclassification. Potential errors associated with the mapping are discussed below.

Table 6.5   User's Accuracy in Study Area, Rondonia 1986–1999

Classified Data	1986	1994	1999
Forest	89.8%	93.5%	96.7%
Secondary forest	45.5%	63.1%	77.4%
Transition	42.9%	75.0%	57.5%
Crops	25.0%	53.6%	67.5%
Pasture	80.0%	77.5%	89.6%
Bare Soil	—	66.7%	28.7%
Water	100.0%	100.0%	93.6%
Overall accuracy	84.6%	88.3%	89.0%
Kappa statistic	0.52	0.69	0.78

6.3.2   Map Accuracy Assessment

The user's accuracy is summarized in Table 6.5. The increase in overall map accuracies for each subsequent year in the analysis was attributed to several factors. First, we used three different sensors (MSS, TM, and ETM+), which introduced increased spatial and spectral resolution of the sensors over time. Second, the 1986 MSS image had clouds that introduced some classification errors. Third, collecting retrospective data was a challenge because interviewees sometimes had difficulty recalling LC and associated LU practices over the study period. In general, retrospective LU information had a higher level of uncertainty than for time periods closer to the date of the interview.

Despite these difficulties, however, overall accuracy was between 85 and 89% for 1986 and 1999, respectively. Accuracy for specific classes ranged between 50 and 90%, achieving ≥ 96% for primary forest in 1999. Some bare soil (1999) and crops (1986) classes were particularly difficult to map and attained accuracies below 30%. The sample size for these particular cover types was relatively small, which may have contributed to this poor outcome. When coupled with the fact that areas of bare soil and crops tend to be small in the study area (≤ 1.0 ha), the lower accuracies were not unexpected for these classes. Error matrices for 1994 and 1999 are presented in Tables 6.6 and 6.7, respectively. The overall accuracy for 1999 was 89.0% (Kappa 0.78). With the exception of bare soil, all the remaining classes had user's accuracies that ranged from 57.5 to 96.7% and producer's accuracies between 66.5 and 100.0%. The overall accuracy for the 1994 land-cover map was 88.3%. In general, accuracy for specific cover types ranges between 50 and 90%, achieving a high of 96.7% for primary forest in 1999. The bare soil (1999) accuracy was below 30%; however, the limited proportion of training sample pixels relative to the total amount of pixels comprising the study area for this specific class may have contributed to this poor outcome.

Table 6.6   Error Matrix for the Land-Cover Map in Study Area, Rondonia 1994

Classified Data	Reference Data								
	Forest	Sec. Forest	Transition	Crops	Pasture	Bare Soil	Water	Total	User's Accuracy
Forest	1218	76	0	5	0	1	3	1303	93.5%
Secondary forest	40	82	0	6	1	1	0	130	63.1%
Transition	0	0	9	3	0	0	0	12	75.0%
Crops	9	15	1	30	0	1	0	56	53.6%
Pasture	15	0	0	6	79	1	1	102	77.5%
Bare soil	4	0	0	5	0	18	0	27	66.7%
Water	0	0	0	0	0	0	25	25	100.0%
Total	1286	173	10	55	80	22	29	1655	
Producer's accuracy	94.7%	47.4%	90.0%	54.6%	98.8%	81.8%	86.2%		

Note: Overall classification accuracy = 88.3%. Kappa statistic = 0.69.

Table 6.7   Error Matrix for the Land-Cover Map in Study Area, Rondonia 1999

| | Reference Data | | | | | | | | User's |
Classified Data	Forest	Sec. Forest	Transition	Crops	Pasture	Bare Soil	Water	Total	Accuracy
Forest	2370	73	0	3	0	0	6	2452	96.7%
Secondary forest	44	233	0	12	8	4	0	301	77.4%
Transition	0	0	54	19	5	16	0	94	57.5%
Crops	0	58	0	206	14	21	6	305	67.5%
Pasture	0	7	0	5	198	9	2	221	89.6%
Bare soil	0	0	0	64	7	29	1	101	28.7%
Water	0	2	0	1	0	0	44	47	93.6%
Total	2414	373	54	310	232	79	59	3521	
Producer's accuracy	98.2%	62.5%	100.0%	66.5%	85.3%	36.7%	74.6%		

Note: Overall classification accuracy = 89.0%. Kappa statistic = 0.78.

The pattern of misclassification and confusion between LC classes is similar for both the 1994 and 1999 error matrices (Table 6.6 and Table 6.7), although different image sensors (TM and ETM+) were used. Confusion between primary and secondary forest was expected because our classification scheme did not separate secondary forest for different successional stages. Some of the polygons delineated in the field as secondary forest exceeded 12 years of regrowth and closely resembled semideciduous primary forest. Accordingly, there was probably some spectral overlap between "old" secondary forest and the semideciduous primary forest. Confusion between secondary forest and crops occurred because many coffee areas were shaded with native species such as rubber tree (*Hevea brasiliensis*), freijó (*Cordia goeldiana*), and Spanish cedar (*Cedrela odorata,*) or included pioneer species such as embaúba (*Cecropia* spp.). Therefore, shaded crops appeared as partially forested areas.

Despite the lack of homogeneity within the "transition" LC class, confusion with other cover types (crops, pasture, and bare soil) was minimal. Confusion most likely occurred because the transition cover type was not particularly unique (Table 6.6 and Table 6.7). The confusion seen in the error matrices between pasture and bare soil and the confusion between bare soil and recently planted coffee areas were expected. Overgrazed pasture had little vegetative matter, allowing these areas easily to be misclassified as bare soil. Also, it was unlikely that spectral reflectance by coffee plants less than 0.5 m tall planted in a 3- × 3-m spacing was detected and discriminated from the surrounding soil background, resulting in confusion between young coffee and bare soil. Water, although spectrally distinct, was easily biased along edge pixels. This was particularly true in the case of small and circuitous watercourses in mixed systems.

6.3.3   Bringing Users into the Map

Initially, the local farmers expressed substantial distrust and skepticism about the mapping project; however, trust was established throughout the mapping process and a good working relationship was established. To best present our findings, we organized community meetings in the areas of the farmer associations involved earlier in the process. Participation in these meetings ranged from as few as six individuals to packed rooms with more than 30 people. These meetings intentionally included the broader community and farmers who had not taken part in the data collection. Each farmer that had provided input during the data collection phase of the study received a color copy of the 1999 LC mapping results. Additional meetings were arranged with agricultural extension agents, leaders of the local rural labor union, municipal officials, and middle school students.

Upon examination, farmers provided verbal confirmation of our estimates and errors. Specific concerns closely resembled the classification errors shown in the accuracy assessment matrices (Table 6.6 and Table 6.7). More than 30 farmers who did not participate in the data collection process compared their estimates of LC for their individual parcels to the statistics generated from the LC maps. In all cases, the general patterns were the same and differences in LC class areas were small. Relevant ideas provided by local farmers were that the map provided a common ground to engage participants in a discussion on environmental awareness and appreciation and that the maps became an instrument of empowerment to local communities. For example, farmers were shocked at the significant changes in LC over time for the whole area (1994 to 1999). This stimulated a debate on the incentives for forest conversion vs. the constraints imposed by the agricultural systems adopted by farmers.

Areas with perennial crops increased dramatically over the years of the study, with coffee becoming the single most important cash crop. Consequently, farmers tended to decrease the amount of land planted with annual subsistence crops such as rice, maize, beans, and cassava. As a result, food security became an issue for some communities. Although new areas would typically come into production within 2 years, the incentives were to expand areas of pasture. An important economic incentive was the dramatic drop in coffee prices worldwide. In the 2001–2002 season, sale prices of coffee in Machadinho D'Oeste were only half of the market value 2 years earlier. Many small farms were not entirely harvested and many farmers reported that they were very inclined to change areas with old coffee trees into pasture. However, the decline in coffee prices motivated enlightened discussions on the economic and environmental dangers of converting most of the land into pasture.

The importance of common forest reserves in the region and the potential and constraints of fostering forest conservation were discussed extensively. There was great appreciation for the fact that the map clearly indicated that the major water resources were within the forest reserves that had not been cleared. Identification of secondary forest along major water streams within the settled areas stimulated a debate on stream bank erosion and nutrient loss into rivers. The general agreement was that farmers went too far in clearing the land and needed to focus efforts on reforesting the areas around the rivers. Farmers voiced the reasons, incentives, and constraints they face in trying to deliberately reforest areas along the water streams. In most of the reported cases of forest recovery, natural regrowth was happening, rather than seeded reforestation. The reported lack of available water in areas in which farmers had irrigated their coffee was a surprise to the researchers.

Two outcomes contributed greatly to farmer empowerment. First, our map offered a synoptic perspective of development patterns that farmers had not entirely realized previously. Farmers felt that having a deeper knowledge of what was happening in their area would enable them to better respond to local needs and contribute to statewide discussions on promoting environmental sustainability. Second, farmers voiced the collective opinion that their participation in the mapping project contributed to better organizing themselves into interest groups. The explicit acknowledgment in the LC map legends of the local associations' contributions was a source of pride within the broader community.

6.4 CONCLUSIONS

Visual inspection and comparison of LC maps with other data sources enabled us to conclude that our efforts provided good estimation of LC change in the study area. The study area changed over the 13-year study period from a typical new colonization area in its early stage, where higher proportions of forests and areas in transition dominate, to one in which these cover types are diminished in area in comparison to the proportional increase in crops and pasture. Statistically based evaluations (error matrices) demonstrated acceptable levels of accuracy with classification errors that were easily explainable and understandable. Participation and input from local farmers was very useful in producing cover maps and proved to be an extremely effective means for collecting classifier training and validation data in areas where other sources were not available. Follow-up meetings with farmers were very constructive for addressing conservation issues with regional and global implications.

Study weaknesses included the intrinsic limitations imposed by the use of the different satellite sensors (i.e., MSS, TM, and ETM+). Also, reference data sizes for some cover classes were relatively small and interviewees expressed greater levels of uncertainty in retrospective data reconstruction than for the current time period. Despite these constraints in data collection, we were confident that they did not represent an extra burden compared to the challenge of obtaining good levels of agreement among remote sensing specialists when using other techniques such as high-resolution videography. The application of an integrated field data collection process would have enhanced the quality of our data. Such an integrated process would comprise simultaneous collection of remotely sensed ground data and household socioeconomic surveys on LU/LC to facilitate direct comparison between data sources.

The level of detail of our classification scheme was similar to that used by other investigators in the region, and our map accuracies compared favorably with their results (Rignot et al., 1997; de Moraes et al., 1998). For local stakeholders, however, our classification scheme was not sufficiently detailed. Stakeholders would most like to clearly distinguish specialty crops such as cacao, coffee, and shade coffee, which would not be practical with the resolution of these data sets.

6.5 SUMMARY

This study assessed LC change in a recent colonization area in the municipalities of Machadinho D'Oeste and Vale do Anari, State of Rondonia, Brazil. Landsat MSS, TM, and ETM+ data were used to create maps of LC conditions for 1986, 1994, and 1999. Images were obtained in July/August (dry season) and field data were collected during August 2000 with the assistance of nine local farm associations and approximately 100 independent farmers. At meetings with the associations, hard copy false-color composites of imagery data with parcel boundaries were presented to individual landowners. Each individual provided historical and contemporary LU for known areas. Polygons were annotated and labeled on stable acetate for each cover type, corresponding to the seven-category classification scheme. Notes were taken during the interview process to indicate the dates of land clearing, cover type, and level of uncertainty expressed by the participants.

Approximately 1000 polygons were field-annotated and random samples were selected for classifier training and map validation. Spectral signature files were generated from training polygons and used in a supervised classification using maximum likelihood classification. Overall accuracy for each year ranged between 85 and 95% (Kappa 0.52–0.78). LC changes were consistent with the trends observed in the study area and reported by others. The participatory process involving local farmers was crucial for achieving the objectives of the study. The specific protocol developed for data collection should be applicable in a wide range of cases and contexts.

The building of trust with the local stakeholders is important with contested issues such as deforestation in the tropics. Systematic data collection among farmers (the primary land users) provided a valuable source of information based on their direct observation in the field and historical data not directly available through other sources. This procedure provided greater confidence for interpreting and understanding classification errors. Finally, the process itself empowered local farmers and provided a forum for discussing land use processes in the region, including challenges to alleviate poverty, increase agrosilvopastoral farming systems, arrest deforestation, and study its implications for developing more effective land use policies.

Including the local stakeholders in the research was a very effective process for evaluating LC change in the region. For stakeholders and researchers, the mapping and reporting process fosters better understanding of the patterns and processes of environmental change in the study area. We foresee that participatory mapping projects such as the one reported in this chapter have the potential to become an important planning device for regional-scale development in Brazil. With greater economic opportunities and stronger institutions at the local level, society is likely to improve the ability to identify and adopt more environmentally sound LU activities.

ACKNOWLEDGMENTS

We acknowledge the Brazilian farmer associations in Machadinho D'Oeste and associates in the Center for Development and Regional Planning (CEDEPLAR) and Center for Remote Sensing (CSR), Federal University of Minas Gerais (UFMG). The Brazilian National Council for Scientific and Technological Research (CNPq), the Teresa Heinz Scholars for Environmental Research program, the Rural Sociological Society (RSS), and the World Wildlife Fund (WWF, Brazil) provided major financial support. At Cornell University, project sponsors included Cornell International Institute for Food, Agriculture, and Development (CIIFAD), Department of Crop and Soil Sciences, Department of Development Sociology and Graduate Field of Development Sociology, Einaudi Center for International Studies, Latin American Studies Program (LASP), Institute for Resource Information Systems (IRIS), and the Population Development Program (PDP). In addition, the Tropical Rain Forest Information Center (TRFIC) at Michigan State University provided the satellite imagery without which this study would not have been possible.

REFERENCES

Alves, D.S., J.L.G. Pereira, C.L. DeSouza, J.V. Soares, and F. Yamaguchi, Characterizing land use change in central Rondonia using Landsat TM imagery, *Int. J. Remote Sens.*, 20, 28–77, 1999.

Asnet, G.P., A.R. Townsend, and M.M.C. Bustamante, Spectrometry of pasture condition and biogeochemistry in the Central Amazon, *Geophysical Research Letters,* 26, 2769–2772, 1999.

Boyd, D.S., G.M. Foody, P.J. Curran, R.M. Lucas, and M. Honzak, An assessment of radiance in Landsat TM middle and thermal infrared wavebands for detection of tropical forest regeneration, *Int. J. Remote Sens.* 17, 249–261, 1996.

Brasil, MIRAD-INCRA e SEPLAN-Projeto RADAMBRASIL, Estudo da Vegetação e do Inventário Florestal, Projeto de Assentamento Machadinho, Glebas 1 e 2, Goiânia, GO, Brazil, 1985.

Crist, E.P., R. Laurin, and R.C. Cicone, Vegetation and Soils Information Contained in Transformed Thematic Mapper Data, in Proceedings of IGARSS' 1986 Symposium, Ecological Society of America Publications, Division, ESA SP-254, 1986.

de Moraes, J.F.L, F. Seyler, C.C. Cerri, and B. Volkoff, Land cover mapping and carbon pool estimates in Rondonia, Brazil, *Int. J. Remote Sens.*, 19, 921–934, 1998.

EMBRAPA/SNLCS, Levantamento de Reconhecimento de Média Intensidade dos Solos e Avaliação da Aptidão Agrícola das Terras em 100.000 Hectares da Gleba Machadinho, no Município de Ariquemes, Rondônia, Boletim de Pesquisa n. 16, EMBRAPA/SNLCS, Rio de Janeiro, RJ, Brazil, 1982.

Fearnside, P.M., The causes of deforestation in the Brazilian Amazon, in *The Geophysiology of Amazonia: Vegetation and Climate Interactions,* Dickenson, R.E., Ed., John Wiley, New York, 1987.

Fearnside, P.M., Amazonian deforestation and global warming: carbon stocks in vegetation replacing Brazil's Amazon forest, *Forest Ecol. Manage.*, 80, 21–34, 1996.

Fearnside, P.M. and W.M. Guimaraes, Carbon uptake by secondary forests in Brazilian Amazon, *Forest Ecol. Manage.,* 80, 35–46, 1996.

Fernandes, E.C.M. and J.C. Matos, Agroforestry strategies for alleviating soil chemical constraints to food and fiber production in the Brazilian Amazon, in *Chemistry of the Amazon: Biodiversity, Natural Products, and Environmental Issues*, Seidl, P.R. et al., Eds., American Chemical Society, Washington, DC, 1995.

Fernandes, E.C.M., Y. Biot., C. Castilla, A.C. Canto, J.C. Matos, S. Garcia, R. Perin, and E. Wandelli, The impact of selective logging and forest conversion for subsistence agriculture and pastures on terrestrial nutrient dynamics in the Amazon, *Ciência Cultura*, 49, 34–47, 1997.

Foody, G.M., G. Palubinskas, R.M. Lucas, P.J. Curran, and M. Honzak, Identifying terrestrial carbon sinks: classification of successional stages in regenerating tropical forest from Landsat TM data, *Remote Sens. Environ.*, 55, 205–216, 1996.

Frohn, R.C., K.C. McGwire, V.H. Dales, and J.E. Estes, Using satellite remote sensing to evaluate a socioeconomic and ecological model of deforestation in Rondonia, Brazil, *Int. J. Remote Sens.*, 17, 3233–3255, 1996.

Huang, C., B. Wylie, L. Yang, C. Homer, and G. Zylstra, Derivation of a tasseled cap transformation based on Landsat 7 at-satellite reflectance, *Int. J. Remote Sens.*, 23, 1741–1748, 2002.

IBGE. Censo Demográfico 1991 – Rondônia, Instituto Brasileiro de Geografia e Estatística (IBGE), Rio de Janeiro, RJ, Brazil, 1994.

IBGE. Censo Demográfico 2000, available at Instituto Brasileiro de Geografia e Estatística (IBGE) Web site, http://www.ibge.gov.br, December 2001.

INPE. Deflorestamento da Amazônia 1995-1997, Instituto Nacional de Pesquisas Espaciais (INPE), São José dos Campos, SP, Brazil, 1998.

Kauth, R.J. and G.S. Thomas, The Tasselled Cap — a graphic description of the spectral-temporal development of agricultural crops as seen by Landsat, in *Symposium on Machine Processing of Remotely Sensed Data*, IEEE, 76CH 1103 – IMPRESO, 1976, pp. 41–51.

Kimes, D.S., R.F. Nelson, W.A. Salas, and D.L. Skole, Mapping secondary tropical forest and forest age from SPOT HRV data, *Int. J. Remote Sens.*, 20, 3625–3640, 1999.

Linden, E., The road to disaster, *Time*, October 16, 2000, pp. 97–98.

Liverman, D., E.F. Moran, R.R. Rindfuss, and P.C. Stern, Eds., *People and Pixels: Linking Remote Sensing and Social Science*, National Academy Press, Washington, DC, 1998.

Lucas, R.M., M. Honzak, G.M. Foody, P.J. Curran, and C. Corves, Characterizing tropical secondary forests using multi-temporal Landsat sensor imagery, *Int. J. Remote Sens.*, 14, 3061–3067, 1993.

Mausel, P., Y. Wu, Y. Li, E. Moran, and E. Brondizio, Spectral identification of successional stages following deforestation in the Amazon, *Geocarto Int.*, 8, 61–71, 1993.

McCracken, S.D., E. Brondizio, D. Nelson, E.F. Moran, A.D. Siqueira, and C. Rodriquez-Pedraza, Remote sensing and GIS at farm level: demography and deforestation in the Brazilian Amazon, *Photogram. Eng. Remote Sens.*, 65, 1311–1320, 1999.

Moran, E.F. and E. Brondizio, Land-use change after deforestation in Amazonia, in *People and Pixels: Linking Remote Sensing and Social Science*, Liverman, D. et al., Eds., National Academy Press, Washington, DC, 1998.

Moran, E.F., E. Brondizio, P. Mausel, and Y. Wu, Integrating Amazonian vegetation, land use, and satellite data, *Bioscience*, 44, 329–338, 1994.

Nepstad, D., D. McGrath, A. Alencar, A.C. Barros, G. Carvalho, M. Santilli, and M. del C. Vera Diaz, Frontier governance in Amazonia, *Science*, 295, 629–631, 2002.

Olmos, F., A.P. de Queiroz Filho, and C.A. Lishoa, As Unidades de Conservação de Rondônia, Rondônia/SEPLAN/PLANAFLORO/PNUD, Porto Velho, RO, Brazil, 1999.

Ostrom, E., Self-Governance and Forest Resources, Occasional Paper 20, Center for International Forestry Research, Jakarta, Indonesia, 1999.

Peralta, P. and P. Mather, An analysis of deforestation patterns in the extractive reserves of Acre, Amazonia, from satellite imagery: a landscape ecological approach, *Int. J. Remote Sens.*, 21, 2555–2570, 2000.

Rignot, E., W. Salas, and D. L. Skole, Mapping deforestation and secondary growth in Rondonia, Brazil, using imaging radar and Thematic Mapper data, *Remote Sens. Environ.*, 59, 167–179, 1997.

Rindfuss, R.R. and P.C. Stern, Linking Remote Sensing and Social Science: The Need and the Challenges, in *People and Pixels: Linking Remote Sensing and Social Science*, Liverman, D. et al., Eds., National Academy Press, Washington, DC, 1998.

Roberts, D.A., G.T. Batista, J.L.G. Pereira, E.K. Waller, and B.W. Nelson, Change Identification Using Multitemporal Spectral Mixture Analysis: Applications in Eastern Amazonia, in *Remote Sensing Change Detection: Environmental Monitoring Methods and Applications*, Lunetta, R.S. and C.D. Elvidge, Eds., Ann Arbor Press, Chelsea, MI, 1998.

Sanchez, P.A., Soil Productivity and Sustainability in Agroforestry Systems, in *Agroforestry: A Decade of Development*, Steppler, H.A. and P.K.R. Nair, Eds., Institute Council for Research in Agroforestry, Nairobi, Kenya, 1987.

Sanchez, P.A. and J.R. Benites, Low input cropping for acid soils of the humid tropics, *Science*, 238, 1521–1527, 1987.

Serrão, E.A.S. and J.M. Toledo, The Search for Sustainability in Amazonian Pastures, in *Alternatives to Deforestation: Steps Toward Sustainable Use of the Amazon Rain Forest*, Anderson, A.B., Ed., Columbia University Press, New York, 1990.

Skole, D.L. and C.J. Tucker, Tropical deforestation and habitat fragmentation in the Amazon: satellite data from 1978–1988, *Science*, 260, 1905–1910, 1993.

Steininger, M.K., Tropical secondary forest regrowth in the Amazon: age, area, and change estimation with Thematic Mapper data, *Int. J. Remote Sens.*, 17, 9–27, 1996.

Sydenstricker-Neto, J., Organizações Locais e Sustentabilidade nos Trópicos Úmidos: Um Estudo Exploratório, Documentos de Trabalho, EMBRAPA-IFPRI, Rio Branco, AC, Brazil, 1997.

Sydenstricker-Neto, J., Parceleiros de Machadinho: História Migratória e as Interações entre a Dinâmica Demográfica e o Ciclo Agrícola em Rondônia, M.A. thesis, State University of Campinas (UNI-CAMP), Campinas, SP, Brazil, 1992.

Sydenstricker-Neto, J. and H.G. Torres, Mobilidade de migrantes: autonomia ou subordinação na Amazônia?, *Revista Brasileira de Estudos de População*, 8, 33–54, 1991.

Szott, L.T., E.C.M. Fernandes, and P.A. Sanchez, Soil-plant interactions in agroforestry systems. *For. Ecol. Manage.*, 45, 127–152, 1991.

Vosti, S.A., J. Witcover, C.L. Carpentier, S.J. Magalhães de Oliveira, and J. Carvalho dos Santos, Intensifying Small-Scale Agriculture in the Western Brazilian Amazon: Issues, Implications and Implementation, in *Tradeoffs or Synergies?: Agricultural Intensification, Economic Development and the Environment*, Lee, D. and C. Barrett, Eds., CAB International, Wallingford, U.K., 2000.

Wood, C.H. and D. Skole, Linking Satellite, Census, and Survey Data to Study Deforestation in the Brazilian Amazon, in *People and Pixels: Linking Remote Sensing and Social Science*, Liverman, D. et al., Eds., National Academy Press, Washington, DC, 1998.

CHAPTER 7

Thematic Accuracy Assessment of Regional Scale Land-Cover Data

Siamak Khorram, Joseph F. Knight, and Halil I. Cakir

CONTENTS

7.1 INTRODUCTION

The Multi-Resolution Land Characteristics (MRLC) consortium, a cooperative effort of several U.S. federal agencies, including the U.S. Geological Survey (USGS) EROS Data Center (EDC) and the U.S. Environmental Protection Agency (EPA), has conducted the National Land Cover Data (NLCD) program. This program used Landsat Thematic Mapper (TM) 30-m resolution imagery as baseline data and successfully produced a consistent and conterminous land-cover (LC) map of the lower 48 states at approximately an Anderson Level II thematic detail. The primary goal of the program was to provide a generalized and regionally consistent LC product for use in a broad range of applications (Lunetta et al., 1998). Each of the 10 U.S. federal geographic regions

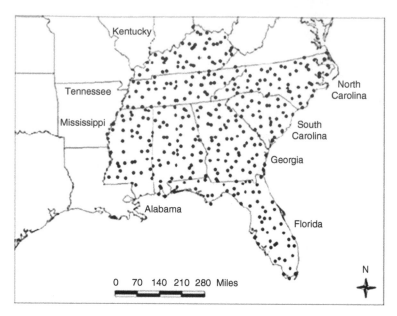

Figure 7.1   Randomly selected photograph center points.

was mapped independently. EPA funded the Center for Earth Observation (CEO) at North Carolina State University (NCSU) to assess the accuracy of the NLCD for federal geographic Region IV.

An accuracy assessment is an integral component of any remote sensing-based mapping project. Thematic accuracy assessment consists of measuring the general and categorical qualities of the data (Khorram et al., 1999). An independent accuracy assessment was implemented for each federal geographic region after LC mapping was completed. The objective for this study was specifically to estimate the overall accuracy and category-specific accuracy of the LC mapping effort. Federal geographic Region IV included the states of Kentucky, Tennessee, Mississippi, Alabama, Georgia, Florida, North Carolina, and South Carolina (Figure 7.1).

7.2 APPROACH

7.2.1   Sampling Design

Quantitative accuracy assessment of regional scale LC maps, produced from remotely sensed data, involves comparing thematic maps with reference data (Congalton, 1991). Since there were no suitable existing reference data that could be used for all federal regions, a practical and statistically sound sampling plan was designed by Zhu et al. (2000) to characterize the accuracy of common and rare classes for the map product using National Aerial Photography Program (NAPP) photographs as the reference data.

The sampling design was developed based on the following criteria: (1) ensure the objectivity of sample selection and validity of statistical inferences drawn from the sample data, (2) distribute sample sites spatially across the region to ensure adequate coverage of the entire region, (3) reduce the variance for estimated accuracy parameters, (4) provide a low-cost approach in terms of budget and time, and (5) be easy to implement and analyze (Zhu et al., 2000).

The sampling was a two-stage design. The first stage, the primary sampling unit (PSU), was the size of a NAPP aerial photograph. One PSU (photo) was randomly selected from a cluster of 128 photographs. These clusters were formed using a geographic frame of 30 × 30 m. Randomly selected PSU locations are shown in Figure 7.1. The second stage was a stratified random sample,

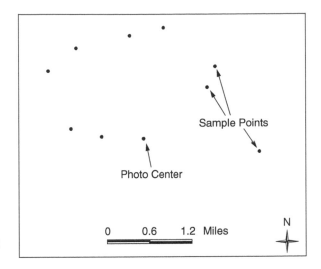

Figure 7.2   Sample sites clustered around
the photograph center.

within the extent of all of the PSUs only, of 100 sample sites per LC class. The selected sites were
referred to as secondary sampling units (SSU). The number of sites per photograph ranged from
1 to approximately 70 (Figure 7.2). The total number of sample sites in the study was 1500 (100
per cover classes), although only 1473 sites were interpreted due to missing NAPP photos. This
sampling approach was chosen by the Eros Data Center (EDC) over a standard random sample to
reduce the cost of purchasing the NAPP photography (Zhu et al., 2000).

7.2.2   Training

Before the NAPP photo interpretation for the sample sites could begin, photo interpreters were
trained to accomplish the goals of the study. To provide consistency among the interpreters, a
comprehensive training program was devised. The program consisted of a full-day training session
and subsequent on the job training. Two experienced aerial photo interpretation and photogram
metry instructors led the formal classroom training sessions. The training sessions included the
following topics: (1) discussion of color theory and photo interpretation techniques, (2) understanding of the class definitions, (3) interpretation of over 100 sample sites of different classes during
the training sessions followed by interactive discussions about potential discrepancies, (4) creation
of sample sites for later reference, and (5) repetition of interpretation practice after the sessions.

The focus was on real-world situations that the interpreters would encounter during the project.
Each participant was presented with over 100 preselected sites and was asked to provide his or her
interpretation of the land cover for these sites. Their interpretations were analyzed and subsequently
discussed to minimize any misconceptions. During the on-the-job portion of the training, each
interpreter was assigned approximately 500 sites to examine. Their progress was monitored daily
for accuracy and proper methodology. The interpreters kept logs of their decisions and the sites for
which they were uncertain about the LC classes. On a weekly basis, their questions were addressed
by the project photo interpretation supervisor. The problem sites (approximately 400) were discussed
until all team members felt comfortable with the class definitions and their consistency in interpretation. Agreement analysis between the three interpreters resulted in an average agreement of 84%.

7.2.3   Photographic Interpretation

7.2.3.1   Interpretation Protocol

The standard protocol used by the photo interpreters was as follows:

- Each interpreter was assigned 500 of the 1500 total sites.
- Interpretation was based on NAPP photographs.
- The sample site locations on the NAPP photos were found by first plotting the sites on TM false-color composite images then finding the same area on the photo by context.
- During the interpretation process, cover type and other related information such as site homogeneity were recorded for later analysis.
- When there was some doubt as to the correct class or there was the possibility that two classes could be considered correct, the interpreters selected an alternate class in addition to the primary class.
- The interpretations were based on the majority of a 3×3 pixel window (Congalton and Green, 1999).

7.2.3.2 Interpretation Procedures

The Landsat TM images were displayed using ERDAS Imagine. By plotting the site locations on the Landsat TM false-color composite images, the interpreters precisely located each site. Then, based on the context from the image, the interpreters located the site on the photographs as best they could. Clearly, some error was inherent in this location process; however, this was the simplest and most cost-effective procedure available. The use of a 3×3 pixel window for interpretation was intended to reduce the effect of location errors.

The interpreters examined each site's characteristics using the aerial photograph and TM image and determined the appropriate LC label for the site according to the classification scheme, then they entered the information into the project database. The following data were entered into the database: site identification number (sample site), coordinates, photography acquisition date, photograph identification code, imagery identification number, primary or dominant LC class, alternate LC class (if any), general site description, unusual observations, general comments, and any temporal site changes between image and photo acquisition dates. The interpreters did not have prior access to the MRLC classification values during the interpretation process.

Individual interpreters analyzed 15% ($n = 75$) of each of the other interpreters' sample sites to create an overlap database to evaluate the performance of the interpreters and the agreement among them. Selection of these 75 sites was done through random sampling. This scheme provided 225 sites that were interpreted by all three interpreters. Agreement analysis using these overlap sites indicated an average agreement of 84% among the three interpreters (Table 7.1).

7.2.3.3 Quality Assurance and Quality Control

Quality assurance (QA) and quality control (QC) procedures were vigorously implemented in the study as designated in the interpretation organization chart (Table 7.2). Discussions among the interpreters and project supervisors during the interpretation process provided an opportunity to discuss the problems that occurred and to resolve problems on the spot.

The QA and QC plan is shown in Figure 7.3. Upon completion of training, a test was performed to determine how similarly the interpreters would call the same sites. The initial results of the analysis revealed that some misunderstandings about class definitions had remained after the training process. As a result, the interpreters were retrained as a group to "calibrate" themselves. This helped to ensure that calls were more consistent among interpreters. Upon satisfactory completion of the retraining, the interpreters were assigned to complete interpretation of the 1500 sample sites.

7.3 RESULTS

7.3.1 Accuracy Estimates

Table 7.3 presents the error matrix for MRLC Level II classes. The numbers across the top and sides of the matrices represent the 15 MRLC classes (Appendix A). Table 7.4 presents the error

Table 7.1   Agreement Analysis Among PIs: Interpreter Call vs. Overlap Consensus for the 225 Overlap Sites

Interpreted Results (rows) vs. Overlap Consensus (columns)

MRLC Class	1.1	2.1	2.2	2.3	3.1	3.2	3.3	4.1	4.2	4.3	8.1	8.2	8.5	9.1	9.2	Tot	%	Corr
1.1	18												1	1		20	0.9	18
2.1	1	21														22	1.0	21
2.2			3	1												4	0.8	3
2.3				9												9	1.0	9
3.1					4							2				6	0.7	4
3.2		1				6										7	0.9	6
3.3							16	1				1				18	0.9	16
4.1							2	14	1	1			1			19	0.7	14
4.2							2		7							9	0.8	7
4.3							3	2		26	1			4		36	0.7	26
8.1											10	1				11	0.9	10
8.2											3	10		1		14	0.7	10
8.5								1					16		2	19	0.8	16
9.1										2				13	1	16	0.8	13
9.2															15	15	1.0	15
Tot	19	22	3	10	4	6	23	18	8	29	14	14	18	19	18	225		
%	0.9	1	0.9	0.9	1	1	0.7	0.8	0.9	0.9	0.7	0.7	0.9	0.7	0.8		0.84	
Corr	18	21	3	9	4	6	16	14	7	26	10	10	16	13	15			188

Table 7.2   Interpretation Team Organization

	Interpreter Organization		
Photo Interpreters	PI #1 (500 pts + 75 pts from PI #2 and 75 pts from PI #3	PI #2 (500 pts + 75 pts from PI #1 and 75 pts from PI #3	PI #3 (500 pts + 75 pts from PI #1 and 75 pts from PI #2
PI supervisor	Random checking for consistency, checking 225 overlapped sites, sites with question from three PIs		
Project supervisor	Checking sites with question from PI supervisor, random checking of overall sites, overall QA/QC		
Project director	Procedure establishment, discussions on issues, random checking, overall QA/QC		

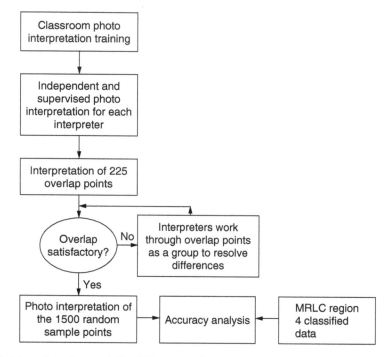

Figure 7.3   Training, photo interpretation (PI), and quality assurance and quality control (QA/QC) procedures.

matrix for MRLC Level I classes. The Level II classes were grouped into the following Level I categories: (1) water, (2) urban or developed, (3) bare surface, (4) agriculture and other grasslands, (5) forest (upland), and (6) wetland (woody or nonwoody). The overall accuracies for the Level I and II classes were 66% and 44%, respectively.

Table 7.3 illustrates the confusion among low-intensity residential, high-intensity residential, and commercial/transportation categories. Many factors may have contributed to the confusion; however, we believe the complex classification scheme used was a dominant factor. For example, the most ambiguous categories were the three urban classes, which were distinguished only by percentage of vegetation. Technically, it was beyond the methods employed in this study to quantify subpixel vegetation content. As a result, many high-intensity residential areas in the classified image were assigned to low-intensity residential and commercial/transportation classes. This occurred because high-intensity residential classes, which had a median percentage of vegetation, were easily confused with lower-intensity and higher-intensity urban development.

Also, many problems were encountered with the interpretation of cropland and pasture/hay since these classes had very similar spectral and spatial patterns that occurred within the same agricultural areas. In addition, cropland was frequently converted to pasture/hay during the interval of two acquisition dates, or vice versa. Confusion also existed within classes of evergreen forest

Table 7.3 Error Matrix for the Level II MRLC Data (15 Classes)

PI Results / MRLC Class	Classified MRLC Data																	
	1.1	2.1	2.2	2.3	3.1	3.2	3.3	4.1	4.2	4.3	8.1	8.2	8.5	9.1	9.2	Tot	%	Corr
1.1	87			3	3	5	2						2	4	2	108	0.8	87
2.1		47	49	22	1		2		2	1	5	1	24	1		155	0.3	47
2.2		1	10	2									4		2	19	0.5	10
2.3		3	22	32	1	5		1	1				4		1	69	0.5	32
3.1		2	3	6	33	18			2			1		1	2	69	0.5	33
3.2				1	3	34										38	0.9	34
3.3	1					13	33	4	2	12	3	4	1	5	1	78	0.4	33
4.1		6		3		8	6	46	3	7	3	7		9		99	0.5	46
4.2	1	1	1	1			7		34	4			2	6	4	61	0.6	34
4.3		24	3	7	2	6	16	29	42	62	9	4	4	16	4	228	0.3	62
8.1		2	1	2	15	4	11	4	1	4	28	18	11	1	2	103	0.3	28
8.2		1		3	11	1	6	3	1	1	37	57	3	2	2	128	0.4	57
8.5		10	11	13	20	3	7	4	1	3	8	4	41	2	3	131	0.3	41
9.1	4	2		1	1	3	8	2	10	4		3	1	43	15	96	0.4	43
9.2	4	1		2	10	1	2	1	1					9	60	91	0.7	60
Tot	98	100	100	98	100	100	100	94	99	98	93	99	97	99	98	1473		
%	0.9	0.5	0.1	0.3	0.3	0.3	0.3	0.5	0.3	0.6	0.3	0.6	0.4	0.4	0.6		0.44	
Corr	87	47	10	32	33	34	33	46	34	62	28	57	41	43	60			647

Table 7.4 Error Matrix for the Level I MRLC Data

			MRLC data							
		1	2	3	4	8	9	Tot	%	Corr
PI Results	1	87	3	10	0	2	6	108	0.81	87
	2	0	188	9	4	38	4	243	0.77	188
	3	1	12	134	21	8	9	185	0.72	134
	4	1	46	45	227	30	39	388	0.59	227
	8	1	43	78	21	207	12	362	0.57	207
	9	8	6	24	18	4	127	187	0.68	127
	Tot	98	298	300	291	289	197	1473		
	%	0.89	0.63	0.45	0.78	0.72	0.64		0.66	
	Corr	87	188	134	227	207	127			970

and mixed forest, deciduous forest and mixed forest, barren ground and other grassland, low-intensity residential and mixed forest, and transitional and all other classes.

The difference between image classification and photo interpretation is that image classification is mostly based on the spectral values of the pixels, whereas photo interpretation incorporates color (tones), pattern recognition, and background context in combination. These issues are inherent in any accuracy assessment project using aerial photos as the reference data (Ramsey et al., 2001). For this project, however, aerial photos were the only reasonable reference data source.

The interpretation process is not the only component of the accuracy assessment process (Congalton and Green, 1999). Additional factors that should be considered are positional and correspondence error. To account for these errors, the following additional criteria for correct classification were considered in this project: (1) primary matches classified pixel, (2) primary or alternate matches classified pixel, (3) primary is most common in classified 3 × 3 pixel areas, (4) primary matches any pixel in a classified 3 × 3 pixel area, (5) primary is most common in classified 3 × 3 pixel area, and (6) primary or alternate matches any pixel in a 3 × 3 pixel area. "Interpreted" refers to the classes chosen during the aerial photo interpretation process, "primary" and "alternate" are the most probable LC classes for a particular site, and "classified" refers to the MRLC classification result for that site. The analysis results for each cover class in six cases are presented in Table 7.5 and Table 7.6. The overall accuracies under various scenarios ranged from 44% to 79.4% (n = 1473) for cases "a" and "f," respectively.

Table 7.5 Summary of Further Accuracy Analysis by Interpreted Cover Class: Number of Sites

Class	No.	Primary PI Matches MRLC	Prim or Alt Matches MRLC	Primary PI Is Mode of 3 × 3	Primary PI Matches Any 3 × 3	Prim or Alt PI Is Mode of 3 × 3	Prim or Alt PI Matches Any 3 × 3
1.1	108	87	95	84	92	94	100
2.1	155	47	69	60	81	124	135
2.2	19	10	11	8	11	15	16
2.3	69	32	39	35	41	44	49
3.1	69	33	35	27	30	34	42
3.2	38	34	36	34	36	35	37
3.3	78	33	44	33	42	40	52
4.1	99	46	55	60	68	79	83
4.2	61	34	39	44	48	52	54
4.3	228	62	98	68	110	148	187
8.1	103	28	39	27	38	46	64
8.2	128	57	82	56	83	83	102
8.5	131	41	61	33	53	56	91
9.1	96	43	53	47	59	68	84
9.2	91	60	68	58	67	67	74
Totals	1473	647	824	674	859	985	1170

Table 7.6   Summary of Further Accuracy Analysis by Interpreted Cover Class: Percentage of Sites for Each Class

Class	Percentage	Primary PI Matches MRLC	Prim or Alt PI Matches MRLC	Primary PI Is Mode of 3 × 3	Primary PI Matches Any 3 × 3	Prim or Alt PI Is Mode of 3 × 3	Prim or Alt PI Matches Any 3 × 3
1.1	100.0	80.6	88.0	77.8	85.2	87.0	92.6
2.1	100.0	30.3	44.5	38.7	52.3	80.0	87.1
2.2	100.0	52.6	57.9	42.1	57.9	78.9	84.2
2.3	100.0	46.4	56.5	50.7	59.4	63.8	71.0
3.1	100.0	47.8	50.7	39.1	43.5	49.3	60.9
3.2	100.0	89.5	94.7	89.5	94.7	92.1	97.4
3.3	100.0	42.3	56.4	42.3	53.8	51.3	66.7
4.1	100.0	46.5	55.6	60.6	68.7	79.8	83.8
4.2	100.0	55.7	63.9	72.1	78.7	85.2	88.5
4.3	100.0	27.2	43.0	29.8	48.2	64.9	82.0
8.1	100.0	27.2	37.9	26.2	36.9	44.7	62.1
8.2	100.0	44.5	64.1	43.8	64.8	64.8	79.7
8.5	100.0	31.3	46.6	25.2	40.5	42.7	69.5
9.1	100.0	44.8	55.2	49.0	61.5	70.8	87.5
9.2	100.0	66.3	73.9	63.0	72.8	72.8	80.4
Total Percentage	100.0	44.0	55.9	45.7	58.3	66.8	79.4

7.3.2   Issues and Problems

7.3.2.1   Heterogeneity

The heterogeneity of many areas caused confusion in assigning an exact class label to the sites. Since the spatial resolution of the Landsat TM data was 30 × 30 m, pixel heterogeneity was a common problem (Plate 7.1a). For example, a site on the image frequently contained a mixture of trees, grassland, and several houses. Thus, the reflectance of the pixel was actually a combination of different reflectance classes within that pixel. This factor contributed to confusion between evergreen forest and mixed forest, deciduous forest and mixed forest, low-intensity residential and other grassland, and transitional and several classes.

7.3.2.2   Acquisition Dates

Temporal discrepancies between photograph and image acquisition dates, if not reconciled, would negatively affect the classification accuracy (Plate 7.1b). For example, to interpret early forest growth areas, the interpreter had to decide whether the site was a transitional or a forested area. If the photograph was acquired before the image (e.g., as much as 6 years earlier), it was clear that those early forest growth sites would show up as forest cover on the satellite image. In this case, the interpreters decided the appropriate cover class based on satellite imagery.

7.3.2.3   Location Errors

Locating the reference site on the photo was sometimes problematic. This frequently occurred when: (1) the LC had changed between the image and photo acquisition dates, (2) there were few clearly identifiable features for positional reference, and (3) the reference site was on the border of two or more classes (boundary pixel problem). When the LC had changed between acquisition dates, locating reference sites was difficult because the features surrounding the reference site were also changed. Similarly, when a reference site fell in an area with few identifiable features for positional reference, the interpreter had to approximate the location of the reference site. For

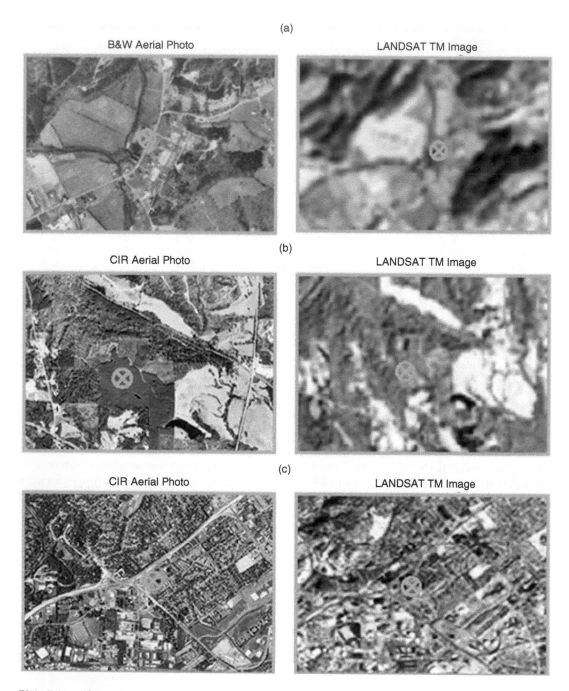

Plate 7.1   (See color insert following page 114.) (a) Heterogeneity problem: reference site consists of several classes. (b) LC class changed between acquisition dates in reference site. (c) Ambiguity of class definitions; it is difficult to differentiate between high-density and commercial class according to definition.

example, when the reference site was on the shadowy side of a mountain, it was impossible to see the reference features except the ridgeline of the mountain; thus, the interpreter was required to locate the reference site based on the approximate distance to and the direction of the ridgeline. The third case was the most common source of confusion in the interpretation process. Reference sites were frequently on the border of two or more classes. In these situations, the interpreter decided between two or more classes by determining which class covered the majority of the 3×3 pixel window.

7.4 FURTHER RESEARCH

The results of this study point to numerous opportunities for further research to improve accuracy assessment methods for regional scale assessments, including: (1) examining the impact of alternate classes in the accuracy assessment, (2) evaluating and analyzing the effect of positional errors on accuracy assessment, (3) collecting field data for the 225 overlapping sample sites to validate the interpretation, and (4) analyzing satellite data with a higher temporal resolution to better identify changes between the acquisition of TM data and NAPP photography (e.g., using NOAA-AVHRR and MODIS data).

ACKNOWLEDGMENTS

The results reported here were generated through an agreement funded by the Environmental Protection Agency (EPA). The views expressed in this report are those of the authors and do not necessarily reflect the views of EPA or any of its subagencies. The authors would like to thank EPA, USGS-EROS Data Center (EDC) for their support and for the assistance given on this project. The authors would also like to thank Heather Cheshire and Linda Babcock of CEO at NCSU for contributing their expertise in photo interpretation and extended help throughout the duration of this project.

REFERENCES

Congalton, R., A review of assessing the accuracy of classifications of remotely sensed data, *Remote Sens. Environ.,* 37, 35–46, 1991.

Congalton, R. and K. Green, A practical look at the sources of confusion in error matrix generation, *Photogram. Eng. Remote Sens.,* 59, 641–644, 1999.

Khorram, S., G.S. Biging, N.R. Chrisman, D.R. Colby, R.G. Congalton, J.E. Dobson, R.L. Ferguson, M.F. Goodchild, J.R. Jensen, and T.H. Mace, Accuracy Assessment of Remote Sensing-Derived Change Detection, monograph, American Society of Photogrammetry and Remote Sensing (ASPRS), Bethesda, MD, 1999.

Lunetta, R S., J.G. Lyon, B. Guidon, and C.D. Elvidge, North American landscape characterization dataset development and data fusion issues, *Photogram. Eng. Remote Sens.,* 64, 821–829, 1998.

Ramsey, E., G. Nelson, and K. Sapkota, Coastal change analysis program implemented in Louisiana, *J. Coastal Res.,* 17, 53–71, 2001.

Zhu, Z.,L. Yang, S.V. Stehman, and R.L. Czaplewski, Accuracy assessment for the U.S. Geological Survey regional land cover mapping program: New York and New Jersey region, *Photogram. Eng. Remote Sens.,* 66, 1425–1438, 2000.

MRLC Classification Scheme and Class Definitions

The MRLC program utilizes a consistent classification scheme for all EPA regions at approximately an Anderson Level II thematic detail. While there are 21 classes in the MRLC system, only 15 were mapped in EPA Region IV. The following classification scheme was applied to the EPA Region IV data set:

1.0 Water: All areas of open water or permanent ice/snow cover.
    1.1 Water: All areas of open water, generally with less than 25% vegetation.

2.0 Developed: Areas characterized by a high percentage of construction materials (e.g., asphalt, concrete, buildings, etc.).
    2.1 Low-intensity residential: Land includes areas with a mixture of constructed materials and vegetation or other cover. Constructed materials account for 30 to 80% of the total area. These areas most commonly include single-family housing areas, especially suburban neighborhoods. Generally, population density values in this class will be lower than in high-intensity residential areas.
    2.2 High-intensity residential: Includes heavily built-up urban centers where people reside. Examples include apartment complexes and row houses. Vegetation occupies less than 25% of the landscape. Constructed materials account for 80 to 100% of the total area. Typically, population densities will be quite high in these areas.
    2.3 High-intensity commercial/industrial/transportation: Includes all highly developed lands not classified as "high-intensity residential," most of which is commercial, industrial, and transportation.

3.0 Barren: Bare rock, sand, silt, gravel, or other earthen material with little or no vegetation regardless of its inherent ability to support life. Vegetation, if present, is more widely spaced and scrubby than that in the vegetated categories.
    3.1 Bare Rock/Sand: Includes areas of bedrock, desert pavement, scarps, talus, slides, volcanic material, glacial debris, beach, and other accumulations of rock and/or sand without vegetative cover.
    3.2 Quarries/strip mines/gravel pits: Areas of extractive mining activities with significant surface expression.
    3.3 Transitional: Areas dynamically changing from one land cover to another, often because of land use activities. Examples include forestlands cleared for timber and may include both freshly cleared areas as well as areas in the earliest stages of forest growth.

4.0 Natural forested upland (nonwet): A class of vegetation dominated by trees generally forming > 25% canopy cover.
    4.1 Deciduous forest: Areas dominated by trees where 75% or more of the tree species shed foliage simultaneously in response to an unfavorable season.
    4.2 Evergreen forest: Areas dominated by trees where 75% or more of the tree species maintain their leaves all year. Canopy is never without green foliage.
    4.3 Mixed forest: Areas dominated by trees where neither deciduous nor evergreen species represent more than 75% of the cover present.

5.0 Herbaceous planted/cultivated: Areas dominated with vegetation that has been planted in its current location by humans and/or is treated with annual tillage, modified conservation tillage, or other intensive management or manipulation. The majority of vegetation in these areas is planted and/or maintained for the production of food, fiber, feed, or seed.
    5.1 Pasture/hay: Grasses, legumes, or grass-legume mixtures planted for livestock grazing or the production of seed or hay crops.

5.2   Row Crops: All areas used for the production of crops such as corn, soybeans, vegetables, tobacco, and cotton.

5.3   Other grasses: Vegetation planted in developed settings for recreation, erosion control, or aesthetic purposes. Examples include parks, lawns, and golf courses.

6.0   Wetlands: Nonwoody or woody vegetation where the substrate is periodically saturated with or covered with water.

6.1   Woody wetlands: Areas of forested or shrubland vegetation where soil or substrate is periodically saturated with or covered with water.

6.2   Emergent herbaceous wetlands: Nonwoody vascular perennial vegetation where the soil or substrate is periodically saturated with or covered with water.

Note: Cover class types 5.0, 6.0, and 7.0 did not occur in federal geographic Region 4.

CHAPTER 8

An Independent Reliability Assessment for the Australian Agricultural Land-Cover Change Project 1990/91–1995

Michele Barson, Vivienne Bordas, Kim Lowell, and Kim Malafant

CONTENTS

8.1 INTRODUCTION

Australia's first National Greenhouse Gas Inventory (NGGI) suggested that agricultural clearing could represent as much as a quarter of Australia's total greenhouse gas emissions (DOEST, 1994). The Agricultural Land Cover Change (ALCC) project was undertaken by the Bureau of Rural Sciences (BRS) and eight Australian state government agencies to document rates of deforestation and reforestation (1990/1991–1995) for the purpose of improving estimates of greenhouse gas emissions (Barson et al., 2000). For this study, woody vegetation was defined as all vegetation, native or exotic, with a height ≥ 2 m and a crown cover density of $\geq 20\%$ (McDonald et al., 1990). This was the definition of "forest" agreed to by state and commonwealth agencies for Australia's National Forest Inventory (NFI) and the definition used for Australia's NGGI (NGGI, 1999). The definition includes vegetation usually referred to as forest (50–100% crown cover), as well as woodlands (20–50% crown cover) and plantations (silvaculture operations), but not open woodlands where crown cover is $< 20\%$. Woodlands occupy 112.0×10^6 ha, followed by forests (43.6×10^6 ha) and plantations (1.0×10^6 ha) (NFI, 1998).

The project documented both increases and decreases in woody vegetation over the period 1990/1991–1995 for the intensive land-use (LU) zone (Figure 8.1). Decreases or clearing were defined as the removal of woody vegetation resulting in a crown cover of $< 20\%$. Increases in woody vegetation (usually due to tree planting) result in a crown cover that exceeds 20%). The

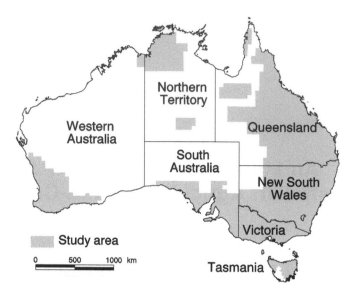

Figure 8.1   Study area for the agricultural land-cover change for the Australian continent 1990/91–1995 project.

intensive LU zone comprises some 288×10^6 ha, representing approximately 38% of the Australian continent, and is where most land clearing has taken place. Outside this zone in the Australian outback, the land-cover (LC) is disturbed but relatively intact (Graetz et al., 1995). The project produced change maps for 156 pairs of Thematic Mapper (TM) scenes, which showed that clearing for agriculture and development activities over the study period totaled approximately 1.2×10^6 ha (308,000 ha/year). The results indicated that more than 80% of the clearing for agriculture was taking place in the state of Queensland (Figure 8.2) and that the annual rates of clearing for the continent were almost 210,000 ha, or 40% lower than the figures compiled for the first NGGI.

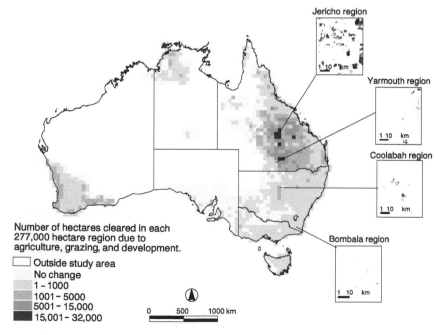

Figure 8.2   Distribution of clearing (ha) of woody vegetation from 1990/91–1995 for agriculture, grazing, and development in Australia.

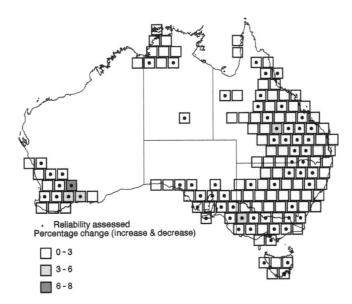

Figure 8.3 Percentage of land-cover change by Thematic Mapper (TM) image (all causes).

Clearing is a sensitive issue politically in Australia because of its impact on biodiversity conservation, greenhouse gas emissions, and land and water salinization. In most states, farmers are now required to apply for a permit before clearing vegetation. Because this study represented the first major operational use of remotely sensed data in Australia, there was substantial interest in the reliability of the LC change estimates.

Traditionally, accuracy assessment of data sets derived from satellite imagery involves comparing them against an independent or reference source of information assumed to be correct (i.e., aerial photographs). However, no suitable data were available for our 288×10^6-ha study area. The limited existing aerial photography had already been used in the quality assurance process, which required that every change pixel identified be checked against another data source (Kitchin and Barson, 1998). Additionally, for most of the 156 TM scene pairs, LC change was a rare event; over 96% of the scenes had less than 3% of their area affected by change (Figure 8.3).

A methodology that did not require a reference data set and could be applied to change data produced using a variety of approaches to image processing and radiometric calibration (Table 8.1) had been developed by Lowell (2001) to evaluate the LC change maps produced for the ALCC project by the state agencies. This chapter reports on the application of Lowell's method and on the reliability of the estimates of change produced for the Australian ALCC project.

Table 8.1 Land-Cover Change Detection Methods Used for Each State

State	Method
NSW	Unsupervised classification of combined 1991 and 1995 images
NT	Band 5 subtraction of 1990 and 1995 images
QLD	Thresholding of band 2, 5, and NDVI difference images
SA	Unsupervised classification (150 classes) of combined 1990 and 1995 images
TAS	Thresholding of NDVI difference data
VIC	Unsupervised classification of combined 1990 and 1995 images to create woody, nonwoody, woody increase, and woody decrease
WA	Combined 1990 and 1995 images and carried out canonical variant analysis based on biogeographic regions to identify suitable indices and bands to classify land-cover change

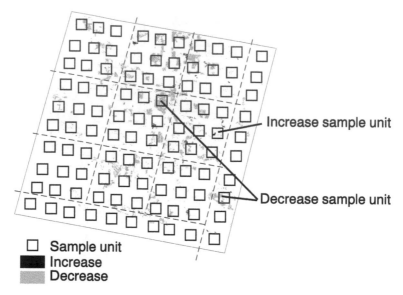

Figure 8.4   Sampling strategy for estimating land-cover change (increase plus decrease in woody vegetation) for a TM (Thematic Mapper) image.

8.2 METHODS

Lowell's method produces an area-based, independent estimate of change for which confidence intervals can be calculated; the estimates of change derived from image analysis are then compared against these confidence intervals. The method uses 500- × 500-pixel sample units (Figure 8.4) selected for each change image in a two-stage procedure to ensure that a minimum number (n) of samples containing change are included. First, sample units are selected containing change areas (increase or decrease), proportional to the amount of change across the image. The remaining sample units are then selected according to a spatially distributed random sample without replacement. Lowell demonstrated that 33 samples were required to obtain stable confidence intervals for TM scenes in which change was relatively rare (i.e., approximately 0.13% of the scene). The n could be reduced to 22 for scenes where change was more common (i.e., approximately 2.5%) (Lowell, 1998). Following sample unit selection, the area of change (increase or decrease in woody vegetation) was estimated for each sample using image enhancement techniques such as unsupervised classification or by displaying band differences or differences among various combinations of bands. Confidence intervals for woody vegetation changes were then calculated for the independent estimates of change made for each scene. These were then compared with the amount of change reported for the scene by the corresponding state agency.

Lowell's method was applied to the ALCC results within a spatial hierarchy to enable analysis of variations in LC change at multiple resolutions (i.e., study area, state, TM scene, and sample unit). Approximate z-score values were calculated using the state estimates of change, the overall sample estimate, and the individual sample results. Significance levels for the z-scores were calculated and compared for both the overall and individual sample unit results. Approximate confidence intervals (95%) for the estimated proportion of change were also compared to the overall state estimate. Dual assessment criteria (Jupp, 1998) were implemented to consider both the overall scene results and the individual sample results; the state change estimates for a scene were only accepted as reliable when both the overall and individual sample z-scores were not significantly different. Preparation of the grid for sample units was automated by writing a program to determine the areal extent of each change image and to establish a grid of nonoverlapping 500- × 500-pixel sample units.

The BRS obtained 156 change images from state agencies and calculated the total change (woody vegetation) for each image. Due to resource limitations, it was decided that only half the scenes in the study area would be assessed. Scenes containing only a small percentage of landmass were excluded, leaving 151 for possible assessment. Within each state, scenes were classified as having "low" or "high" levels of change. A weighting program that took into account the amount of change in each scene, the individual states' overall contribution to Australia's LC change, and the methods used for change detection was used to select scenes for sampling. A total of 67 scenes were selected for reliability assessment (Figure 8.3).

Analysts who had not previously been involved in the project were selected through a competitive tendering process to prepare the independent sample-based estimates of change for comparison with the results produced by state agencies. The analysts were provided with coregistered TM images for the 67 scenes, pixel coordinates for the upper left-hand corners for sample units within scenes, and 1990/1991 LC maps showing the distribution of woody vegetation, but no LC change information. They verified the image coregistration, and after some preliminary testing they calculated the normalized vegetation index (NDVI) for each image then subtracted the NDVI images and displayed the difference image (Jensen, 1996).

The NDVI difference image for each sample unit was examined at a range of threshold values to determine the location of positive (increases in woody vegetation) and negative (decreases) differences. These areas were checked in detail against the 1990/1991 and 1995 images and the upper and lower thresholds for increases and decreases were recorded. A final classification of change was performed using selected thresholds, and the change areas were checked against the LC image. This ensured that only areas that were woody in 1990/1991 could be identified as a "decrease" and those that were not woody in 1990/1991 as an "increase" in woody cover. The analysis provided an estimate of the number of pixels of increase and decrease for each sample unit in the image.

For quality assurance (QA) purposes, four to six sample units from half the images being assessed were randomly sampled. Change for these sample units was assessed as described above, but by different operators. Differences in interpretation were discussed and evaluated statistically using a paired *t*-test. A sample unit failed QA if the average of the differences found by two operators was not the same. In this case the main operator reexamined all the sample units for the image. The analysts provided BRS with a spreadsheet for each image containing the sample locations and the number of pixels of increase and decrease for each sample in the image, plus notes on any other areas of possible change identified. The BRS implemented an automatic analysis to evaluate the differences between the state estimate of change and those provided by the consultants. For scenes where the state's and the analyst's estimates of change differed substantially, the BRS investigated the possible reasons. The approximate spatial distribution of the change in the state change map was also examined to determine whether a lack of acceptance was due to highly localized changes difficult to sample effectively using the current method.

The investigation of lack of acceptance included inspecting the state's sources of information used for initial checking of the change (i.e., aerial photography, other satellite imagery, or field data) (Kitchin and Barson, 1998). Discrepancies were forwarded to respective states for advice on likely reasons for such differences. If no reason could be identified (e.g., where severe drought had led to leaf drop so that spectrally the area appeared to have been cleared, but ground inspection showed that it had not), the scene was reprocessed by the analysts.

8.3 RESULTS

In the first assessment, 60 of 67 scenes met the acceptance criteria described above. The seven noncompliant scenes were forwarded to the states for comment, and a new set of 500- × 500-pixel sample units was generated for reprocessing these scenes. On reprocessing, five additional scenes

Table 8.2   Distribution of Scenes for Independent Assessment

State	Total Number of Scenes in Study Area	Scenes Assessed	Scenes Reprocessed	Scenes Meeting Acceptance Criteria
New South Wales including Australian Capital Territory	37	10	2	9
Northern Territory	10	3	0	3
Queensland	53	22	0	22
South Australia	17	11	3	10
Tasmania	5	3	1	3
Victoria	16	9	0	9
Western Australia	18	9	1	9
Total	156	67	7	65

met the acceptance criteria (Table 8.2). Of the remaining two scenes, one contained a significant proportion of change due to fire on rocky hillsides that was difficult to map from the image without additional photo interpretation. The second also contained changes due to fire and the loss of native vegetation followed by plantation establishment. These changes were difficult to detect without local knowledge and ancillary data.

Further analyses of change maps were undertaken to estimate the variability in overall state and continental estimates of change, and change estimates within each state. The analyst and state results provide a spatial hierarchy in which change proportions and variability could be estimated (Figure 8.5). The overall mean proportions and variability were estimated for the change scenes. Approximate 95% confidence intervals for the means were calculated at each of the levels and were used to identify any significant differences between the estimates at each of the levels. In most cases, the mean change proportions estimated from the consultant's process provided values well within the 95% confidence interval estimated by the states (Table 8.3). The only exceptions were two scenes from Queensland with mean change estimates in excess of the confidence interval estimated by the state. Although the analyst's process provided lower mean change proportion estimates than those of the state, the two estimates of variability were generally similar. The continental estimates were within the 95% confidence interval, although the state estimate of continental change was 1.3% vs. the analyst's lower estimate of 0.9%. The variability of the state estimate was lower (0.2%) than that of the estimate provided by the consultant's process (0.3%). Table 8.4 summarizes the mean and variability of change for the spatial hierarchy. It shows that the variability estimates from the analyst's results are greater than those from the state at comparable levels. The ranges of means for the change proportions were consistently lower for the analyst's results, but not statistically different. Our results indicated that the states' results were the most accurate, as was evidenced by the relatively small confidence intervals.

Of the 67 scenes evaluated, 90% were determined acceptable after initial processing and 97% after additional processing. This high level of acceptance provided confidence in the results of the ALCC project. The total potential error in LC change estimates across Australia is shown in Table

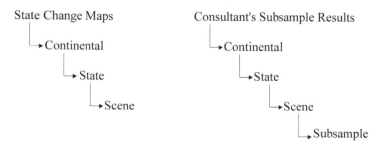

Figure 8.5   Estimation hierarchy from the state's and analyst's results sets.

Table 8.3　Comparative Analysis for State's and Analyst's Sample Unit Change Maps

State	Source	Mean	Standard Deviation	95% Confidence Interval
New South Wales	State	0.003864	0.007294	0–0.018160
	Analyst	0.001831	0.008726	0–0.018935
Northern Territory	State	0.000644	0.000478	0–0.001579
	Analyst	0	0	0
Queensland	State	0.011593	0.012670	0–0.036427
	Analyst	0.011075	0.025121	0–0.060313
South Australia	State	0.012995	0.014141	0–0.040712
	Analyst	0.005625	0.027887	0–0.060284
Victoria	State	0.013600	0.019658	0–0.052130
	Analyst	0.006055	0.029289	0–0.063462
Tasmania	State	0.039160	0.026792	0–0.091672
	Analyst	0.014050	0.026977	0–0.066925
Western Australia	State	0.032650	0.026897	0–0.085369
	Analyst	0.020183	0.061761	0–0.141235
Continental estimate	State	0.012898	0.018069	0–0.048313
	Analyst	0.008903	0.032335	0–0.072279

Note: Differences between state's and analyst's estimates were not significant ($p = 0.05$) for any state shown.

Table 8.4　Comparative Analysis for State's and Analyst's Sample Unit Change Maps

Source	Level	Mean Range	Standard Deviation Range
State change maps	Continental	0.012898	0.018069
	State	0.000644–0.039160	0.000478–0.026897
	Scene	0–0.100215	
Analysts	Continental	0.008903	0.032335
	State	0.001831–0.020183	0.008726–0.061761
	Scene	0.000041–0.068090	0.000163–0.138222
	Subsample	0–0.57344	

Table 8.5　Potential Error in Land-Cover Change Estimates for Australia if 90% or 97% of Images Are Reliable and the Remaining 10% to 3% of Images Have Various Amounts of Error

Error per Image	90% Correct	97% Correct
10%	1%	0.3%
20%	2%	0.6%
30%	3%	0.9%

8.5. If the interpretation that 10% of the change scenes failed is correct (even though an additional 7% passed with reprocessing) and the error on the failing change scenes is as high as 30%, the difference in LC change is only 3%. It would be surprising if the error were as high as this, since no scenes failed for the state with the largest amount (> 80%) of clearing (Queensland). Thus, the error was likely to be distributed among the states having the least amount of change.

8.4　DISCUSSION AND CONCLUSIONS

The goal of this study was to provide an independent evaluation of the reliability of the estimates made by state agencies of LC change in Australia from 1990/1991–1995. Traditional approaches for assessing the accuracy of LC product derived from remote sensor data were determined to be inappropriate for an assessment of our LC change products because these methods require statistically sufficient class representation (n) and relatively homogeneous distribution (CGC, 1994; Congalton and Mcleod, 1994). In our study LC change was a relatively rare event and tended to be concentrated in relatively few areas. Reference data to support a traditional accuracy assessment approach were not available.

The area-based method developed by Lowell (2001) was implemented to provide an independent estimate of change for which confidence intervals were calculated. State estimates of change were then compared against these confidence intervals to provided a means of evaluating the accuracy of the state-produced LC change products. State estimates were within the established 95% confidence intervals for 60 of the 67 scenes initially tested. The seven scenes that did not meet the acceptance criteria were reprocessed and retested, and five were subsequently accepted. LC change rates were underestimated by the analysts or overestimated by the states for the two scenes not accepted. The method overcame the difficulties caused by the lack of suitable reference data. This is likely to be a common difficulty in large-area studies of LC change. Suitable reference data will rarely be available to match the multiple dates for LC change studies. Even when multiple-date data are available, obtaining a "true" change map will be difficult, since the overlay of multiple-date LC is likely to introduce error (Lowell, 2001).

We used an area-based sampling unit rather than a discrete sample-based approach because of the relative rarity of woody vegetation change and the difficulty (and cost) of sampling enough points across a change map to support a rigorous statistical assessment. Based on extensive testing of the sample unit size, Lowell (2001) determined that the 500- × 500-pixel sample unit provided stable estimates of the confidence intervals after relatively few sample units had been examined. The present study demonstrated that a 500-pixel sample unit was a practical size for evaluation. When sample unit location had been automated, one operator could evaluate a change map with 33 samples in approximately 10 h. The area-based reliability method provided a cost-effective evaluation of the results of the ALCC project and represented only 3.5% of the total project budget.

Overall, the assessment demonstrated that the process of detecting LC change from TM data provided repeatable and reliable results. Different change techniques and approaches to radiometric calibration among individual states did not negatively affect results. Because LC change was a relatively rare event, the area-based methodology had a considerable advantage over more traditional point-based evaluation methods that require a large number of points (n) to support a rigorous statistical analysis. The method is particularly useful when suitable reference data for testing the change estimates are unavailable.

Digital data sets and the final report are available on CD-ROM. Copies can be obtained from the first two authors or downloaded (http://adl.brs.gov.au/ADLsearch/).

8.5 SUMMARY

Australia's first NGGI identified that land clearing could be contributing as much as 25% of Australia's total greenhouse gas emissions. These figures were regarded as very uncertain, and a collaborative project was undertaken with eight state agencies using TM imagery and other data to document the rates of change in woody vegetation from 1990/1991–1995. The reliability of this project's results was assessed using a method developed by Lowell (2001) for this purpose. Traditional methods of accuracy assessment were impractical given the large size of the study area, the relative rarity of the change detected, and the lack of an appropriate reference data set. Lowell's method was implemented to provide an independent estimate of change against which state agency estimates were compared. The reliability assessment demonstrated that the process of detecting land-cover change from TM imagery was repeatable and provided consistent results across states.

This approach may be useful in other environments where reference data suitable for checking land-cover change are unavailable.

ACKNOWLEDGMENTS

The authors gratefully acknowledge the substantial efforts of the following state agencies that participated in the project: Agriculture Western Australia; New South Wales Land Information Centre; Department of Information Technology and Management; Northern Territory Department of Lands, Planning and Environment; Queensland Department of Natural Resources; South Australian Department of Primary Industries and Resources; Tasmanian Department of Primary Industry, Water and Environment; Victorian Department of Natural Resources and Environment; and the Western Australian Department of Land Administration.

The authors would like to thank David Jupp (CSIRO Earth Observation Centre) for his contributions to the development of the independent LC change assessment method and the consultants, the Royal Melbourne Institute of Technology University's Geospatial Science Initiative and Geoimage, who undertook the independent assessment. David Jupp, Eric Lambin (Catholic University, Louvain, Belgium), and the Australian Greenhouse Office are thanked for comments on an earlier manuscript. The Australian government and state government agencies jointly funded this project.

REFERENCES

Barson, M.M., L.A. Randall, and V.M. Bordas, Land Cover Change in Australia: Results of the Collaborative Bureau of Rural Sciences – State Agencies' Project on Remote Sensing of Land Cover Change, Bureau of Rural Sciences, Canberra, Australia, 2000.

Congalton, R. and R. Mcleod, Change Detection Accuracy Assessment on the NOAA Chesapeake Bay Pilot Study, in Proceedings of the First International Symposium on the Spatial Accuracy of Natural Resource Data Bases, American Society for Photogrammetry and Remote Sensing, Bethesda, MD, 1994, pp. 78–87.

CGC (Computer Graphics Center), Accuracy Assessment of Land Cover Change Detection, CGC Report No. 101, North Carolina State University, Raleigh 1994.

DOEST (Department of Environment, Sport and Territories), National Greenhouse Gas Inventory 1988 and 1990, National Greenhouse Gas Inventory Committee, Canberra, Australia, 1994.

Graetz, R.D., M.A. Wilson, and S.K. Campbell, Landcover Disturbance over the Australian Continent: A Contemporary Assessment, Biodiversity Series, Paper No. 7, Biodiversity Unit, Department of Environment, Sport and Territories, Canberra, Australia, 1995.

Jensen, J.J., Introductory Digital Image Processing: A Remote Sensing Perspective, 2nd ed., Prentice Hall, Englewood Cliffs, NJ, 1996.

Jupp, D.L.B., Report on the Development of an Independent Accuracy Assessment Methodology for the Remote Sensing of Agricultural Land Cover Change 1990–1995 Project for the Australian Continent, Bureau of Resource Sciences Report, Canberra, Australia, 1998.

Kitchin, M. and M. Barson, Monitoring Land Cover Change: Specifications for the Agricultural Land Cover Change 1990–1995 project, version 4, Bureau of Rural Sciences, Canberra, Australia, 1998.

Lowell, K., Development of an Independent Accuracy Assessment Methodology for the Remote Sensing of Agricultural Land Cover Change 1990–1995 Project for the Australian Continent, report to the Bureau of Resource Sciences, Canberra, Australia, 1998.

Lowell, K., An area based accuracy assessment methodology for digital change maps, Int. J. Remote Sens., 22, 3571–3596. 2001.

McDonald, R.C., R.F. Isbell, J.G. Speight, J. Walker, and M.S. Hopkins, Vegetation, in Australian Soil and Land Survey: Field Handbook, 2nd ed., Inkata Press, Melbourne, Australia, 1990.

NFI (National Forest Inventory), Australia's State of the Forests Report 1998, Bureau of Rural Sciences, Canberra, Australia, 1998.

NGGI (National Greenhouse Gas Inventory), National Greenhouse Gas Inventory Land Use Change and
    Forestry Sector 1990–1997, in *Workbook 4.2 and Supplementary Methodology*, Australian Greenhouse
    Office, Canberra, Australia, 1999.

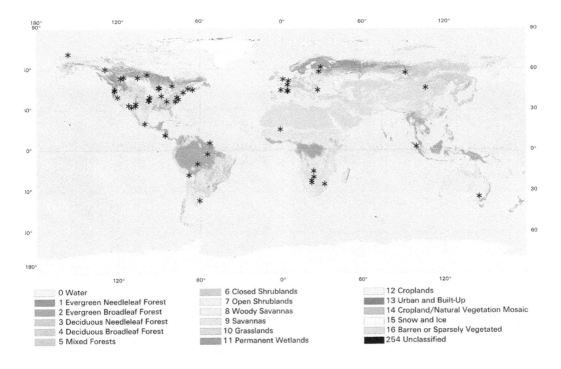

0 Water	6 Closed Shrublands	12 Croplands
1 Evergreen Needleleaf Forest	7 Open Shrublands	13 Urban and Built-Up
2 Evergreen Broadleaf Forest	8 Woody Savannas	14 Cropland/Natural Vegetation Mosaic
3 Deciduous Needleleaf Forest	9 Savannas	15 Snow and Ice
4 Deciduous Broadleaf Forest	10 Grasslands	16 Barren or Sparsely Vegetated
5 Mixed Forests	11 Permanent Wetlands	254 Unclassified

PLATE 3.1 CEOS land-cover product evaluation core site locations.

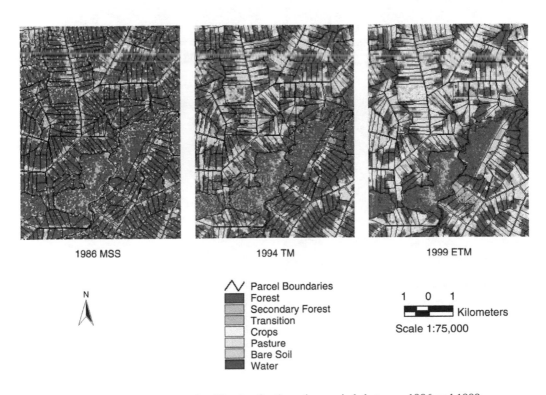

1986 MSS   1994 TM   1999 ETM

N

Parcel Boundaries
Forest
Secondary Forest
Transition
Crops
Pasture
Bare Soil
Water

1   0   1
            Kilometers
Scale 1:75,000

PLATE 6.1 Land-cover classification for three time periods between 1986 and 1999.

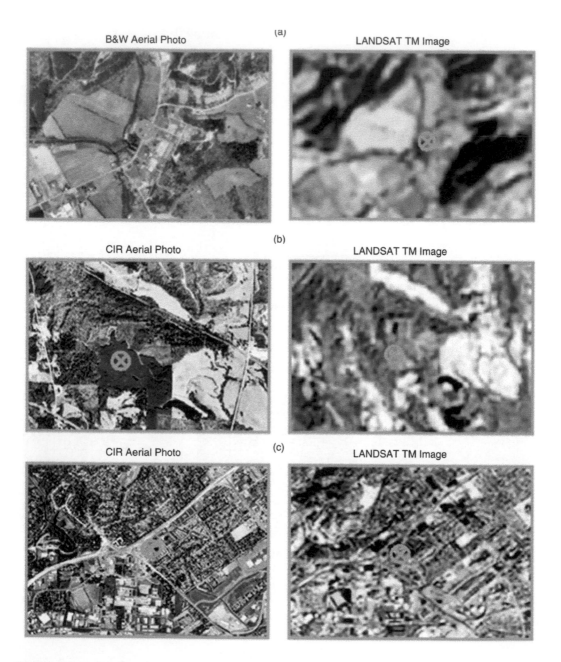

PLATE 7.1 (a) Heterogeneity problem: reference site consists of several classes. (b) LC class changed between acquisition dates in reference site. (c) Ambiguity of class definitions; it is difficult to differentiate between high-density and commercial class according to definition.

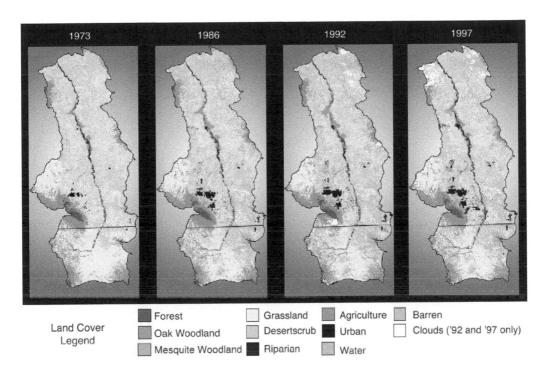

1973	1986	1992	1997

Land Cover Legend

- Forest
- Oak Woodland
- Mesquite Woodland
- Grassland
- Desertscrub
- Riparian
- Agriculture
- Urban
- Water
- Barren
- Clouds ('92 and '97 only)

PLATE 9.1 1973, 1986, 1992 and 1997 land-cover maps of the upper San Pedro River watershed with key to classes.

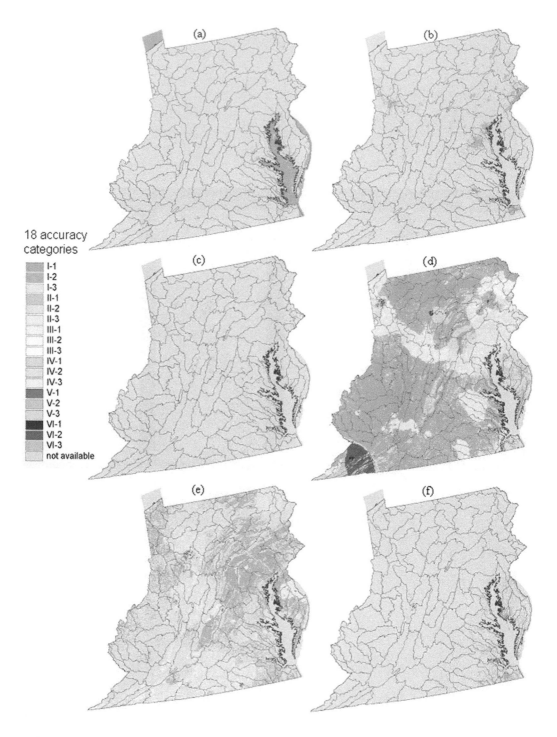

PLATE 13.1 Fuzzy accuracy maps of (a) water, (b) developed, (c) barren, (d) forested upland, (e) herbaceous planted/cultivated, and (f) wetlands.

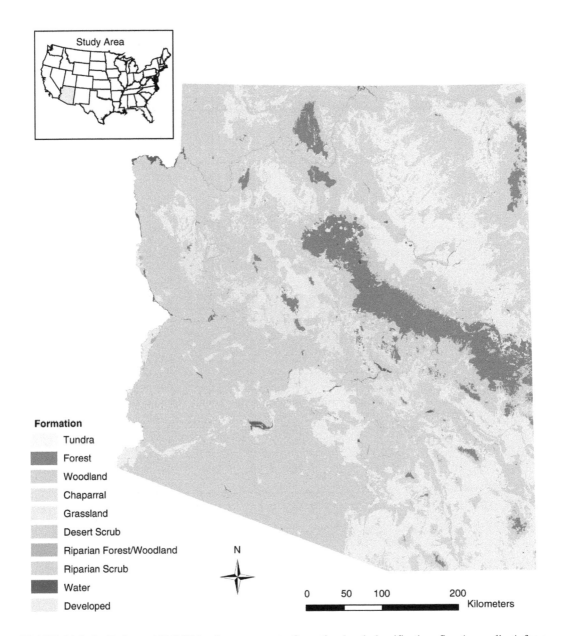

Formation

- Tundra
- Forest
- Woodland
- Chaparral
- Grassland
- Desert Scrub
- Riparian Forest/Woodland
- Riparian Scrub
- Water
- Developed

N

0   50   100   200
                        Kilometers

PLATE 14.1 Preliminary AZ-GAP land-cover map to formation level classification. See Appendix A for a complete list of all cover classes. The preliminary map contained 58,170 polygons describing 105 vegetation types (Appendix A).

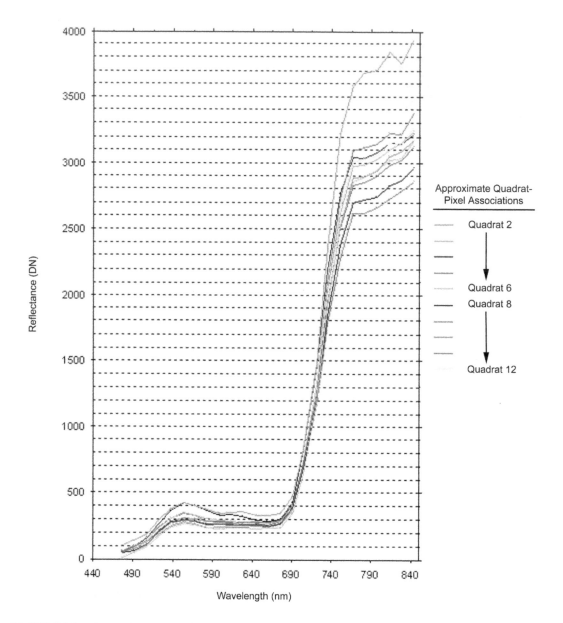

PLATE 18.1 Comparison of *Phragmites australis* among 10 field-sampled quadrats using spectral reflectance of PROBE-1 data (480 nm–840 nm). Pixel locations were in the approximate location of quadrats in the northernmost *Phragmites* stand at Pointe Mouillee.

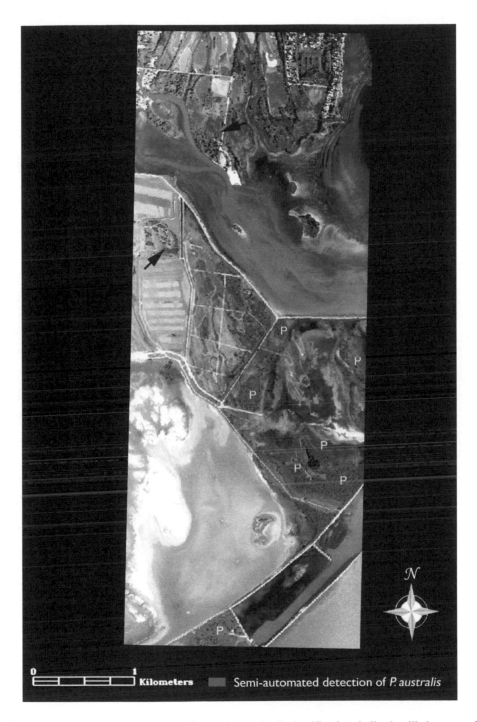

PLATE 18.2 Results of a Spectral Angle Mapper (supervised) classification, indicating likely areas of relatively homogeneous stands of *Phragmites australis* (solid blue) and field-based ecological data. Black arrows show field-sampled patches of *Phragmites*. Areas of mapped *Phragmites* are overlaid on a natural-color PROBE-1 image of Pointe Mouillee wetland complex (August 29, 2001). Yellow "P" indicates location of generally known areas of *Phragmites*, as determined from 1999 aerial photographs.

PLATE 19.1 An approximately 15-ha portion of the Dead Run sub-watershed showing (a) truth vector file (roads and rooftops) overlain on a USGS DOQQ and (b) rasterized 3-m reference GRID overlain on the 30-m subpixel estimate grid.

CHAPTER 9

Assessing the Accuracy of Satellite-Derived Land-Cover Classification Using Historical Aerial Photography, Digital Orthophoto Quadrangles, and Airborne Video Data

Susan M. Skirvin, William G. Kepner, Stuart E. Marsh, Samuel E. Drake, John K. Maingi, Curtis M. Edmonds, Christopher J. Watts, and David R. Williams

CONTENTS

9.1 INTRODUCTION

There is intense interest among federal agencies, states, and the public to evaluate environmental conditions on community, watershed, regional, and national scales. Advances in computer technology, geographic information systems (GIS), and the use of remotely sensed image data have provided the first opportunity to assess ecological resource conditions on a number of scales and to determine cross-scale relationships between landscape composition and pattern, fundamental ecological processes, and ecological goods and services. Providing quantifiable information on the thematic and spatial accuracy of land-cover (LC) data derived from remotely sensed sources is a fundamental step in achieving goals related to performing large spatial assessments using space-based technologies.

Remotely sensed imagery obtained from Earth-observing satellites now spans three decades, making possible the mapping of LC across large regions by the classification of satellite images. However, the accuracy of these derived maps must be known as a condition of the classification. Theoretically, the best reference data against which to evaluate classifications are those collected on the ground at or near the time of satellite overpass. However, such data are rarely available for retrospective multitemporal studies, thus mandating the use of alternative data sources. Accordingly, the U.S. Environmental Protection Agency (EPA) has established a priority research area for the development and implementation of methods to document the accuracy of classified LC and land characteristics databases (Jones et al., 2000).

To meet the ever-growing need to generate reliable LC products from current and historical satellite remote sensing data, the accuracy of derived products must be assessed using methods that are both effective and efficient. Therefore, our objective was to demonstrate the viability of utilizing new high-resolution digital orthophotography along with other airborne data as an effective substitute when historical ground-sampled data were not available. The achievement of consistent accuracy assessment results using these diverse sources of reference data would indicate that these techniques could be more widely applied in retrospective LC studies.

In this study, classification accuracies for four separate LC maps of the San Pedro River watershed in southeastern Arizona and northeastern Sonora, Mexico (Figure 9.1) were evaluated using historical aerial photography, digital orthophoto quadrangles, and high-resolution airborne video. Landsat Multispectral Scanner (MSS) data (60-m pixels) were classified for the years 1973, 1986, and 1992. Lastly, 1997 Landsat Thematic Mapper (TM) data (30-m pixels) were resampled to 60 m to match the MSS resolution and classified. All data were analyzed at the Instituto del Medio Ambiente y el Desarrollo Sustentable del Estado de Sonora (IMADES) in Hermosillo, Mexico. Map accuracy was assessed by Lockheed-Martin (Las Vegas, Nevada) for 1973 and 1986 and at the University of Arizona (Tucson, Arizona) for 1992 and 1997. This study incorporated previous accuracy assessment methods developed for the San Pedro watershed by Skirvin et al. (2000) and Maingi et al. (2002).

9.2 BACKGROUND

9.2.1 Upper San Pedro Watershed Study Area

The study location comprised the upper watershed of the San Pedro River, which originates in Sonora, Mexico, and flows north into southeastern Arizona. Covering approximately 7600 km^2 (5800 km^2 in Arizona and 1800 km^2 in Sonora, Mexico), this area represents the transition between

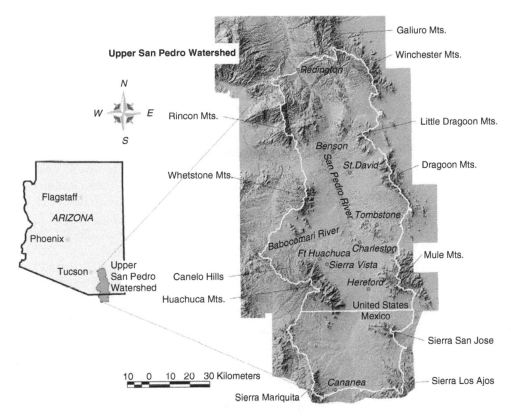

Figure 9.1   Location of the upper San Pedro River watershed study area with shaded relief map.

the Sonoran and Chihuahuan deserts, and topography, climate, and vegetation vary substantially across the watershed. Elevation ranges from 900 to 2900 m and annual rainfall ranges from 300 to 750 mm. Biome types include desertscrub, grasslands, oak woodland-savannah, mesquite woodland, riparian forest, and conifer forest, with limited areas of irrigated agriculture. Urban areas, including several small towns and the rapidly growing U.S. city of Sierra Vista, are fringed by low-density development that also occurs far from population centers. Numerous geospatial data sets covering the upper San Pedro watershed can be viewed and downloaded at the U.S. Environmental Protection Agency San Pedro Web site (USEPA, 2000).

9.2.2   Reference Data Sources for Accuracy Assessment

Aerial photography has long served in the creation of LC maps, both as a mapping base and more recently as a source of higher-resolution reference data for comparison with maps produced by classification of satellite imagery. Coverage for the conterminous U.S. at a scale of 1:40,000 is available through the National Aerial Photography Program (NAPP) and is scheduled for update on a 10-year, repeating cycle. Digital orthophoto quarter quadrangles (DOQQs) are produced from the 1:40,000-scale NAPP or equivalent high-altitude aerial photography that has been orthorectified using digital elevation models (DEMs) and ground control points of known location. A DOQQ image pixel represents 1 m² on the ground, permitting detection of landscape features as small as approximately 2 m in diameter. However, the image analyst may need site visits and/or supplementary higher-resolution images to visually calibrate for DOQQ-based LC interpretation.

Marsh et al. (1994) described the utility of airborne video data as a cost-effective means to acquire significant numbers of reference data samples for classification accuracy assessment. In that study, very similar classification accuracies were derived from airborne video reference data and from aerial color 35-mm reference photography acquired under the same conditions. With the addition of Global Positioning System (GPS) coordinate data encoded directly onto the videotape for georeferencing, sample points can be rapidly located for interpretation during playback.

9.2.3 Reporting Accuracy Assessment Results

The current standard for reporting results of classification accuracy assessment focuses on the error or confusion matrix, which summarizes the comparison of map class labels with reference data labels. Some easily computed summary statistics for the error matrix include overall map accuracy, proportion correct by classes (user and producer accuracy), and errors of omission and commission. Additional summary statistics usually include a Kappa (Khat) coefficient that adjusts the overall proportion correct for the possibility of chance agreement (Congalton et al., 1983; Rosenfield and Fitzpatrick-Lins, 1986; Congalton and Green, 1999). Although Kappa is widely used, some authors have criticized its characterization of actual map accuracy (Foody, 1992). Ma and Redmond (1995) proposed some alternatives to the Kappa coefficient, including a Tau statistic that is more readily computed and easier to interpret than Kappa. Stehman (1997) reviewed a variety of summary statistics and concluded that overall map accuracy and user and producer accuracies have direct probabilistic interpretations for a given map, whereas other summary statistics must be used with caution. The error matrix itself is recognized as the most important accuracy assessment result when accompanied by descriptions of classification protocols, accuracy assessment design, source of reference data, and confidence in reference sample labels (Stehman and Czaplewski, 1998; Congalton and Green, 1999; Foody, 2002).

9.3 METHODS

Four LC maps for the upper San Pedro River Watershed (Plate 9.1) were generated using 1973, 1986, and 1992 North American Landscape Characterization (NALC) project MSS data (Lunetta et al., 1993) and the 1997 TM data. All images were coregistered and georeferenced to a 60- × 60-m Universal Transverse Mercator (UTM) ground coordinate grid with a nominal geometric precision of 1 to 1.5 pixels (60 to 90 m).

9.3.1 Image Classification

The same LC classes ($n = 10$) were used to develop all four maps (Table 9.1). Vegetation cover classes represented very broad biome-level categories of biological organization, similar to the ecological formation levels as described in the classification system for biotic communities of North America (Brown et al., 1979). The classes included forest, oak woodland, mesquite woodland, grassland, desertscrub, riparian, agriculture, urban, water, and barren and were selected after direct consultation with the major land managers and stakeholder groups within the San Pedro watershed in Arizona and Mexico (Kepner et al., 2000). Most of the watershed was covered by grassland, desert scrub, and mesquite and oak woodland (Table 9.2).

The classification process for each data set began with an unsupervised classification using the green, red, and near-infrared spectral bands to produce a map with 60 spectrally distinct classes. The choice of 60 classes was based on previous experience with NALC data that usually gave a satisfactory trade-off between the total number of classes and the number of mixed classes. In this context, it proved helpful to define a set of 21 intermediate classes, which were easier to relate to the spectral information. For example, the barren class contained bare rock, chalk deposits, mines, tailing ponds, etc., that had unique spectral signatures. Each class was then displayed over the false-color image and assigned to one of the LC categories or to a mixed class.

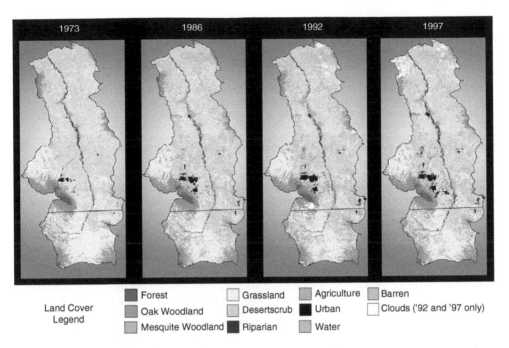

Plate 9.1   (See color insert following page 114.) 1973, 1986, 1992 and 1997 land-cover maps of the upper San Pedro River watershed with key to classes.

Interactive manipulation of spectral signatures for each class permitted many of the mixed classes to be resolved. The remaining mixed classes were separated into different categories using a variety of ancillary information sources, such as topographic maps produced by the Mexican National Institute of Statistics, Geography and Information (INEGI) (1:50,000 scale) and the U.S. Geological Survey (1:24,000 scale). The ancillary information used depended on the image being analyzed; for example, classification of the 1992 image relied heavily on field visits to establish ground control. Five 3-day site visits were conducted from September 1997 to June 1998 to enable analysts to collect specific LC data with the aid of GPS equipment.

9.3.2   Sampling Design

Because available reference data only partially covered the study area, pixels within each map were not equally likely to be selected for sampling; thus, a trade-off between practical constraints and statistical rigor was necessary (Congalton and Green, 1999). Sample points were selected using a stratified random sampling design, stratified by LC area for each of the four accuracy assessments. Reference data covering the Mexican portion of the study area were not available. The number of sample points was calculated using the following equation based on binomial probability theory (Fitzpatrick-Lins, 1981):

$$N = \frac{Z^2 pq}{E^2}$$

where N = number of samples, p = expected or calculated accuracy (%), $q = 100 - p$, E = allowable error, and Z = standard normal deviate for the 95% two-tail confidence level = 1.96.

For the lowest expected map accuracy of 60% with an allowable error of 5%, 369 sample points were required. Under area-stratified sampling, rare classes of small total area (i.e., water and barren) would not be sampled sufficiently to detect classification errors, so the minimum sample size was

Table 9.1   Land-Cover Class Descriptions for the Upper San Pedro Watershed

Forest	Vegetative communities comprised principally of trees potentially over 10 m in height and frequently characterized by closed or multilayered canopies. Species in this category are evergreen (with the exception of aspen), largely coniferous (e.g., ponderosa pine, pinyon pine), and restricted to the upper elevations of mountains that arise off the desert floor.
Oak Woodland	Vegetative communities dominated by evergreen trees (*Quercus* spp.) with a mean height usually between 6 and 15 m. Tree canopy is usually open or interrupted and singularly layered. This cover type often grades into forests at its upper boundary and into semiarid grassland below.
Mesquite Woodland	Vegetative communities dominated by leguminous trees whose crowns cover 15% or more of the ground, often resulting in dense thickets. Historically maintained maximum development on alluvium of old dissected flood plains; now present without proximity to major watercourses. Winter deciduous and generally found at elevations below 1200 m.
Grassland	Vegetative communities dominated by perennial and annual grasses with occasional herbaceous species present. Generally grass height is under 1 m and they occur at elevations between 1100 and 1700 m, sometimes as high as 1900 m. This is a landscape largely dominated by perennial bunch grasses separated by intervening bare ground or low-growing sod grasses and annual grasses with a less-interrupted canopy. Semiarid grasslands are mostly positioned in elevation between evergreen woodland above and desertscrub below.
Desertscrub	Vegetative communities comprised of short shrubs with sparse foliage and small cacti that occur between 700 and 1500 m in elevation. Within the San Pedro river basin this community is often dominated by one of at least three species (i.e., creosotebush, tarbush, and whitethorn acacia). Significant areas of barren ground devoid of perennial vegetation often separate individual plants. Many desertscrub species are drought deciduous.
Riparian	Vegetative communities adjacent to perennial and intermittent stream reaches. Trees can potentially exceed an overstory height of 10 m and are frequently characterized by closed or multilayered canopies depending on regeneration. Species within the San Pedro basin are largely dominated by two species: cottonwood and Goodding willow. Riparian species are largely winter deciduous.
Agriculture	Crops actively cultivated and irrigated. In the San Pedro River basin these are primarily found along the upper terraces of the riparian corridor and are dominated by hay and alfalfa. They are minimally represented in overall extent (less than 3%) within the basin and are irrigated by ground and pivot-sprinkler systems.
Urban (Low and High Density)	This is a land-use dominated by small ejidos (farming villages or communes), retirement homes, or residential neighborhoods (Sierra Vista). Heavy industry is represented by a single open-pit copper mining district near the headwaters of the San Pedro River near Cananea, Sonora (Mexico).
Water	Sparse free-standing water is available in the watershed. This category would be mostly represented by perennial reaches of the San Pedro and Babocomari rivers with some attached pools or repressos (earthen reservoirs), tailings ponds near Cananea, ponds near recreational sites such as parks and golf courses, and sewage treatment ponds east of the city of Sierra Vista, Arizona.
Barren	A cover class represented by large rock outcropping or active and abandoned mines (including tailings) that are largely absent of above-ground vegetation.

set to 20 where available (van Genderen and Lock, 1977). Work by Congalton (1991) and Congalton and Green (1999) suggests that sample sizes derived from multinomial theory are appropriate for comparing class accuracies, with a minimum sample size of 50 per class; however, this goal was not attainable for rare classes in this study.

After evaluation of selected sample points in each reference data set, an error matrix was constructed, comparing map class labels to reference data labels for each LC classification. Overall map accuracy and class-specific user and producer accuracies were calculated for each class. A Khat (Cohen's Kappa) and Tau (Ma and Redmond, 1995) were computed for the four error matrices, followed by a significant difference test (Z-statistic) based on Khat values (Congalton and Green, 1999).

9.3.3   Historical Aerial Photography

Reference data for the 1973 and 1986 LC maps were developed using aerial photography stereo pairs covering the Arizona portion of the study area (1:40,000 scale). A team, including a photo interpreter, an image processing specialist, a GIS specialist, and a statistician, conducted accuracy assessments. A preliminary study was conducted, using data collected during a field trip to the study area, to evaluate the effectiveness and accuracy of using aerial photographs to discriminate grassland, desertscrub, and mesquite woodland classes. These classes were particularly difficult to distinguish on the aerial photographs.

9.3.3.1 Image Collection, Preparation, and Site Selection

Landsat MSS data registration and other data integrity issues were reviewed for the 1973 and 1986 maps. These efforts included checking projection parameters and visual alignment using GIS data layers (i.e., roads, streams, digital raster graphics, and digital elevation models). Random sample points were generated using DOQQs acquired in 1992 (1:25,000 scale), and individual sample points were located on the aerial photographs using the DOQQs for accurate placement. A 180- × 180-m interpretation grid was generated and overlaid onto the LC maps.

Two mutually exclusive sets of sample points were generated for both 1973 and 1986 maps. The second set of sample points served as a pool of substitute points when no aerial photographs were available for a sample point in the first set. Whenever possible, pixels selected as sample sites represented the center of a 3 × 3 pixel window representing a homogeneous cover type. For rare classes (e.g., water), pixel sample points were chosen with at least six pixels in the window belonging to the same class. A total of 813 reference samples were used to assess the 1973 ($n =$ 429) and 1986 ($n = 384$) maps. Multiple dates of aerial photographs were used in assessment: June 1971 and April 1972 (1973 map) and June 1983, June 1984, and September 1984 (1986 map).

9.3.3.2 Photograph Interpretation and Assessment

Photointerpreter training included using a subset of the generated sample points identified during visits to the San Pedro watershed locations as interpretation keys. To avoid bias, photointerpreters did not know what classifications had been assigned to sample points on the digital LC maps. To locate the randomly chosen sample sites on the aerial photographs, the site locations were first displayed on the DOQQ. Interpreters could then visually transfer the location of each site to the appropriate photograph by matching identical spatial data such as roads, vegetation patterns, rock outcrops, or other suitable features visible on the DOQQ and on the photograph. Each transferred sample point was examined on stereoscopic photographs and identified using the definitions shown in Table 9.1. LC categories for each sample point were recorded on a spreadsheet. A comment column on the spreadsheet allowed the interpreter to enter any notes about the certainty or ambiguity of the classification. The senior photointerpreter checked the accuracy of 10% of the sample point locations and 15% of the spreadsheet entries to ensure completeness and consistency. All LC class interpretations noted by a photointerpreter as "difficult" were classified by consensus opinion of all the interpreters.

9.3.4 Digital Orthophoto Quadrangles

Approximately 60 panchromatic DOQQs acquired in 1992 for the U.S. portion of the study area were available as reference data to evaluate the 1992 results. To obtain a precise geographic matching between the DOQQs and the satellite-derived map, the 1992 source MSS image data were geometrically registered to an orthorectified 1997 TM scene, and the resulting transformation parameters were applied to the 1992 thematic map.

9.3.4.1 Interpreter Calibration

To effectively visualize conditions represented by the LC class descriptions (Table 9.1), University of Arizona and IMADES team members participated in a field visit to numerous sites in the San Pedro watershed study area, including areas that were intermediate between classes. The analyst performing the 1992 assessment also reviewed high-resolution color airborne video data for comparison with the appearance of LC classes in the DOQQs. The video data were acquired over the watershed in 1995 and vegetation in selected frames at 1:200 scale was identified to species or species groups (Drake, 2000). Image "chips" were extracted from the DOQQs as an aid to LC class recognition in the reference data (Maingi et al., 2002).

9.3.4.2 Sample Point Selection

Generation of sample points from LC maps relied on a window majority rule. A window kernel of 3 × 3 pixels was moved across each cover class and resulted in selection of a sample point if a majority of six of the nine pixels belonged to the same class. This ensured that points were extracted from areas of relatively homogenous LC. A 180- × 180-m DOQQ sample size was used to match the 3 × 3 pixel map window and a map class was assigned and recorded for the DOQQ sample. A total of 457 points were sampled to assess the 1992 map.

9.3.5 Airborne Videography

Accuracy assessment of the 1997 LC map was performed using airborne color video data encoded with GPS time and latitude and longitude coordinates. The video data were acquired on May 2 through May 5, 1997, and were therefore nearly coincident with the June Landsat TM scene. There were 11 h of continuously recorded videography of the San Pedro Watershed for the area north of the U.S.–Mexico border, acquired at a flying height of 600 m above ground level. The nadir-looking video camera used a motorized 15× zoom lens that was computer controlled to cycle every 12 sec during acquisition, with a full-zoom view held for 3 sec. The swath width at wide angle was about 750 m and was approximately 50 m at full zoom. At full zoom, the ground pixel size was about 7.0 cm and the frame was approximately 1:200-scale when displayed on a 13-inch monitor. Although the nominal accuracy of the encoded GPS coordinates was only 100 m, ground sampling revealed that average positional accuracy was closer to 40 m (McClaran et al., 1999; Drake, 2000). The video footage was acquired by flying north–south transects spaced 5 km apart and the total flight coverage encompassed a distance of nearly 2000 km.

9.3.5.1 Video and GIS Data Preparation

The encoded GPS time and geographic coordinate data were extracted from the video into a spreadsheet for each flight line. Coordinate data from the spreadsheets were used to create GIS point coverages of frames from each flight line. Individual frames of the video data were identified during viewing by a time display showing hours, minutes, and seconds, in addition to a counter that numbered the 30 frames recorded per second. The time display information was included as an attribute to the GIS point coverages, which were inspected for erroneous coordinate or time data indicated by points that fell off the flight lines or were out of time sequence; such points were deleted.

9.3.5.2 Video Sample Point Selection

To minimize the likelihood of video sample points falling on boundaries between cover classes, selection of random sample points along the video flight lines was restricted to relatively homogeneous areas within classes. This was accomplished by applying a 3 × 3 diversity or variety filter to the 1997 map, which replaced the center pixel in a moving window by the number of different data file values (cover classes) present in the window. Pixels assigned the value of one therefore

Table 9.2   Upper San Pedro Watershed Land-Cover Classes: Absolute and Relative Areas; Representative Values from 1997 Land-Cover Classification

Land-Cover Class	Area (ha)	Proportion of Total Area (%)
Grassland	263,475	36
Desertscrub	229,571	31
Woodland Mesquite	101,559	14
Woodland Oak	90,540	12
Urban	16,562	2
Agriculture	14,530	2
Riparian	9,217	1
Forest	7,193	1
Barren	6,814	1
Water	417	<0.1
Total	739,878	100

represented centers of 180- × 180-m homogeneous areas on the map. Background, clouds, and cloud-shadowed pixels were excluded to prevent the selection of pixels that fell at the edge of the map, within openings in clouds, or in cloud-shadowed areas where the adjacent cover classes were not known.

Video flight line coverages were overlaid on the map of homogeneous cover, and a subset of frames falling on homogeneous areas ($n = 4,567$) was drawn from all study area frames ($n = 18,104$). The map class under each subset frame was added as an attribute to the "candidate frames" GIS point coverage for stratification purposes.

9.3.5.3   Random Frame Selection and Evaluation

Video sample points were drawn randomly from the homogeneous subset, stratified by map class area, and were distributed throughout the Arizona portion of the study area. The water class was excluded from analysis for lack of adequate reference data ($n = 6$) and was not presented in the final error matrix. A surplus of approximately 15% over the calculated minimum number of frames needed for each cover class was selected. The videography interpreter was provided with spreadsheet records containing the videotape library identifier, latitude, and longitude for each sample frame, along with GPS time for frame location on the tape. A cover class was assigned to each sample point and recorded in the spreadsheet.

Although the accuracy of video frame interpretation was not assessed in this study, it is expected to be very high. Drake (1996) reported that LC identification of similar airborne videography at the more detailed biotic community level averaged 80% accuracy after only 3 h of interpreter training. The interpreter for this study had substantial prior experience in both video frame interpretation and ground sampling for videography accuracy assessment in this region.

9.4 RESULTS

9.4.1   Aerial Photography Method

Results of accuracy assessment are presented in Table 9.3 (1973) and Table 9.4 (1986). Overall map accuracies were similar at 70% for 1973 and 68% for 1986. Khat and Tau statistics were also similar at 0.62 and 0.59 (Khat), and 0.66 and 0.65 (Tau) for 1973 and 1986, respectively. The user's and producer's accuracies were similar to overall accuracy for all except the mesquite woodland

Table 9.3 Error Matrix Comparing Aerial Photo Interpretation and 1973 Digital Land-Cover Classification, with Producer's and User's Accuracy by Class

1973 Land-Cover Class	Reference (Aerial Photo Interpretation Class)										
	1	2	3	4	5	6	7	8	9	10	Grand total
1	19	1	0	0	0	0	0	0	0	0	20
2	1	33	0	3	0	0	0	0	0	0	37
3	0	1	16	1	0	2	0	0	0	0	20
4	0	0	13	92	21	0	0	0	1	1	128
5	0	0	14	11	96	0	0	0	0	1	122
6	0	0	3	0	2	15	0	0	0	0	20
7	0	0	3	0	7	1	10	0	0	1	22
8	0	0	0	2	5	0	0	13	0	0	20
9	0	0	4	3	6	0	1	0	3	3	20
10	0	0	0	2	15	0	0	0	1	2	20
Grand Total	20	35	53	114	152	18	11	13	5	8	429

Land-Cover Class	1973 Map Total	Photointerpreter Total	Number Correct	Producer's Accuracy (%)	User's Accuracy (%)
1. Forest	20	20	19	95	95
2. Woodland Oak	37	35	33	94	89
3. Woodland Mesquite	20	53	16	30	80
4. Grassland	128	114	92	81	72
5. Desertscrub	122	152	96	63	79
6. Riparian Forest	20	18	15	83	75
7. Agriculture	22	11	10	91	45
8. Urban	20	13	13	100	65
9. Water	20	5	3	60	15
10. Barren	20	8	2	25	10
Total	429	429	299		

Note: Overall accuracy = 70%; Tau = 0.66; Cohen's Kappa (Khat) = 0.62; standard error = 0.027.

and barren classes, which showed substantially less than average accuracies in both years. The water class in 1973 had very low accuracies of 25% (producer's) and 10% (user's) and could not be assessed for 1986.

9.4.2 Digital Orthophoto Quadrangle Method

Accuracy assessment results are summarized in Table 9.5. Overall accuracy was about 75%, with Khat of 0.70 and Tau of 0.72. The producer's accuracy was 100% for four classes (forest, urban, water, and barren), indicating that all pixels examined in the DOQQs for these classes were correctly labeled in the 1992 map. The user's accuracy was also high for forest and water classes but was substantially less for urban and barren classes at 44 and 55%, respectively. Accuracies of mesquite woodland and grassland classes were lower than those for other classes.

9.4.3 Airborne Videography Method

Overall 1997 map accuracy was 72%, with Khat of 65% and Tau of 68% (Table 9.6). A detailed examination of results by cover class shows substantial variability in classification accuracy, with producer's accuracies ranging from 54 to 100% and user's accuracies from 13 to 100%. For most classes the two measures were roughly comparable and fell within the range of 60 to 90%. Exceptions were the mesquite woodland class with accuracies around 50% and agriculture and barren classes with relatively high producer's accuracies (71 to 100%) but lower user's accuracies (13 to 21%).

Table 9.4   Error Matrix Comparing Aerial Photo Interpretation and 1986 Land-Cover Classification, with Producer's and User's Accuracy by Class

1986 Land-Cover Classes	Reference (Aerial Photo Interpretation Class)									
	1	2	3	4	5	6	7	8	10	Grand Total
1	19	1	0	0	0	0	0	0	0	20
2	3	35	0	1	0	0	0	0	0	39
3	0	0	17	3	19	0	1	0	2	42
4	0	0	12	77	12	0	1	0	2	104
5	0	0	8	13	74	0	0	0	0	95
6	0	0	0	1	1	19	2	0	0	23
7	0	0	1	4	3	2	9	1	0	20
8	0	0	0	5	3	0	0	13	0	21
10	0	0	3	10	7	0	0	0	0	20
Grand Total	22	36	41	114	119	21	13	14	4	384

Land-Cover Class	1986 Map Total	Photointerpreter Total	Number Correct	Producer's Accuracy (%)	User's Accuracy (%)
1. Forest	20	22	19	86	95
2. Woodland Oak	39	36	35	97	90
3. Woodland Mesquite	42	41	17	42	41
4. Grassland	104	114	77	68	74
5. Desertscrub	95	119	74	62	78
6. Riparian Forest	23	21	19	91	83
7. Agriculture	20	13	9	69	45
8. Urban	21	14	13	93	62
10. Barren	20	4	0	0	0
Total	384	384	263		

Note: Overall accuracy = 68%; Tau = 0.65; Cohen's Kappa (Khat) = 0.61; standard error = 0.029.

9.5 DISCUSSION

9.5.1   Map Accuracies

Statistics describing map accuracy were very similar among the four dates tested regardless of differences in assessment methods and reference data. Overall map accuracies ranged from 67 to 75% and Tau values from 0.65 to 0.72. There were no statistically significant differences among Khat values (0.61 to 0.70) for all possible date comparisons.

One aspect of sampling that differed among the assessments was the application of homogeneity standards to the context of map sample points. Selection was made from the center of uniform 3 × 3 pixel windows for the 1973 and 1986 assessments, with an exception for rare cover classes requiring only a majority of five or more pixels to match the center pixel. All sample points were selected from uniform 3 × 3 windows in the 1997 assessment. In contrast, for the 1992 assessment, a map class label was assigned as the majority of six or more pixels within a 3 × 3 window centered on the sample point. Although a positive bias may have been introduced by sampling only in homogeneous areas (Hammond and Verbyla, 1996), this effect was not apparent in results presented here.

9.5.2   Class Confusion

For all dates evaluated the producer's and user's accuracies tended to be similar to the overall classification accuracies and ranged between 61 and 100%. Generally low classification accuracies were expected in a spatially heterogeneous setting such as the San Pedro watershed, where cover types were distributed in a patchy fashion across the landscape due to climatic and edaphic effects

Table 9.5   Results of DOQQ-Based Accuracy Assessment of 1992 Land-Cover Classification: Error Matrix and Producer's and User's Accuracy by Class

1992 Land-Cover Classes	Reference (Digital Orthophoto Quads)										
	1	2	3	4	5	6	7	8	9	10	Grand Total
1	22	2	0	0	0	0	0	0	0	0	24
2	0	44	0	3	1	0	0	0	0	0	48
3	0	2	40	9	10	1	0	0	0	0	62
4	0	6	12	68	17	0	0	0	0	0	103
5	0	1	8	11	89	0	0	0	0	0	109
6	0	0	0	0	0	20	3	0	0	0	23
7	0	0	1	0	0	4	18	0	0	0	23
8	0	0	2	1	10	0	1	11	0	0	25
9	0	0	1	0	0	0	0	0	19	0	20
10	0	0	0	7	2	0	0	0	0	11	20
Grand Total	22	55	64	99	129	25	22	11	19	11	457

Land-Cover Class	1992 Map Total	DOQQ Total	Number Correct	Producer's Accuracy (%)	User's Accuracy (%)
1. Forest	24	22	22	100	92
2. Woodland Oak	48	55	44	80	92
3. Woodland Mesquite	62	64	40	63	65
4. Grassland	103	99	68	69	66
5. Desertscrub	109	129	89	69	82
6. Riparian Forest	23	25	20	80	87
7. Agriculture	23	22	18	82	78
8. Urban	25	11	11	100	44
9. Water	20	19	19	100	95
10. Barren	20	11	11	100	55
Total	457	457	342		

Note: Overall accuracy = 75%; Tau = 0.72; Cohen's Kappa (Khat) = 0.70; standard error = 0.025.

and land-use practices. Classes mapped with lower than average accuracy included the small-area agriculture, urban, water, and barren classes and the widespread mesquite woodland class. Factors likely to have contributed to class confusions included: (1) LC changes between the dates of image and reference data (especially for the 1973 and 1986 maps), (2) high spatial variability within classes (including areas dominated by soil background reflectance), (3) variable interpretations of class definitions by independent assessment teams, and (4) errors in reference data interpretation. Geometric misregistration did not appear to be a factor in the results presented here.

The agriculture class had higher producer than user accuracies for all dates and was most frequently confused with riparian, desertscrub, and mesquite woodland classes. The spatial distribution of agricultural areas in the watershed essentially outlined the riparian corridors, contributing to mixed pixel spectral response and classification confusion. There may have been difficulty in distinguishing fallow and abandoned agricultural fields from adjacent desertscrub and mesquite woodland, since the spectral response of these cover types was generally dominated by soil background.

The urban class included low-density settlement on both sides of the border. Low-density development was difficult to distinguish from surrounding cover types even at the DOQQ scale, suggesting the possibility of error in both maps and reference data. The accelerating pace of development in the watershed, particularly in Arizona, may have contributed to cover changes occurring between the dates of imagery and reference data.

The water class had the smallest area and was likely to have changed between the dates of images and reference data, due to the ephemeral nature of most surface water in this semiarid environment. For example, the 1973 NALC scene was acquired after a high-rainfall, El Niño–Southern Oscillation (ENSO) event during the winter of 1972–73 and portrayed wetter conditions than

Table 9.6 Results of Video-Based Accuracy Assessment of the 1997 Land-Cover Classification: Error Matrix and User's and Producer's Accuracy by Class

1997 Land-Cover Classes	Reference (Video Frame Data)									
	1	2	3	4	5	6	7	8	10	Grand Total
1	20	4	0	0	0	0	0	0	0	24
2	2	50	0	3	0	0	0	0	0	55
3	0	1	27	13	12	2	0	1	0	56
4	0	8	16	113	21	0	0	1	0	159
5	0	4	4	12	115	0	0	2	0	137
6	0	0	0	0	0	21	2	1	0	24
7	0	0	1	0	15	2	5	1	0	24
8	0	0	0	0	0	0	0	24	0	24
10	0	0	2	0	19	0	0	0	3	24
Grand Total	22	67	50	141	182	25	7	30	3	527

Land-Cover Class	1997 Map Total	Video Total	Number Correct	Producer's Accuracy (%)	User's Accuracy (%)
1. Forest	24	22	20	91	83
2. Woodland Oak	55	67	50	75	91
3. Woodland Mesquite	56	50	27	54	48
4. Grassland	159	141	113	80	71
5. Desertscrub	137	182	115	63	84
6. Riparian	24	25	21	84	88
7. Agriculture	24	7	5	71	21
8. Urban	24	30	24	80	100
9. Water	N/A	N/A	N/A	N/A	N/A
10. Barren	24	3	3	100	13
Total	527	527	378		

Note: Overall accuracy = 72%; Tau = 0.68; Cohen's Kappa (Khat) = 0.65; standard error = 0.024.

reference aerial photography acquired in 1971 and 1972 (Easterling et al., 1996; NOAA, 2001). The water class was not evaluated in 1986 and 1997 assessments due to insufficient representation in reference data.

The barren class was mapped with poor accuracy overall, including 0% correct in 1986. This class was most often confused with mesquite woodland, grassland, and desertscrub. These classes generally have sparse vegetation cover, with many image pixels dominated by soil or rock spectral responses, and were difficult to distinguish from truly barren areas at the MSS 60-m pixel size. A total of 38% of samples interpreted as barren on reference aerial photography from 1971 and 1972 were mapped as water in 1973; this was probably due to the interannual variations in precipitation mentioned above.

The mesquite woodland class may be interpreted as an indicator of landscape change in the San Pedro Watershed (Kepner et al., 2000, 2002). Conversion of many grassland areas to shrub dominance during the last 120 years is well documented for this region (Bahre, 1991, 1995; Wilson et al., 2001), and these change detection results were of potential interest to many researchers. However, both user and producer accuracies of all four dates were generally low for mesquite woodland (30 and 80%, respectively, for 1973 and 40 to 65% for other years). Class confusions included all but the forest class, with especially large errors in the grassland and desertscrub classes. This result may substantially reflect both the spatially and temporally transitional nature of the class and differences in interpretation among the groups performing image classification and accuracy assessment. Additionally, it was likely that neither the spectral nor the spatial resolution of MSS imagery was adequate to distinguish the mesquite woodland class in a heterogeneous semiarid environment, where most pixels are mixtures of green and woody vegetation, standing litter, and soils of varying brightness (Asner et al., 2000).

9.5.3   Future Research

Assessment of future LC classifications for the upper San Pedro area should incorporate some measure of the reference data variability, perhaps also allowing a secondary class label (Zhu et al., 2000; Yang et al., 2001). This may help to clarify the results for some cover classes. For example, the low accuracies and class confusions associated with the mesquite woodland class may have been due, in large part, to its gradational nature. If the interpreter would have been able to quantify the confidence associated with reference point interpretations, there would not have been a need to select sample points from homogeneous map areas, thus reducing the possibility of a positive accuracy bias (Foody, 2002). Another useful tool for future San Pedro LC work is the map of all sample points used in the accuracy assessment. Each point was attributed with geographic coordinates and both map and reference data labels (Skirvin et al., 2000). These data could be applied to generate a geographic representation of the continuous spatial distribution of LC errors (Kyriakidis and Dungan, 2001) to highlight especially difficult areas that should be field checked or otherwise handled in the future.

9.6   CONCLUSIONS

The results discussed in this chapter indicate that historical aerial photography, DOQQ data, and high-resolution airborne video data can be used successfully to perform classification accuracy assessment on LC maps derived from historical satellite data. Archived aerial photographs may be the only reference data available for retrospective analysis before 1992. However, their resolution (1:40,000 scale for NAPP data) often makes this task difficult. Successful use of DOQQ data requires precise geometric registration of the LC map to allow the overlay of orthorectified DOQQs. The use of georeferenced high-resolution airborne videography as a proxy for actual ground sampling in accuracy assessment provided the best method for current reference data development in the San Pedro watershed. The advantages include: (1) cost-effective collection of a statistically meaningful number of sample points, (2) effective control of coordinate locational error, and (3) variable-scale videography that permits the identification of specific plant species or communities of interest. Additionally, the videography provides a clear depiction of cultural features and land-use activities. The main limitation of this method is that data are collected along predetermined flight paths, thus constraining the sampling frame design.

9.7   SUMMARY

Because the rapidly growing archives of satellite remote sensing imagery now span decades, there is increasing interest in the study of long-term regional LC change across multiple image dates. However, temporally coincident ground-sampled data may not be available to perform an independent accuracy assessment of the image-derived LC map products. This study explored the feasibility of utilizing historical aerial photography, DOQQs, and high-resolution airborne color video data to assess the accuracy of satellite-derived LC maps for the upper San Pedro River watershed in southeastern Arizona and northeastern Sonora, Mexico. Satellite image data included Landsat Multi-Spectral Scanner (MSS) and Landsat Thematic Mapper (TM) data acquired over an approximately 25-year period. Four LC classifications were performed using three dates of MSS imagery (1973, 1986, and 1992) and one TM image (1997). The TM imagery was aggraded from 30 to 60 m to match the coarser MSS pixel size.

A stratified random sampling design was incorporated with samples apportioned by LC area, using a minimum sample size of $n = 20$ for rare classes. Results indicated similar map accuracies were obtained using the three alternative methods. Aerial photography provided reference data to assess the 1973 and 1986 LC maps with overall classification accuracies of 70% (1973) and 67% (1986). Assignments of class labels to sample points on 1992 reference DOQQs were verified by comparison with higher-resolution airborne video data, with overall 1992 map classification accuracy of 75%. Accuracy assessment of the 1997 products used contemporaneous airborne color video data and resulted in an overall map accuracy of 72%. There was no evidence of positive bias in accuracy resulting from use of homogeneous vs. heterogeneous pixel contexts in sampling the LC maps.

The use of historical aerial photography, high-resolution DOQQs, and airborne videography as a proxy for actual ground sampling for satellite image classification accuracy has merit. Selection of a reference data set for this study depended on the date of image acquisition. For example, prior to 1992, historical aerial photographs were the only data available. DOQQs covered the period since initiation of the high-resolution NAPP in 1992, and high-resolution airborne videography provided a cost-effective means of acquiring many reference sample points near the time of image acquisition. Problems that were difficult to avoid included inadequate sampling of rare classes and reconciling cover changes between acquisition dates of aerial photography or DOQQs and satellite image data. Other issues, including the need for consistent geometric rectification and criteria for mutually exclusive and reproducible LC class descriptions, need special attention when satellite image classification and subsequent LC map accuracy assessment are performed by different teams.

ACKNOWLEDGMENTS

The U.S. Environmental Protection Agency, Office of Research and Development provided funding for this work. The authors wish to thank participants from U.S. EPA, Lockheed Martin Environmental Services, Instituto del Medio Ambiente y el Desarrollo Sustentable del Estado de Sonora (IMADES), and the Arizona Remote Sensing Center at the University of Arizona for their assistance.

REFERENCES

Asner, G.P., C.A. Wessman, C.A. Bateson, and J.L. Privette, Impact of tissue, canopy, and landscape factors on the hyperspectral reflectance variability of arid ecosystems, *Remote Sens. Environ.*, 74, 69–84, 2000.

Bahre, C.J., *A Legacy of Change*, The University of Arizona Press, Tucson, 1991.

Bahre, C.J., Human impacts on the grasslands of southeastern Arizona, in *The Desert Grassland*, McClaran, M.P. and T.R. Van Devender, Eds., The University of Arizona Press, Tucson, 1995.

Brown, D.E., C.H. Lowe, and C.P. Pase, A digitized classification system for the biotic communities of North America, with community (series) and association examples for the Southwest, *J. Arizona-Nevada Acad. Sci.*, 14 (Suppl. 1), 1–16, 1979.

Congalton, R., A review of assessing the accuracy of classifications of remotely sensed data, *Remote Sens. Environ.*, 37, 35–46, 1991.

Congalton, R.G. and K. Green, *Assessing the Accuracy of Remotely Sensed Data: Principles and Practices*, CRC Press, Boca Raton, FL, 1999.

Congalton, R.G., R.G. Oderwald, and R.A. Mead, Assessing Landsat classification accuracy using discrete multivariate statistical techniques, *Photogram. Eng. Remote Sens.*, 49, 1671–1678, 1983.

Drake, S.E., Climate-Correlative Modeling of Phytogeography at the Watershed Scale, Ph.D. dissertation, University of Arizona, Tucson, 2000.

Drake, S.E., Visual interpretation of vegetation classes from airborne videography: an evaluation of observer proficiency with minimal training, *Photogram. Eng. Remote Sens.*, 62, 969–978, 1996.

Easterling, D.R., T.R. Karl, E.H. Mason, P.Y. Hughes, D.P. Bowman, R.C. Daniels, and T.A. Boden, United States Historical Climatology Network (U.S. HCN) Monthly Temperature and Precipitation Data, Revision 3, Carbon Dioxide Information Analysis Center, Oak Ridge National Laboratory, Oak Ridge, TN, 1996.

Fitzpatrick-Lins, K., Comparison of sampling procedures and data analysis for a land-use and land-cover map. *Photogram. Eng. Remote Sens.*, 47, 343–351, 1981.

Foody, G.M., On the compensation for chance agreement in image classification accuracy assessment, *Photogram. Eng. Remote Sens.*, 58, 1459–1460, 1992.

Foody, G.M., Status of land cover accuracy assessment, *Remote Sens. Environ.*, 80, 185–201, 2002.

Hammond, T.O. and D.L. Verbyla, Optimistic bias in classification accuracy assessment, *Int. J. Remote Sens.*, 17, 1261–1266, 1996.

Jones, K.B., L.R. Williams, A.M. Pitchford, E.T. Slonecker, J.D. Wickham, R.V. O'Neill, D. Garofalo, and W.G. Kepner, A National Assessment of Landscape Change and Impacts to Aquatic Resources: A 10-Year Strategic Plan for the Landscape Sciences Program, EPA/600/R-00/001, U.S. Environmental Protection Agency, Office of Research and Development, Washington, DC, 2000.

Kepner, W.G., C.J. Watts, and C.M. Edmonds, Remote Sensing and Geographic Information Systems for Decision Analysis in Public Resource Administration: A Case Study of 25 Years of Landscape Change in a Southwestern Watershed, EPA/600/R-02/039, U.S. Environmental Protection Agency, Office of Research and Development, Washington, DC, 2002.

Kepner, W.G., C.J. Watts, C.M. Edmonds, J.K. Maingi, S.E. Marsh, and G. Luna, A landscape approach for detecting and evaluating change in a semi-arid environment. *Environ. Monit. Assess.*, 64, 179–195, 2000.

Kyriakidis, P.C. and J.L. Dungan, A geostatistical approach for mapping thematic classification accuracy and evaluating the impact of inaccurate spatial data on ecological model predictions, *Environ. Ecol. Stat.*, 8, 311–330, 2001.

Lunetta, R.L., J.G. Lyon, J.A. Sturdevant, J.L. Dwyer, C.D. Elvidge, L.K. Fenstermaker, D. Yuan, S.R. Hoffer, and R. Werrackoon, North American Landscape Characterization: Research Plan, EPA/600/R-93/135, U.S. Environmental Protection Agency, Las Vegas, NV, 1993.

Ma, Z. and R.L. Redmond, Tau coefficients for accuracy assessment of classification of remote sensing data, *Photogram. Eng. Remote Sens.*, 61, 435–439, 1995.

Maingi, J.K., S.E. Marsh, W.G. Kepner, and C.M. Edmonds, An Accuracy Assessment of 1992 Landsat-MSS Derived Land Cover for the Upper San Pedro Watershed (U.S./Mexico), EPA/600/R-02/040, U.S. Environmental Protection Agency, Office of Research and Development, Washington, DC, 2002.

Marsh, S.E., J.L. Walsh, and C. Sobrevila, Evaluation of airborne video data for land-cover classification accuracy assessment in an isolated Brazilian forest, *Remote Sens. Environ.*, 48, 61–69, 1994.

McClaran, M., S.E. Marsh, D. Meko, S.M. Skirvin, and S.E. Drake. Evaluation of the Effects of Global Climate Change on the San Pedro Watershed: Final Report, Cooperative Agreement No. A950-A1-0012 between the University of Arizona and the U.S. Geological Survey, Biological Resource Division, 1999.

NOAA (National Oceanic and Atmospheric Administration), Climate Prediction Center: ENSO Impacts on the U.S.: Previous Events, Web page [accessed 22 October 2002], available at http://www.cpc.ncep.noaa.gov/products/analysis_monitoring/ensostuff/ensoyears.html.

Rosenfield, G.H. and K. Fitzpatrick-Lins, A coefficient of agreement as a measure of thematic classification accuracy, *Photogram. Eng. Remote Sens.*, 52, 223–227, 1986.

Skirvin, S.M., S.E. Drake, J.K. Maingi, S.E. Marsh, and W.G. Kepner, An Accuracy Assessment of 1997 Landsat Thematic Mapper Derived Land Cover for the Upper San Pedro Watershed (U.S./Mexico), EPA/600/R-00/097, U.S. Environmental Protection Agency, Office of Research and Development, Washington, DC, 2000.

Stehman, S.V., Selecting and interpreting measures of thematic classification accuracy, *Remote Sens. Environ.*, 62, 77–89, 1997.

Stehman, S.V. and R.L. Czaplewski, Design and analysis for thematic map accuracy assessment: fundamental principles, *Remote Sens. Environ.*, 64, 331–344, 1998.

USEPA (U.S. Environmental Protection Agency), Upper San Pedro River, Web page [accessed 17 October 2002]. Available at http://www.epa.gov/nerlesd1/land-sci/html2/sanpedro_home.html.

van Genderen, J.L. and B.F. Lock, Testing land use map accuracy, *Photogram. Eng. Remote Sens.*, 43, 1135–1137, 1977.

Wilson, T.B., R.H. Webb, and T.L. Thompson, Mechanisms of Range Expansion and Removal of Mesquite in Desert Grasslands of the Southwestern United States, General Technical Report RMRS-GTR-81, U.S. Forest Service, Rocky Mountain Research Station, 2001.

Yang, L., S.V. Stehman, J.H. Smith, and J.D. Wickham, Thematic accuracy of MRLC land cover for the eastern United States, *Remote Sens. Environ.*, 76, 418–422, 2001.

Zhu, Z., L. Yang, S.V. Stehman, and R.L. Czaplewski, Accuracy assessment for the U.S. Geological Survey regional land-cover mapping program: New York and New Jersey region, *Photogram. Eng. Remote Sens.*, 66, 1425–1435, 2000.

Using Classification Consistency in Interscene Overlap Areas to Model Spatial Variations in Land-Cover Accuracy over Large Geographic Regions

Bert Guindon and Curtis M. Edmonds

CONTENTS

10.1 INTRODUCTION

Over the past decade a number of programs have been undertaken to create definitive data sets of processed satellite imagery that encompass national and global coverage at specific acquisition epochs. Initial initiatives included the Multi-Resolution Land Characteristics (MRLC), the North American Landscape Characterization (NALC), and the GEOCover programs (Loveland and Shaw, 1996; Sohl and Dwyer, 1998; Dykstra et al., 2000). Subsequent initiatives have been spawned to generate information layers from these data sets, including the National Land Cover Data (NLCD) layer (Vogelmann et al., 2001). It is recognized that a quantitative assessment to characterize product accuracies is needed to support their acceptance and application by the general scientific community (Zhu et al., 2000). An "ideal" accuracy assessment methodology for large-area products would meet the following objectives: it would (1) provide an estimation of classification confidence, (2) effectively characterize spatial variations in accuracy, (3) have the ability to be implemented coincident with the classification process (feedback mechanism), (4) be consistent and repeatable, and (5) be sufficiently robust in design to support subsequent change detection assessments.

The most common approach to classification assessment is through the analysis of confusion matrices (Congalton, 1991). In this approach product classifications for a statistically robust number of samples (n) are compared with "reference" data derived from an independent source (e.g., interpretation of aerial photography). The cost of "reference" data acquisition represents a significant challenge. This results in numerous limitations, which include: (1) only a small fraction of the area of interest is used in the assessment process, (2) the content of a single confusion matrix is used to characterize the accuracy of diverse areas (Zhu et al., 2000); (3) rare classes are frequently underrepresented (n), and (4) accuracy characterization is limited to "macroscopic" levels (i.e., overall product and individual class levels).

Cost and logistics preclude highly detailed accuracy characterization based solely on conventional ground reference data, and therefore one must investigate complementary, albeit indirect, methods of accuracy assessment. This chapter describes an assessment strategy based on classification consistency. For most land resources satellites (e.g., Landsat), extensive image overlap occurs between scenes from adjacent World Reference System (WRS) frames. For a given adjacent path/row pair, each scene provides a quasi-independent classification estimate of those pixels resident in the overlap region. Intuitively, we would expect the level of classification agreement, hereafter referred to as classification consistency, to be indicative of the absolute levels of classification accuracy (i.e., high levels of consistency should be associated with high levels of classification accuracy).

The objectives here are to (1) establish a statistical link between classification consistency and both user's and producer's accuracies, (2) develop an integrated accuracy assessment strategy to quantify classification consistency and hence infer classification confidence, and (3) illustrate and assess this approach using synoptic land-cover (LC) products.

10.2 LINK BETWEEN CLASSIFICATION CONSISTENCY AND ACCURACY

To develop the statistical relationship between classification consistency for user's and producer's accuracies, consider the case of two adjacent scenes, hereafter referred to as scenes number 1 and 2. If each scene is independently classified to a common scheme, the overlap region can be used to quantify the classification consistency. For example, the consistency of class A in scene number 1 can be written as:

$$C_{1A} = \sum_{T=1}^{M} N_T P_{1TA} P_{2TA} \left/ \left(\sum_{T=1}^{M} N_T P_{1TA} \right) \right. \tag{10.1}$$

where C_{1A} = the consistency, defined as the fraction of overlap pixels classed as A in scene number 1 that are also classed as A in scene number 2, M = the number of classes, P_{kTA} = the probability that a pixel of true class T is labeled as class A in scene number k, and N_T = number of true class T pixels in the overlap region. Note that P_{kTT} is the producer accuracy of class T in scene k.

The user accuracy for scene number 1_A will be equal to the ratio of the number of correctly classified class A pixels to the total number labeled as A:

$$Q_{1A} = N_A P_{1AA} \left/ \left(\sum_{T=1}^{M} N_T P_{1TA} \right) \right. \tag{10.2}$$

The restricted two-class scenario (i.e., classes A and B) provides useful insights for those classes within a larger class mix whose labeling accuracy is limited primarily by pairwise class confusion. In this case, Equation (10.1) reduces to:

$$C_{1A} = [f\ P_{1AA}\ P_{2AA} + P_{1BA}\ P_{2BA}\ ]/[f\ P_{1AA} + P_{1BA}\ ] \tag{10.3}$$

where f is the ratio of numbers of true class A to true class B pixels. That is:

$$f = N_A/N_B \tag{10.4}$$

It can be seen that consistency is a function not only of the producer accuracies but also the relative class proportions. Similarly, user accuracy can be expressed as a function of producer accuracy and f. For example:

$$Q_{1A} = f\ P_{1AA}/[f\ P_{1AA} + P_{1BA}\ ] \tag{10.5}$$

If the two classifications are derived from similar data sources (e.g., scenes from the same sensor), each scene will typically exhibit similar producer accuracies (i.e., $P_{1AA} = P_{2AA} = P_{AA}$, etc.). In this instance, consistency and user accuracy will be the same for each scene:

$$C_{1A} = C_{2A} = C_A = [f\ P_{AA}^2 + P_{BA}^2]/[f\ P_{AA} + P_{BA}] \tag{10.6}$$

and

$$Q_{1A} = Q_{2A} = Q_A = f\ P_{AA}/[f\ P_{AA} + P_{BA}] \tag{10.7}$$

We have examined the relationships of consistency and user's accuracy as functions of producer's accuracy and f for a range of parameters applicable to the Laurentian Great Lakes region in which LC has been classed as either forest or nonforest. Producer's accuracies in the range 0.5 to 1 need only be considered since 0.5 corresponds to random class assignment. Also, for this level of stratification, we would expect high producer's accuracy performance (e.g., > 0.8 with Landsat Multispectral Scanner (MSS) data). Finally, in the Great Lakes region, f varies dramatically from approximately 0.1 in the agricultural south to 10 in the north for forested land and vice versa for unforested land.

Figure 10.1 and Figure 10.2 illustrate the relationships of consistency and user accuracy with producer's accuracy, respectively, for f values ranging from 0.1 to 10 and a nominal class B producer's accuracy of 0.8. These results are typical of a range of realistic cases. From an inspection of these plots we can draw a number of conclusions: (1) both consistency and user's accuracy increase monotonically with producer's accuracy, suggesting that consistency is an indicator of classification accuracy performance and (2) consistency and user's accuracy exhibit similar sensitivities to f. We hypothesize that consistency can be employed as a "surrogate" of user's accuracy to monitor variations in accuracy at scene-level spatial scales.

10.3 USING CONSISTENCY WITHIN A CLASSIFICATION METHODOLOGY

Our approach for applying consistency measures is dependent on the specific algorithms and methodologies employed for our study area. The following discussion addresses key aspects of our Great Lakes LC methodology and how they incorporate consistency and address our accuracy objectives. Figure 10.3 illustrates the overall data processing flow.

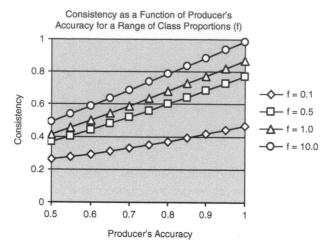

Figure 10.1 Relationship of classification consistency as a function of producer's accuracy for a range of class proportions (f). The four cases shown span the range of forested and nonforested class proportions encountered in scenes of the Laurentian Great Lakes watershed.

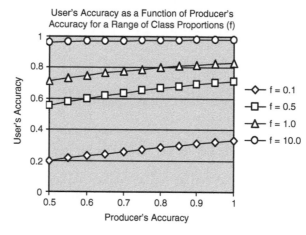

Figure 10.2 User's accuracy as a function of producer's accuracy for a range of class proportions (f). The four cases shown spanned the range of forested and nonforested class proportions encountered in scenes of the Laurentian Great Lakes watershed.

- Each Landsat scene is independently classified and composited with other scenes to generate a final large-area LC product. This approach was labor intensive and is suitable primarily for synoptic mapping (i.e., categorization into a few broad classes). However, it did have a number of important practical advantages: image information content could be thoroughly exploited, and consistency analyses were undertaken on each scene by comparing its classification with those of its nearest four neighbours (cross- and along-track). Thus, regional variations in classification accuracy, arising from interscene quality differences and spatial diversity in class proportions, were monitored at the scene level.
- Scene classification was achieved through unsupervised spectral clustering (K-means algorithm, 150 clusters), followed by cluster labeling. For synoptic mapping (i.e., < 10 classes), each class was described by a number of clusters (5–50). Cluster-based classification had some important ramifications for accuracy considerations, including: (a) the true "unit of classification" was the cluster, since it was at this level that label decision-making occurs; (b) since each class was represented by a number of clusters, we did not expect that the labeling of each cluster would be equally reliable; and (c) if consistency was evaluated at the cluster level and not at the "conventional"

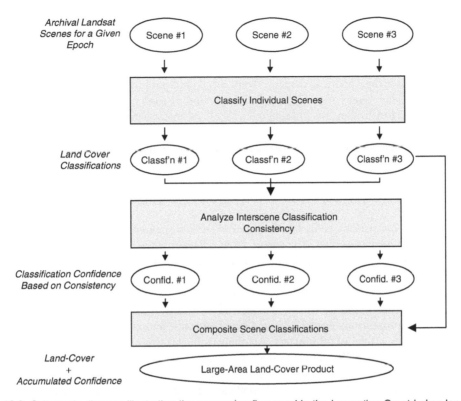

Figure 10.3 Schematic diagram illustrating the processing flow used in the Laurentian Great Lakes land-cover mapping initiative. Classification consistency was used both to check individual scene classifications and in the classification and compositing process to rationalize multiple classifications in overlap regions and to generate a classification confidence layer.

class level, it provided a better model of "microscopic" aspects of user's accuracy and an accuracy estimate closer to the individual pixel level than conventional class-level assessment methods.

- Accuracy assessment was undertaken during the LC product generation process. Interscene classification comparison identified potentially mislabeled clusters, since these exhibited low classification consistency levels. The statistical foundation for "grading" cluster label quality is described elsewhere (Guindon and Edmonds, 2002). Suspect clusters were then revisited and relabeled before the scene classifications were composited into the final product.

- Consistency played a pivotal role in the classification compositing process. Consistency can be viewed as an indicator of the "confidence" that can be assigned to the accuracy of the class label. For overlap regions, relative consistency was used to select the most likely correct classification if two or more scenes predicted conflicting class labels. Additionally, net consistency or confidence was accumulated during compositing, leading to a confidence overlay sampled at the pixel level for the final product. This layer encapsulated (1) parent cluster confidence, (2) the spatial distribution of available image data, and (3) interscene information agreement where multiple scene coverage was available. As such, it provided a valuable ancillary product both for accuracy assessment and to support postproduction interpretation activities.

10.4 GREAT LAKES RESULTS

The classification and accuracy assessment methodologies outlined above were implemented using QUAD-LACC (Guindon, 2002). Here we will illustrate example outputs relevant to the accuracy components. These processing examples were drawn from the creation of two synoptic LC products of the mid-1980s and early 1990s NALC epochs. Each was sampled at 6" (longitude) by 4" (latitude), or approximately 140 m, and included four general cover classes (i.e., water, forest, urban or developed, and other nonforest land). For illustrative purposes, here we stratified the cover classes into two categories (i.e., forest vs. nonforest).

A total of 5300 reference sites were identified within extended regions of thematically homogeneous cover based on supporting evidence from aerial photography interpretation and topographic map inspection. They represented the spectral dispersion of each class. Since each Landsat scene encompassed 100 to 150 sites, the classification labels of pixels within five-by-five-pixel neighborhoods of each site were analyzed to derive estimates of producer's accuracy. These estimates were optimistic, since pixels near interclass boundaries were not included, and should not be viewed as a measure of accuracy in the absolute sense.

10.4.1 Variation of Consistency among Clusters of a Given Class

Classification consistency analysis was undertaken on a scene once it and its immediate neighboring scenes had been classified. The scenes from adjacent paths were most important since they provided the greatest overlap and were not temporally correlated to the central scene. Using QUAD-LACC, consistency evaluations were performed at the cluster level with each cluster assigned an integral consistency measure of 0.0 to 10.0 corresponding to a range of classification agreement of 0.0 to 100%. As an example, we use the case of scene 16/29 from the 1990s epoch. The LC of this scene was approximately equally divided between forest and nonforest classes, with the forest class encompassing a total of 52 clusters. An analysis of the two cross-track overlap regions (i.e., with scenes 15/29 and 17/29) indicated that 76.4% of 710,610 overlap pixels classed as forested in scene 16/29 were also labeled as forest in one of the cross-track neighboring scenes, leading to an overall class measure of 8.0. For the hypothesis that all clusters were equivalent in terms of consistency, we estimated the approximate dispersion in cluster consistency measures from binomial theory (Thomas and Allcock, 1984). Assuming equal pixel populations per cluster, the predicted 1-sigma spread in consistency among clusters should be only ± 0.05% (i.e., practically all clusters should exhibit a consistency measure of 8.0). Figure 10.4 shows the spread in observed consistency measures for the clusters of scene 16/29. Note that the histogram contained 104 entries, since each overlap region provided an independent measure estimate for each cluster. The observed distribution was much broader than predicted by the binomial model, indicating that there is a significant spread in classification quality among clusters and, hence, added accuracy information was available at the cluster level.

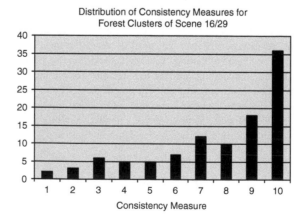

Figure 10.4 Histogram of consistency levels for forest clusters of scene 16/29. The dispersion among values is indicative of the broad differences in classification "quality" among member clusters within a given class.

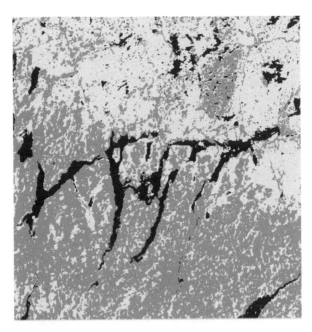

Figure 10.5 Land-cover classification of a portion of scene 17/29. Three classes are shown: water (dark), non-forest land (medium grey), and forest (white).

10.4.2 Aspects of Scene-Based Consistency Overlays

Once a final classification was obtained for a given scene, a "confidence" overlay was produced wherein the confidence value of each pixel corresponded to the consistency level of its parent cluster. Example results are shown in Figure 10.5 for the 1980s path/row scene 17/29. Figure 10.5 illustrates the three primary classes (i.e., water [dark], nonforested land [medium grey], and forested [white]). Figure 10.6 shows an enlargement of the confidence layer of the central portion of Figure 10.5. The confidence range 0.0 to 10.0 was presented as a grey-level scale from black to white. The following points are worthy of note: (1) Water was easily recognized, and hence the central portions of most water bodies exhibited a high, uniform confidence level. (2) Pixels along interclass boundaries, such as the edges of lakes or forest patches, tended to be of low confidence (Figure 10.6). They are members of clusters containing primarily "mixed" pixels and therefore have a low accuracy. (3) Forested areas exhibited a slightly higher average confidence than nonforested areas. This is related to the fact that this scene has more forest than nonforest cover. Consequently, the population of pixels classed as forest will contain a relatively lower proportion of commission errors, resulting in a corresponding higher level of interscene classification consistency.

10.4.3 Aspects of the Accumulated Confidence Layer

Figure 10.7 and Figure 10.8 illustrate a portion of a three-class LC product and accompanying confidence overlay respectively. The interscene overlap regions are readily distinguishable in Figure 10.8 by their higher levels of accumulated confidence. In these regions significant confidence variations still arise, either from conflicting classifications or information loss in one of the constituent scenes because of cloud contamination (e.g., in central Michigan). Finally, in Figure 10.7 there are data gaps, appearing as nearly horizontal black lines, that arise because of along-track data loss during the preprocessing steps of resolution reduction and haze removal (Guindon and Zhang, 2002).

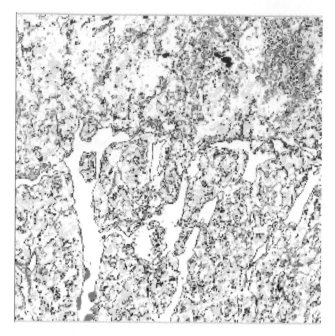

Figure 10.6 Confidence overlay, derived from cluster consistency analyses, for the central quadrant of the classification where brightness is proportional to confidence.

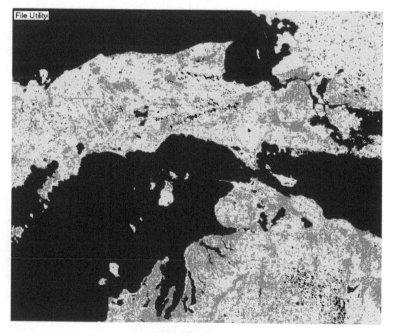

Figure 10.7 Three-class (water [dark], nonforest [medium grey], and forest [white]) land-cover product of the central portion of the Laurentian Great Lakes watershed.

Figure 10.8 Accumulated confidence layer for the classification where brightness is proportional to confidence. On the consistency scale described in the text, numerical values range from 0.0 to 10.0 in nonoverlapping regions and 0.0 to 40.0 in regions where up to four individual scenes contribute classification estimates.

10.4.4 Relationship of Accumulated Confidence and User's Accuracy

The data set of 5300 five-by-five-pixel reference sites was used to investigate the relationship between accumulated confidence and user's accuracy. For each pixel, the appropriate reference class was compared to the assigned class of the final LC product.

Confusion matrices were generated for pixels grouped according to number of contributing scenes and accumulated confidence. User's accuracies of both the forested and nonforested classes were computed for each matrix. Figure 10.9 shows the relationship of user's accuracy vs. accumu-

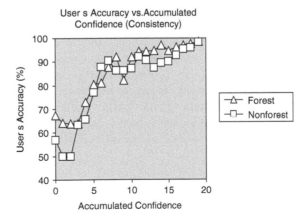

Figure 10.9 Plot of user's accuracy vs. accumulated confidence for forested and nonforested reference sites located in areas where two scene classifications were available. The results indicated that classification confidence based on consistency monotonically increases with increasing user's accuracy and therefore is a useful indicator of the latter.

lated confidence for those pixels whose classification was determined based on two scenes. The monotonic relationship between these variables confirms the earlier statistical arguments that consistency is a legitimate "surrogate" of user's accuracy.

10.5 CONCLUSIONS

Multiple-scene LC products can be expected to exhibit significant internal variations in user accuracy. Detailed characterization of this variability was not feasible using conventional ground reference sampling because of cost and logistics. However, the level of interscene classification consistency provided an indirect "surrogate" measure and was used to gauge local accuracy. This alternative approach was especially attractive for application with Landsat-based maps since extensive overlap areas exist for adjacent orbital paths located in nonequatorial latitudes.

Consistency measures were effectively employed using a number of processing steps. First, assessments were evaluated at the cluster level, thereby providing an estimation of performance at the level of the labeling unit rather than only at the class level. Then, by analyzing the consistency during the product generation phase, detection and correction of incorrectly labeled clusters was accomplished prior to the creation of the final product, thereby improving its quality. Finally, within the interscene overlap regions, consistency served as a "compositing" criterion to select an optimum label and could be accumulated to encapsulate the added confidence associated with multiple independent class estimations.

10.6 SUMMARY

During the past decade, a number of initiatives have been undertaken to create systematic national and global data sets of processed satellite imagery. An important application of these data is the derivation of large geographic area (i.e., multiscene) LC products. These products exhibit internal variations in information quality for two principal reasons. First, they have been assembled from a multitemporal mix of satellite scenes acquired under differing seasonal and atmospheric conditions. Second, intraproduct landscape diversity will lead to spatially varying levels of class commission errors. Detailed modeling of these variations with conventional ground truth is prohibitively expensive, and hence an alternative accuracy assessment method must be sought.

In this chapter we presented a method for confidence estimation based on the analysis of classification consistency in regions of overlapping coverage between Landsat scenes from adjacent orbital paths and rows. A LC mapping methodology has been developed that exploits consistency evaluation to (1) improve scene-based classification performance, (2) support the integration of scene classifications through compositing, (3) provide a detailed confidence characterization of the final product, and (4) conduct postgeneration accuracy assessment. This methodology was implemented within a prototype mapping system, QUAD-LACC, to derive synoptic LC products of the Laurentian Great Lakes watershed. It should be noted that others researchers have suggested using overlap regions to assess the accuracy of landscape metrics (Brown et al., 2000).

REFERENCES

Brown, D.G., J.D. Duh, and S.A. Drzysga, Estimating error in the analysis of forest fragmentation change using North American landscape characterization (NALC) data, *Remote Sens. Environ.*, 71, 106–117, 2000.

Congalton, R.G., A review of assessing the accuracy of classifications of remotely sensed data, *Remote Sens. Environ.*, 37, 35–46, 1991.

Dykstra, J.D., M.C. Place, and R.A. Mitchell, GEOCOVER-ORTHO: Creation of a Seamless Geodetically Accurate, Digital Base Map of the Entire Earth's Land Mass Using Landsat Multispectral Data, in Proceedings of the ASPRS 2000 Conference, Washington, DC, 2000.

Guindon, B. QUAD-LACC: A Proto-type System to Generate and Interpret Satellite-Derived Land Cover Products, in Proceedings of the IGARSS2002 Symposium, Toronto, Ontario, June 24–28, 2002, pp. 1459–1461.

Guindon, B. and C.M. Edmonds, Large-area land cover mapping through scene-based classification compositing, *Photogram. Eng. Remote Sens.*, 68, 589–596, 2002.

Guindon, B. and Y. Zhang, Robust Haze Reduction: An Integral Processing Component in Satellite-based Land Cover Mapping, in Proceedings of the Joint International Symposium on Geospatial Theory, Processing and Applications, Ottawa, Ontario, July 8–12, 2002.

Loveland, T.R. and D. M. Shaw. Multiresolution land characterization: building collaborative partnerships, in *GAP Analysis: A Landscape Approach to Biodiversity Planning*, (J.M. Scott and F. Davis, Editors), ASPRS, Bethesda, MD, 1996, pp. 83–89.

Sohl, T.L. and J.L. Dwyer, North American landscape characterization project: the production of a continental scale three-decade Landsat data set, *Geocarto Int.*, 13, 43–51, 1998.

Thomas, I.L. and C.M. Allcock, Determining the confidence level for a classification, *Photogram. Eng. Remote Sens.*, 50, 1491–1496, 1984.

Vogelmann, J.E., S.M. Howard, L. Yang, C.R. Larson, B.K. Wylie, and N. Van Driel, Completion of the 1990s national land cover data set for the conterminous United States from Landsat Thematic Mapper data and ancillary data sources, *Photogram. Eng. Remote Sens.*, 67, 650–662, 2001.

Zhu, Z., L. Yang, S.V. Stehman, and R.L. Czaplewski, Accuracy assessment for the U.S. Geological Survey regional land-cover mapping program: New York and New Jersey region, *Photogram. Eng. Remote Sens.*, 66, 1425–1435, 2000.

CHAPTER 11

Geostatistical Mapping of Thematic Classification Uncertainty

Phaedon C. Kyriakidis, Xiaohang Liu, and Michael F. Goodchild

CONTENTS

11.1 INTRODUCTION

Thematic data derived from remotely sensed imagery lie at the heart of a plethora of environmental models at local, regional, and global scales. Accurate thematic classifications are therefore becoming increasingly essential for realistic model predictions in many disciplines. Remotely sensed information and resulting classifications, however, are not error free, but carry the imprint of a suite of data acquisition, storage, transformation, and representation errors and uncertainties (Zhang and Goodchild, 2002). The increased interest in characterizing the accuracy of thematic classification has promoted the practice of computing and reporting a set of different, yet complementary, accuracy statistics all derived from the confusion matrix (Congalton, 1991; Stehman, 1997; Congalton and Green, 1999; Foody, 2002). Based on these accuracy statistics, users of remotely sensed imagery can evaluate the appropriateness of different maps on their particular application and subsequently decide to retain one classification vs. another.

Accuracy statistics, however, express different aspects of classification quality and consequently appeal differently to different people, a fact that hinders the use of a single measure of classification accuracy (Congalton, 1991; Stehman, 1997; Foody, 2002). Recent efforts to provide several measures of map accuracy based on map value (Stehman, 1999) constitute a first attempt to address this problem, but in practice map accuracy is still communicated in the form of confusion-matrix-based accuracy statistics. The confusion matrix, and all derived accuracy statistics, however, is a regional (location-independent) measure of classification accuracy: it does not pertain to any pixel or subregion of the study area. For example, user's accuracy denotes the probability that any pixel classified as forest is actually forest on the ground. In this case, all pixels classified as forest have the same probability of belonging to that class on the ground, a fact that does not allow identification of pixels or subregions (of the same class) that warrant additional sampling. A new sampling campaign based on this type of accuracy statistic would just place more samples at pixels allocated to the class with the lower user's accuracy measure, irrespective of the location of these pixels and their proximity to known (training) pixels. In other words, confusion-matrix-based accuracy assessment has no explicit spatial resolution; it only has explicit class resolution.

In this chapter, we capitalize on the fact that conventional (hard) class allocation is typically based on the probability of class occurrence at each particular pixel calculated during the classification procedure. Maps of such posterior probability values portray the spatial distribution of classification quality and are extremely useful supplements to traditional accuracy statistics (Foody et al., 1992). As opposed to confusion-matrix-based accuracy assessment, such maps could identify pixels of the same category where additional sampling is warranted, based precisely on a measure of uncertainty regarding class occurrence at each particular pixel.

Evidently, the above classification uncertainty maps will depend on the classification algorithm adopted. Conventional classifiers typically use the information brought by reflectance values (feature vector) collocated at the particular pixel where classification is performed. In some cases, however, classes are not easily differentiated in the spectral (feature) space, due to either sensor noise or to the inherently similar spectral responses of certain classes. Improvements to the above classification procedures could be introduced in a variety of ways, including geographical stratification, classifier operations, postclassification sorting, and layered classification (Hutchinson, 1982; Jensen, 1996; Atkinson and Lewis, 2000). The above methods enhance the classification procedure by introducing, explicitly or implicitly, contextual information (Tso and Mather, 2001). Within this contextual classification framework, one of the most widely used avenues of incorporating ancillary information is that of pixel-specific prior probabilities (Strahler, 1980; Switzer et al., 1982).

Along these lines, we propose a simple, yet efficient, method for modeling pixel-specific context information using geostatistics (Isaaks and Srivastava, 1989; Cressie, 1993; Goovaerts, 1997). Specifically, we adopt indicator kriging to estimate the conditional probability that a pixel belongs to a specific class, given the nearby training pixels and a model of the spatial correlation for each class (Journel, 1983; Solow, 1986; van der Meer, 1996). These context-based probabilities are then combined with conditional probabilities of class occurrence derived from a conventional (noncontextual) classification via Bayes' rule to yield posterior probabilities that account for both spectral and spatial information. Steele (2000) and Steele and Redmond (2001) used a similar approach based on Bayesian integration of spectral and spatial information, the latter being derived using the nearest neighbor spatial classifier. In this work, we also use Bayes' rule to merge spatial and spectral information, but we use the indicator kriging classifier that incorporates texture information via the indicator covariance of each class. De Bruin (2000) and Goovaerts (2002) also adopted similar approaches using indicator kriging but did not link them to contextual classification. This research extends the above approaches in a formal contextual classification framework and illustrates their use for mapping thematic classification uncertainty.

Once posterior probabilities of class occurrence are derived at each pixel, they can be converted to classification accuracy values. In this chapter, we distinguish between classification uncertainty and classification accuracy: a measure of classification uncertainty, such as the posterior probability of class occurrence, at a particular pixel does not pertain to the allocated class label at that pixel, whereas a measure of classification accuracy pertains precisely to the particular class label allocated at that pixel. We propose a simple procedure for converting posterior probability values to classification accuracy values, and we illustrate its application in the case study section of this chapter using a realistically simulated data set.

11.2 METHODS

Let $C(\mathbf{u})$ denote a categorical random variable (RV) at a pixel with 2D coordinate vector $\mathbf{u} = (u_1, u_2)$ within a study area A. The RV $C(\mathbf{u})$ can take K mutually exclusive and exhaustive outcomes (realizations): $\{c(\mathbf{u}) = c_k, k = 1, \ldots, K\}$, which might correspond to K alternative land-cover types. In this chapter, we do not consider fuzzy classes, i.e., we assume that each pixel $\mathbf{u}$ is composed only of a single class and do not consider the case of mixed pixels.

Let $p_k[c(\mathbf{u})] = Prob\{C(\mathbf{u}) = c_k\}$ denote the probability mass function (PMF) modeling uncertainty about the k-th class c_k at location $\mathbf{u}$. In the absence of any relevant information, this probability $p_k[c(\mathbf{u})]$ is deemed constant within the study area A, i.e., $p_k^*[c(\mathbf{u})] = p_k$. For the set of K classes, these K probabilities are typically estimated from the class proportions based on a set of G training samples $\mathbf{c}_g = [c(\mathbf{u}_g), g = 1, \ldots, G]'$ within the study area A, as $p_k^* = \dfrac{1}{G} \sum_{g=1}^{G} i_k(\mathbf{u}_g)$,

where $i_k(\mathbf{u}_g) = 1$ if pixel $\mathbf{u}_g$ belongs to the k-th class, 0 if not (superscript ' denotes transposition). In a Bayesian classification framework of remotely sensed imagery, these K probabilities $\{p_k, k = 1, \ldots, K\}$ are termed *prior probabilities*, because they are derived before the remote sensing information is accounted for.

11.2.1 Classification Based on Remotely Sensed Data

Traditional classification algorithms, such as the maximum likelihood (ML) algorithm, update the prior probability p_k of each class by accounting for local information at each pixel $\mathbf{u}$ derived from reflectance data recorded in various spectral bands. Given a vector $\mathbf{x}(\mathbf{u}) = [x_1(\mathbf{u}), \ldots, x_B(\mathbf{u})]'$ of reflectance values at a pixel $\mathbf{u}$ in the study area, an estimate of the conditional (or posterior) probability $p_k[c(\mathbf{u}) | \mathbf{x}(\mathbf{u})] = Prob\{C(\mathbf{u}) = c_k | \mathbf{x}(\mathbf{u})\}$ for a pixel $\mathbf{u}$ to belong to the k-th class can be derived via Bayes' rule as:

$$p_k^*[c(\mathbf{u}) | \mathbf{x}(\mathbf{u})] = Prob^*\{C(\mathbf{u}) = c_k | \mathbf{x}(\mathbf{u})\} = \frac{p^*[\mathbf{x}(\mathbf{u}) | c(\mathbf{u}) = c_k] \cdot p_k^*}{p^*[\mathbf{x}(\mathbf{u})]} \tag{11.1}$$

where $p^*[\mathbf{x}(\mathbf{u}) | c(\mathbf{u}) = c_k] = Prob^*\{X_1(\mathbf{u}) = x_1(\mathbf{u}), \ldots, X_B(\mathbf{u}) = x_B(\mathbf{u}) | c(\mathbf{u}) = c_k\}$ denotes the class-conditional multivariate likelihood function, that is, the PDF for the particular spectral combination $\mathbf{x}(\mathbf{u}) = [x_1(\mathbf{u}), \ldots, x_B(\mathbf{u})]'$ to occur at pixel $\mathbf{u}$, given that the pixel belongs to class k. In the denominator, $p^*[\mathbf{x}(\mathbf{u})] = Prob^*\{X_1(\mathbf{u}) = x_1(\mathbf{u}), \ldots, X_B(\mathbf{u}) = x_B(\mathbf{u})\}$ denotes the unconditional (marginal) PDF for the same spectral combination $\mathbf{x}(\mathbf{u})$ to occur at the same pixel. For a particular pixel $\mathbf{u}$, this latter marginal PDF is just a normalizing constant (a scalar). It is common to all K classes (i.e., it does not affect the allocation decision), and it is typically computed as

$$p^*[\mathbf{x}(\mathbf{u})] = \sum_{k=1}^{K} p^*[\mathbf{x}(\mathbf{u}) \mid c(\mathbf{u}) = c_k] \cdot p_k^*, \text{ to ensure that the sum of the resulting } K \text{ conditional}$$

probabilities $\{p_k^*[c(\mathbf{u}) \mid \mathbf{x}(\mathbf{u})], k = 1, ..., K\}$ is 1. The final step in the classification procedure is typically the allocation of pixel $\mathbf{u}$ to the class c_m with the largest conditional probability:

$$p_m^*[c(\mathbf{u}) \mid \mathbf{x}(\mathbf{u})] = \max_k \{p_k^*[c(\mathbf{u}) \mid \mathbf{x}(\mathbf{u})], k = 1, ..., K\}, \text{ which is termed } maximum \ a \ posteriori \ \text{(MAP)}$$

selection.

In the case of Gaussian maximum likelihood (GML), the likelihood function is B-variate Gaussian and fully specified in terms of the $(B \times 1)$ class-conditional multivariate mean vector $\mathbf{m}_k = [E\{X_b(\mathbf{u}) \mid c(\mathbf{u}) = c_k\}, b = 1, ..., B]'$ and the $(B \times B)$ variance-covariance matrix $\boldsymbol{\Sigma}_k = [\mathrm{Cov}\{X_b(\mathbf{u}), X_{b'}(\mathbf{u}) \mid c(\mathbf{u}) = c_k\}, b = 1, ..., B, b' = 1, ..., B]$ of reflectance values. The exact form of the likelihood function then becomes:

$$p^*[\mathbf{x}(\mathbf{u}) \mid c(\mathbf{u}) = c_k] = (2p)^{-B/2} \cdot |\boldsymbol{\Sigma}_k|^{-1/2} \cdot \exp\left(-[\mathbf{x}(\mathbf{u}) - \mathbf{m}_k]' \cdot \boldsymbol{\Sigma}_k^{-1} \cdot [\mathbf{x}(\mathbf{u}) - \mathbf{m}_k]/2\right) \qquad (11.2)$$

where $|\boldsymbol{\Sigma}_k|$ and $\boldsymbol{\Sigma}_k^{-1}$ denote, respectively, the determinant and inverse of the class-conditional variance-covariance matrix $\boldsymbol{\Sigma}_k$.

In many cases, there exists ancillary information that is not accounted for in the classification procedure by conventional classifiers. One approach to account for this ancillary information is that of local prior probabilities, whereby the prior probabilities p_k^* are replaced with, say, elevation-dependent probabilities $p_k^*[c(\mathbf{u}) \mid e(\mathbf{u})]$, where $e(\mathbf{u})$ denotes the elevation or slope value at pixel $\mathbf{u}$. Such probabilities are location-dependent due to the spatial distribution of elevation or slope.

In the absence of ancillary information, the spatial correlation of each class (which can be modeled from a representative set of training samples) provides important information that should be accounted for in the classification procedure. Fragmented classifications, for example, might be incompatible with the spatial correlation of classes inferred from the training pixels. This characteristic can be expressed in probabilistic terms via the notion that a pixel $\mathbf{u}$ is more likely to be classified in class k than in class k', i.e., $p_k[c(\mathbf{u}) \mid \mathbf{x}(\mathbf{u})] > p_{k'}[c(\mathbf{u}) \mid \mathbf{x}(\mathbf{u})]$, if the information in the neighborhood of that pixel indicates the presence of a k-class neighborhood. This notion of context is typically incorporated in the remote sensing literature via Markov random field models (MRFs); see, for example, Li (2001) or Tso and Mather (2001) for details.

11.2.2 Geostatistical Modeling of Context

In this chapter, we propose an alternative procedure for modeling context based on indicator geostatistics, which provides another way for arriving at local prior probabilities $p_k^*[c(\mathbf{u}) \mid \mathbf{c}_g]$ given the set of G class labels $\mathbf{c}_g = [c(\mathbf{u}_g), g = 1, ..., G]'$; see, for example, Goovaerts (1997). Contrary to the MRF approach, the geostatistical alternative: (1) does not rely on a formal parametric model, (2) is much simpler to explain and implement in practice, (3) can incorporate complex spatial correlation models that could also include large-scale (low-frequency) spatial variability, and (4) provides a formal way of integrating other ancillary sources of information to yield more realistic local prior probabilities.

Indicator geostatistics (Journel, 1983; Solow, 1986) is based on a simple, yet effective, measure of spatial correlation: the covariance $\sigma_k(\mathbf{h})$ between any two indicators $i_k(\mathbf{u})$ and $i_k(\mathbf{u} + \mathbf{h})$ of the same class separated by a distance vector $\mathbf{h}$, and is defined as:

$$\sigma_k(\mathbf{h}) = E\{I_k(\mathbf{u}+\mathbf{h})\cdot I_k(\mathbf{u})\} - E\{I_k(\mathbf{u}+\mathbf{h})\}\cdot E\{I_k(\mathbf{u})\}$$

$$= Prob\{I_k(\mathbf{u}+\mathbf{h})=1, I_k(\mathbf{u})=1\} - Prob\{I_k(\mathbf{u}+\mathbf{h})=1\}\cdot Prob\{I_k(\mathbf{u})=1\} \tag{11.3}$$

The indicator covariance $\sigma_k(\mathbf{h})$ quantifies the frequency of occurrence of any two pixels of the same category k, found $\mathbf{h}$ distance units apart. Intuitively, as the modulus of vector $\mathbf{h}$ becomes larger, that frequency of occurrence would decrease. Note that the indicator covariance is related to the bivariate probability $Prob\{I_k(\mathbf{u}+\mathbf{h})=1, I_k(\mathbf{u})=1\}$ of two pixels of the same k-th category being $\mathbf{h}$ distance units apart, and is thus related to joint count statistics. For an application of joint count statistics in remote sensing accuracy assessment, the reader is referred to Congalton (1988).

Under second-order stationarity, the sample indicator covariance $\sigma_k^*(\mathbf{h})$ of the k-th category for a separation vector $\mathbf{h}$ is inferred as:

$$\sigma_k^*(\mathbf{h}) = \frac{1}{G(\mathbf{h})}\sum_{g=1}^{G(\mathbf{h})} i_k(\mathbf{u}_g+\mathbf{h})\cdot i_k(\mathbf{u}_g) - p_k^2 \tag{11.4}$$

where $G(\mathbf{h})$ denotes the number of training samples separated by $\mathbf{h}$.

A plot of the modulus $|\mathbf{h}_l|$ (in the isotropic case) of several vectors $\{\mathbf{h}_l, l=1,\ldots,L\}$ vs. the corresponding covariance values $\{\sigma_k^*(\mathbf{h}_l), l=1,\ldots,L\}$ constitutes the sample covariance function. Parametric and positive definite covariance models $\Sigma_k = \{\sigma_k(\mathbf{h}), \forall\mathbf{h}\}$ for any arbitrary vector $\mathbf{h}$ are then fitted to the sample covariance functions. The parameters of these functions (e.g., covariance function type, relative nugget, or range) might be different from one category to another, indicating different spatial patterns of, say, land-cover types. For a particular separation vector $\mathbf{h}$, the corresponding model-derived indicator covariance is denoted as $\sigma_k(\mathbf{h})$.

The spatial information of the training pixels is encoded partially in the indicator covariance model $\sigma_k(\mathbf{h})$ for the k-th category and partially in their actual location and class label. In Fourier analysis jargon, the covariance model $\sigma_k(\mathbf{h})$ provides amplitude information (i.e., textural information), whereas the actual locations of the training samples and their class labels provide phase information (i.e., location information). Taken together, locations and covariance of training pixels provide contextual information that can be used in the classification procedure.

Ordinary indicator kriging (OIK) is a nonparametric approximation to the conditional PMF $p_k[c(\mathbf{u})|\mathbf{c}_g] = Prob\{C(\mathbf{u})=c_k|\mathbf{c}_g\}$ for the k-th class to occur at pixel $\mathbf{u}$, given the spatial information encapsulated in the G training samples $\mathbf{c}_g = [c(\mathbf{u}_g), g=1,\ldots,G]'$; see Van der Meer (1996), and Goovaerts (1997) for details. The OIK estimate $p_k^*[c(\mathbf{u})|\mathbf{c}_g]$ for the conditional PMF $p_k[c(\mathbf{u})|\mathbf{c}_g]$ that the k-th class prevails at pixel $\mathbf{u}$ is expressed as a weighted linear combination of the $G(\mathbf{u})$ sample indicators $\mathbf{i}^k = [i_k(\mathbf{u}_g), g=1,\ldots,G(\mathbf{u})]'$ for the same k-th class found in a neighborhood $N(\mathbf{u})$ centered at pixel $\mathbf{u}$:

$$p_k^*[c(\mathbf{u})|\mathbf{c}_g] \approx p_k^*[c(\mathbf{u})|\mathbf{i}^k] = Prob^*\{C(\mathbf{u})=c_k|\mathbf{i}^k\} = \sum_{g=1}^{G(\mathbf{u})} w_k(\mathbf{u}_g)\cdot i_k(\mathbf{u}_g) \tag{11.5}$$

under the constraint $\displaystyle\sum_{g=1}^{G(\mathbf{u})} w_k(\mathbf{u}_g)=1$; this latter constraint allows for local, within-neighborhood $N(\mathbf{u})$, departures of the class proportion from the prior (constant) proportion p_k. In the previous equation, $w_k(\mathbf{u}_g)$ denotes the weight assigned to the g-th training sample indicator of the k-th category $i_k(\mathbf{u}_g)$ for estimation of $p_k[c(\mathbf{u})|\mathbf{c}_g]$ for the same k-th category at pixel $\mathbf{u}$. The size of the neighborhood $N(\mathbf{u})$ is typically identified to the range of correlation of the indicator covariance model Σ_k.

When modeling context at pixel $\mathbf{u}$ via the local conditional probability $p_k^*[c(\mathbf{u})\,|\,\mathbf{c}_g]$, the $G(\mathbf{u})$ weights $\{w_k(\mathbf{u}_g), g = 1,\ldots, G(\mathbf{u})\}$ for the k-th category indicators are derived per solution of the (ordinary indicator kriging) system of equations:

$$\sum_{g'=1}^{G(\mathbf{u})} w_k(\mathbf{u}_{g'}) \cdot \sigma_k(\mathbf{u}_{g'} - \mathbf{u}_g) + \psi_k = \sigma_k(\mathbf{u} - \mathbf{u}_g), \quad g = 1,\ldots, G(\mathbf{u})$$

$$\sum_{g'=1}^{G(\mathbf{u})} w_k(\mathbf{u}_{g'}) = 1$$

(11.6)

where ψ_k denotes the Lagrange multiplier that is linked to the constraint on the weights; see Goovaerts (1997) for details. The solution of the above system yields a set of $G(\mathbf{u})$ weights that account for: (1) any spatial redundancy in the training samples by reducing the influence of clusters and (2) the spatial correlation between each sample indicator $i_k(\mathbf{u}_g)$ of the k-th category and the unknown indicator $i_k(\mathbf{u})$ for the same category.

A favorable property of OIK is its data exactitude: at any training pixel, the estimated probability $p_k^*[c(\mathbf{u})\,|\,\mathbf{c}_g]$ identifies the corresponding observed indicator; for example, $p_k^*[c(\mathbf{u}_g)\,|\,\mathbf{c}_g] = i_k(\mathbf{u}_g)$. This feature is not shared by traditional spatial classifiers, such as the nearest neighbor classifier (Steele et al., 2001), which allow for misclassification at the training locations. On the other hand, at a pixel $\mathbf{u}$ that lies further away from the training locations than the correlation length of the indicator covariance model Σ_k, the estimated OIK probability is very similar to the corresponding prior class proportion (i.e., $p_k^*[c(\mathbf{u})\,|\,\mathbf{c}_g] = p_k$). In short, the only information exploited by IK is the class labels at the training sample locations and their spatial correlation. Near training locations, IK is faithful to the observed class labels, whereas away from these locations IK has no other information apart from the K prior (constant) class proportions $\{p_k, k = 1,\ldots, K\}$.

11.2.3 Combining Spectral and Contextual Information

Once the two conditional probabilities $p_k^*[c(\mathbf{u})\,|\,\mathbf{x}(\mathbf{u})]$ and $p_k^*[c(\mathbf{u})\,|\,\mathbf{c}_g]$ are derived from spectral and spatial information, respectively, the goal is to fuse these probabilities into an updated estimate of the conditional probability $p_k^*[c(\mathbf{u})\,|\,\mathbf{x}(\mathbf{u}), \mathbf{c}_g] = \text{Prob}\{C(\mathbf{u}) = c_k\,|\,\mathbf{x}(\mathbf{u}), \mathbf{c}_g\}$, which accounts for both information sources. In what follows, we will drop the superscript * from the notation for simplicity, but the reader should bear in mind that all quantities involved are estimated probabilities. In accordance with Bayesian terminology, we will refer to the individual source conditional probabilities, $p_k^*[c(\mathbf{u})\,|\,\mathbf{x}(\mathbf{u})]$ and $p_k^*[c(\mathbf{u})\,|\,\mathbf{c}_g]$, as preposterior probabilities and retain the qualifier posterior only for the final conditional probability $p_k^*[c(\mathbf{u})\,|\,\mathbf{x}(\mathbf{u}), \mathbf{c}_g]$ that accounts for both information sources.

Bayesian updating of the individual source preposterior probabilities for, say, the k-th class is accomplished by writing the posterior probability $p_k[c(\mathbf{u})\,|\,\mathbf{x}(\mathbf{u}), \mathbf{c}_g]$ in terms of the prior probability p_k and the joint likelihood function $p[\mathbf{x}(\mathbf{u}), \mathbf{c}_g\,|\,c(\mathbf{u}) = c_k]$:

$$p_k[c(\mathbf{u})\,|\,\mathbf{x}(\mathbf{u}), \mathbf{c}_g] = \text{Prob}\ \{C(\mathbf{u}) = c_k\,|\,\mathbf{x}(\mathbf{u}), \mathbf{c}_g\} = \frac{p\ [\mathbf{x}(\mathbf{u}), \mathbf{c}_g\,|\,c(\mathbf{u}) = c_k] \cdot p_k}{p\ [\mathbf{x}(\mathbf{u}), \mathbf{c}_g]}$$

(11.7)

where $p[\mathbf{x}(\mathbf{u}), \mathbf{c}_g\,|\,c(\mathbf{u}) = c_k] = \text{Prob}\{X(\mathbf{u}_1) = x(\mathbf{u}_1),\ldots, X(\mathbf{u}_B) = x(\mathbf{u}_B), C(\mathbf{u}_1) = c_{k_1},\ldots, C(\mathbf{u}_G) = c_{k_G}\,|\,c(\mathbf{u}) = c_k\}$ denotes the probability that the particular combination of B reflectance values and G sample class labels occurs at pixel $\mathbf{u}$ and its neighborhood (for simplicity, G and $G(\mathbf{u})$ are not differentiated notation-wise). In the denominator, $p[\mathbf{x}(\mathbf{u}), \mathbf{c}_g]$ denotes the marginal (unconditional) probability, which can be expressed in terms of the entries of the numerator using the law of total probability.

Assuming class-conditional independence between the spatial and spectral information, that is, $p[\mathbf{x}(\mathbf{u}), \mathbf{c}_g \mid c(\mathbf{u}) = c_k] = p[\mathbf{x}(\mathbf{u}) \mid c(\mathbf{u}) = c_k] \cdot p[\mathbf{c}_g \mid c(\mathbf{u}) = c_k]$, one can write:

$$p_k[c(\mathbf{u}) \mid \mathbf{x}(\mathbf{u}), \mathbf{c}_g] = \frac{p[\mathbf{x}(\mathbf{u}) \mid c(\mathbf{u}) = c_k] \cdot p[\mathbf{c}_g \mid c(\mathbf{u}) = c_k] \cdot p_k}{p[\mathbf{x}(\mathbf{u}), \mathbf{c}_g]} \tag{11.8}$$

Class-conditional independence implies that the actual class $c(\mathbf{u}) = c_k$ at pixel $\mathbf{u}$ suffices to model the spectral information independently from the spatial information, and vice versa. Although conditional independence is rarely checked in practice, it has been extensively used in the literature because it renders the computation of the conditional probability tractable. It appears in evidential reasoning theory (Bonham-Carter, 1994), in multisource fusion (Benediktsson et al., 1990; Benediktsson and Swain, 1992), and in spatial statistics (Cressie, 1993). The consequence of this assumption is that one can combine spectrally derived and spatially derived probabilities without accounting for the interaction of spectral and spatial information.

Using Bayes' rule, one arrives at the final form of posterior probability under conditional independence (Lee et al., 1987; Benediktsson and Swain, 1992):

$$p_k[c(\mathbf{u}) \mid \mathbf{x}(\mathbf{u}), \mathbf{c}_g] = \frac{\dfrac{p_k[c(\mathbf{u}) \mid \mathbf{x}(\mathbf{u})] \cdot p_k[c(\mathbf{u}) \mid \mathbf{c}_g]}{p_k}}{\dfrac{p_k[c(\mathbf{u}) \mid \mathbf{x}(\mathbf{u})] \cdot p_k[c(\mathbf{u}) \mid \mathbf{c}_g]}{p_k} + \dfrac{p_k[\overline{c(\mathbf{u})} \mid \mathbf{x}(\mathbf{u})] \cdot p_k[\overline{c(\mathbf{u})} \mid \mathbf{c}_g]}{p_k}} \tag{11.9}$$

where $c(\overline{\mathbf{u}}) = \overline{c_k}$ denotes the complement event of the k-th class and $\overline{p_k}$ denotes the prior probability for that event. In the case of three mutually exclusive and exhaustive classes, forest, shrub, and rangeland, for example, if the k-th class corresponds to forest then the complement event is the absence of forest (i.e., presence of either shrub or rangeland), and the probability for that complement event is the sum of the shrub and rangeland probabilities.

In words, the final posterior probability $p_k[c(\mathbf{u}) \mid \mathbf{x}(\mathbf{u}), \mathbf{c}_g]$ that accounts for both sources of information (spectral and spatial) under conditional independence is a simple product of the spectra-based conditional probability $p_k[c(\mathbf{u}) \mid \mathbf{x}(\mathbf{u})]$ and the space-based conditional probability $p_k[c(\mathbf{u}) \mid \mathbf{c}_g]$ divided by the prior class probability p_k. Each resulting probability $p_k[c(\mathbf{u}) \mid \mathbf{x}(\mathbf{u}), \mathbf{c}_g]$ is finally standardized by the sum $\sum_{k=1}^{K} p_k[c(\mathbf{u}) \mid \mathbf{x}(\mathbf{u}), \mathbf{c}_g]$ of all resulting probabilities over all K classes to ensure a unit sum.

A more intuitive version of the above fusion equation is easily obtained as:

$$p_k[c(\mathbf{u}) \mid \mathbf{x}(\mathbf{u}), \mathbf{c}_g] \propto \frac{p_k[c(\mathbf{u}) \mid \mathbf{x}(\mathbf{u})]}{p_k} \cdot \frac{p_k[c(\mathbf{u}) \mid \mathbf{c}_g]}{p_k} \cdot p_k \tag{11.10}$$

where the proportionality constant is still the sum $\sum_{k=1}^{K} p_k[c(\mathbf{u}) \mid \mathbf{x}(\mathbf{u}), \mathbf{c}_g]$ of all resulting probabilities, which ensures that they sum to 1.

This version of the posterior probability equation entails that the ratio $p_k[c(\mathbf{u}) \mid \mathbf{x}(\mathbf{u}), \mathbf{c}_g] / p_k$ of the final posterior probability $p_k[c(\mathbf{u}) \mid \mathbf{x}(\mathbf{u}), \mathbf{c}_g]$ to the prior probability p_k is simply the product of the ratio $p_k[c(\mathbf{u}) \mid \mathbf{x}(\mathbf{u})] / p_k$ of the spectrally derived preposterior probability $p_k[c(\mathbf{u}) \mid \mathbf{x}(\mathbf{u})]$

to the prior probability p_k times the ratio $p_k[c(\mathbf{u})\,|\,\mathbf{c}_g]/\,p_k$ of the derived preposterior probability $p_k[c(\mathbf{u})\,|\,\mathbf{c}_g]$ to the prior probability p_k. Note that this is a congenial assumption whose consequences have not received much attention in the remote sensing literature (and in other disciplines). Under this assumption, the final posterior probability $p_k[c(\mathbf{u})\,|\,\mathbf{x}(\mathbf{u}),\mathbf{c}_g]$ can be seen as a modulation of the prior probability p_k by two factors: the first factor $p_k[c(\mathbf{u})\,|\,\mathbf{x}(\mathbf{u})]/\,p_k$ quantifies the influence of remote sensing, while the second factor $p_k[c(\mathbf{u})\,|\,\mathbf{c}_g]/\,p_k$ quantifies the influence of the spatial information.

Note that, in the above formulation, both information sources are deemed equally reliable, which need not be the case in practice. Although individual source preposterior probabilities in the fusion Equation 11.9 can be discounted via the use of reliability exponents (Benediktsson and Swain, 1992; Tso and Mather, 2001), this avenue is not explored in this chapter due to space limitations.

11.2.4 Mapping Thematic Classification Accuracy

The set of K posterior probabilities of class occurrence $\{p_{k'}[c(\mathbf{u})\,|\,\mathbf{x}(\mathbf{u}),\mathbf{c}_g],k'=1,\dots,K\}$ derived at a particular pixel $\mathbf{u}$ can be readily converted into a classification accuracy value $a(\mathbf{u})$. If pixel $\mathbf{u}$ is allocated to, say, category c_k, then a measure of accuracy associated with this particular class allocation is simply $a(\mathbf{u})=p_{k'=k}[c(\mathbf{u})\,|\,x(\mathbf{u}),\mathbf{c}_g]$, whereas a measure of inaccuracy (error) associated with this allocation is $1-a(\mathbf{u})=1-p_{k'=k}[c(\mathbf{u})\,|\,x(\mathbf{u}),\mathbf{c}_g]$. If such posterior probabilities are available at each pixel $\mathbf{u}$, any classified map product can be readily accompanied by a map (of the same dimensions) that depicts the spatial distribution of classification accuracy.

The accuracy value at each pixel $\mathbf{u}$ is a sole function of the K posterior probabilities available at that pixel; different probability values will therefore yield different accuracy values at the same pixel. Evidently, the more realistic the set of posterior probabilities at a particular pixel $\mathbf{u}$, the more realistic the accuracy value at that pixel. Consider for example, the set of K preposterior probabilities $\{p_{k'}[c(\mathbf{u})\,|\,\mathbf{x}(\mathbf{u})],k'=1,\dots,K\}$ derived from a conventional maximum likelihood classifier (Section 11.2.1) and the set of K posterior probabilities $\{p_{k'}[c(\mathbf{u})\,|\,\mathbf{x}(\mathbf{u}),\mathbf{c}_g],k'=1,\dots,K\}$ derived from the proposed fusion of spectral and spatial information (Section 11.2.3). These two sets of probability values will yield two different accuracy measures $a_c(\mathbf{u})$ and $a_f(\mathbf{u})$ at the same pixel $\mathbf{u}$ (subscripts c and f distinguish the use of conventional vs. fusion-based probabilities). It is argued that the use of contextual information for deriving the latter posterior probabilities yields a more realistic accuracy map than that typically constructed using the former preposterior probabilities derived from a conventional classifier (Foody et al., 1992).

11.2.5 Generation of Simulated TM Reflectance Values

This section describes a procedure used in the case study (Section 11.3) to realistically simulate a reference classification and the corresponding set of six TM spectral bands. Availability of an exhaustive reference classification allows computation of accuracy statistics without the added complication of a particular sampling design.

Starting from raw TM imagery, a subscene is classified into L clusters using the Iterative Self-Organizing Data Analysis Technique (ISODATA) clustering algorithm (Jensen, 1996). These L clusters are assigned into K known classes. To reduce the degree of fragmentation in the resulting classified map, the classification is smoothed using MAP selection within a window around each pixel $\mathbf{u}$ (Deutsch, 1998). The resulting land-cover (LC) map is regarded as the exhaustive reference classification.

Based on this reference classification, the class-conditional joint PDF of the six TM bands is modeled as multivariate Gaussian with mean and covariance derived from raw TM bands. Let $\mathbf{m}^o_{\mathbf{X}|k}$ and $\Sigma^o_{\mathbf{X}|k}$ denote the (6×1) vector of class-conditional mean and the (6×6) matrix of class-conditional (co)variances of the raw reflectance values in the k-th class. Let $\mathbf{m}_{\mathbf{X}}$ and Σ denote the (6×1) mean vector and (6×6) covariance matrix, respectively, of the above K class-conditional mean vectors $\left\{\mathbf{m}^o_{\mathbf{X}|k}, k = 1, \ldots, K\right\}$. A set of K simulated (6×1) vectors $\left\{\mathbf{m}_{\mathbf{X}|k}, k = 1, \ldots, K\right\}$ of class-conditional means are generated from a six-variate Gaussian distribution with mean $\mathbf{m}_{\mathbf{X}}$ and covariance $\boldsymbol{\Sigma}$. In the case study, simulated class-conditional mean vectors $\left\{\mathbf{m}_{\mathbf{X}|k}, k = 1, \ldots, K\right\}$ were used instead of their original counterparts $\left\{\mathbf{m}^o_{\mathbf{X}|k}, k = 1, \ldots, K\right\}$ in order to introduce class confusion. Simulated reflectance values are then generated for each pixel in the reference classification from the appropriate class-conditional distribution, which is assumed Gaussian with mean $\mathbf{m}_{\mathbf{X}|k}$, and covariance $\boldsymbol{\Sigma}^o_{\mathbf{X}|k}$. For example, if a pixel in the reference classification has LC forest $(k = 1)$, six simulated reflectance values are simulated at that pixel from a Gaussian distribution with mean $\mathbf{m}_{\mathbf{X}|1}$ and covariance $\boldsymbol{\Sigma}^o_{\mathbf{X}|1}$. A similar procedure for generating synthetic satellite imagery (but without the simulation of class-conditional mean values $\left\{\mathbf{m}_{\mathbf{X}|k}, k = 1, \ldots, K\right\}$) was adopted by Swain et al. (1981) and Haralick and Joo (1986). The simulated reflectance values are further degraded by introducing white noise generated by a six-variate Gaussian distribution with mean $\mathbf{0}$ and (co)variance $0.2\,\boldsymbol{\Sigma}$; this entails that the simulated noise is correlated from one spectral band to another.

Independent simulation of reflectance values from one pixel to another implies the nonrealistic feature of low spatial correlation in the simulated reflectance values. In the case study, in order to enhance spatial correlation as well as positional error, typical of real images, a motion blur filter with a horizontal motion of 21 pixels in the $-45°$ direction was applied to each band to simulate the linear motion of a camera. The resulting reflectance values were further degraded by addition of a realization of an independent multivariate white noise process, which implies correlated noise from one spectral band to another. This latter realization was generated using a multivariate Gaussian distribution with mean $\mathbf{0}$ and (co)variance $0.05\,\boldsymbol{\Sigma}$. To avoid edge effects introduced by the motion blur filter, the results of Gaussian maximum likelihood classification, as well as those for indicator kriging, were reported on a smaller (cropped) subscene.

The last step in the simulated TM data generation consists of a band-by-band histogram transformation: the histogram of reflectance values for each spectral band in the simulated image is transformed to the histogram of the original TM reflectance values for that band through histogram equalization. The purpose of this transformation is to force the simulated TM imagery to have the same histogram as that of the original TM imagery, as well as similar covariance among bands. The (transformed) simulated reflectance values are finally rounded to preserve the integer digital nature of the data.

11.3 RESULTS

To illustrate the proposed methodology for fusing spatial and spectral information for mapping thematic classification uncertainty, a case study was conducted using simulated imagery based on a Landsat Thematic Mapper subscene from path 41/row 27 in western Montana, and the procedure described in Section 11.2.5. The TM imagery, collected on September 27, 1993, was supplied by the U.S. Geological Survey's (USGS) Earth Resources Observation Systems (EROS) Data Center and is one of a set from the Multi-Resolution Land Characteristics (MRLC) program (Vogelmann et al., 1998). The study site consisted of a subscene covering a portion of the Lolo National Forest $(541 \times 414$ pixels). The original 30-m TM data served as the basis for generating the simulated TM imagery used in this case study.

The subscene was classified into $L = 150$ clusters using the ISODATA algorithm, and these L clusters were assigned to $K = 3$ classes: forest $(k = 1)$, shrub $(k = 2)$, and rangeland $(k = 3)$. The resulting classification was smoothed using MAP selection within a 5×5 window around each pixel $\mathbf{u}$. The resulting LC map is regarded as the exhaustive reference classification (unavailable in practice). A small subset $(G = 314)$ of the 541×414 pixels (0.14% of the total population) was selected as training pixels through stratified random sampling. The sample and reference class proportions of forest, shrub, and rangeland were $p_1 = 0.65$, $p_2 = 0.21$, and $p_3 = 0.14$, respec-

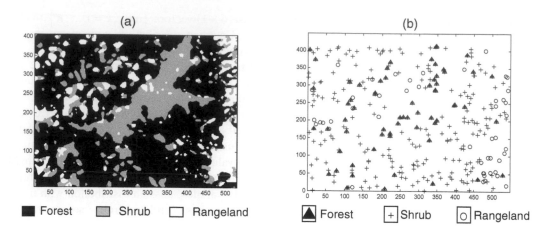

Figure 11.1 Reference classification (a) and 314 training pixels (b) selected via stratified random sampling.

Table 11.1 Parameters of the Three Indicator Covariance Models, σ_1, σ_2, σ_3, for Forest, Shrub, and Rangeland, Respectively

		Sill		Range	
	Nugget	(1)	(2)	(1)	(2)
Forest	0.02	0.61	0.37	30	120
Shrub	0.03	0.59	0.38	25	100
Rangeland	0.01	0.75	0.75	22	400

Note: All indicator covariances were modeled using a nugget contribution and two exponential covariance structures with respective sills and practical ranges: sill(1), sill(2), range(1), and range(2). Sill values are expressed as a percentage of the total variance: $p_k(1 - p_k) = 0.23$, 0.17, 0.12, for forest, shrub, and rangeland, respectively; range values are expressed in numbers of pixels.

tively. The remaining unsampled reference pixels were used as validation data for assessing the accuracy of the different methods. The cropped (ranging from 7 to 530 and from 9 to 406 pixels) reference classification and the $G = 314$ training samples used in this study are shown in Figure 11.1a and Figure 11.1b.

The class labels and the corresponding simulated reflectance values at the training sample locations were used to derive statistical parameters: the class-conditional means $\mathbf{m}_{X|1}, \mathbf{m}_{X|2}, \mathbf{m}_{X|3}$ and the class-conditional (co)variances $\boldsymbol{\Sigma}_{X|1}^o, \boldsymbol{\Sigma}_{X|2}^o, \boldsymbol{\Sigma}_{X|3}^o$ for forest, shrub, and rangeland, respectively. The class labels of the training pixels were also used to infer the three indicator covariance models, $\sigma_1, \sigma_2, \sigma_3$, for forest, shrub, and rangeland, respectively (Equation 11.5). All indicator covariance models (not shown) were isotropic, and their parameters are tabulated in Table 11.1. The forest and shrub indicator covariance models, σ_1, σ_2, consisted of a nugget component (2 to 3% of the total variance), a small-scale structure of practical range 25 to 30 pixels (59 to 61% of the total variance), and a larger-scale structure of practical range 100 to 120 pixels (37 to 38% of the total variance). The rangeland indicator covariance model, σ_3, consisted of a nugget component (1% of the total variance), a small-scale structure of practical range 22 pixels (75% of the total variance), and one larger-scale structure of practical range 400 pixels (24% of the total variance). These covariance model parameters imply that forest and shrub have a very similar spatial correlation that differs slightly from that of rangeland. The latter class has more pronounced small-scale variability, and less large-scale variability, which is also of longer range than that of forest and shrub. For further details regarding the interpretation of variogram and covariance functions computed from remotely sensed imagery, see Woodcock et al. (1988).

11.3.1 Spectral and Spatial Classifications

Using the class-conditional means $\mathbf{m}_{X|1}, \mathbf{m}_{X|2}, \mathbf{m}_{X|3}$ and (co)variances $\Sigma_{X|1}^o, \Sigma_{X|2}^o, \Sigma_{X|3}^o$, three Gaussian likelihood functions were established for any vector $\mathbf{x(u)}$ of reflectance values at any pixel $\mathbf{u}$ not in the training set (Equation 11.1). The three Gaussian likelihood functions were subsequently inverted (Equation 11.2) to compute the three spectrally derived preposterior probabilities, $p_1[c(\mathbf{u}) | \mathbf{x(u)}]$, $p_2[c(\mathbf{u}) | \mathbf{x(u)}]$, and $p_3[c(\mathbf{u}) | \mathbf{x(u)}]$, for forest, shrub, and rangeland, respectively. These GML preposterior probabilities are shown in Figure 11.2a–c. Note (1) the high degree of noise in the probabilities, (2) the confusion of shrub and rangeland (probabilities close to 0.5), and (3) the motion-like appearance that entails diffuse class boundaries. The corresponding MAP selection at each pixel $\mathbf{u}$ is shown in Figure 11.2d. Note again the high degree of fragmentation in the classified map. The overall classification accuracy (evaluated against the reference classification) was 0.73 (Kappa = 0.44), indicating a rather severe misclassification.

Arguably, in the presence of noise, the original spectral vector could have been replaced by a vector of the same dimensions whose entries are averages of reflectance values within a (typically 3×3) neighborhood around each pixel (Switzer, 1980). This, however, amounts to implicitly introducing contextual information into the classification procedure: spatial variability in the reflectance values is suppressed via a form of low-pass filter to introduce more spatial correlation, and thus produce less fragmented classification maps. In the absence of noise-free data, any such filtering procedure is rather arbitrary: there is no reason to use a 3×3 vs. a 5×5 filter, for example. In this chapter, we propose a method for introducing that notion of compactness in classification via a model of spatial correlation inferred from the training pixels themselves.

Ordinary indicator kriging (OIK) (Equation 11.5 and Equation 11.6) was performed using the three sets of G training class indicators and their corresponding indicator covariance models to compute the space-derived preposterior probabilities $p_1[c(\mathbf{u}) | \mathbf{c}_g]$, $p_2[c(\mathbf{u}) | \mathbf{c}_g]$, $p_3[c(\mathbf{u}) | \mathbf{c}_g]$ for forest, shrub, and rangeland, respectively. These OIK preposterior probabilities are shown in Figure 11.3a–c. Note the very smooth spatial patterns and the absence of clear boundaries, as opposed to those found in the spectrally derived posterior probabilities of Figure 11.2. Note also that the training sample class labels are reproduced at the training locations, per the data-exactitude property of OIK. The corresponding MAP selection at each pixel $\mathbf{u}$ is shown in Figure 11.3d. The overall classification accuracy is 0.73 (Kappa = 0.44), the same as that computed from the spectrally derived classification, indicating the same level of severe misclassification for the spatially derived classification.

11.3.2 Merging Spectral and Contextual Information

Bayesian fusion (Equation 11.9), was performed to combine the individually derived spectral and spatial preposterior probabilities into posterior probabilities $p_1[c(\mathbf{u}) | x(\mathbf{u}), \mathbf{c}_g]$, $p_2[c(\mathbf{u}) | x(\mathbf{u}), \mathbf{c}_g]$, and $p_3[c(\mathbf{u}) | x(\mathbf{u}), \mathbf{c}_g]$, for forest, shrub, and rangeland, respectively; these posterior probabilities account for both information sources and are shown in Figure 11.4a–c. Compared to the spectrally derived preposterior probabilities of Figure 11.2, the latter posterior probabilities have smoother spatial patterns and much less noise. Compared to the spacially derived preposterior probabilities of Figure 11.3, the latter posterior probabilities have more variable patterns and indicate clearer boundaries. The corresponding MAP selection at each pixel $\mathbf{u}$ is shown in Figure 11.4d. The overall classification accuracy increased to 0.80 and the Kappa coefficient to

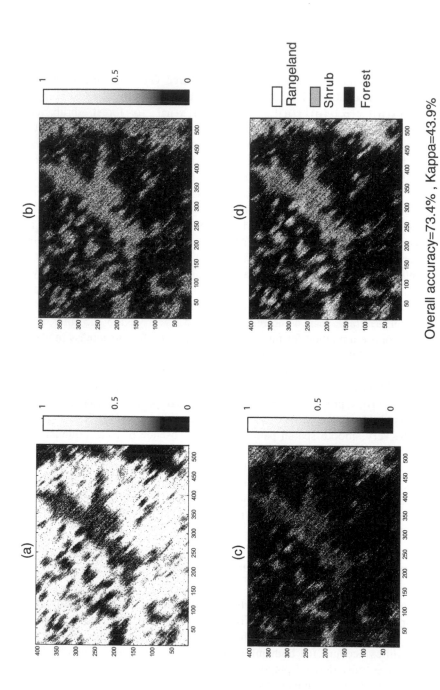

Figure 11.2 Conditional probabilities for forest (a), shrub (b), and rangeland (c), based on Gaussian maximum likelihood (GML), and corresponding MAP selection (d).

Overall accuracy=73.4% , Kappa=43.9%

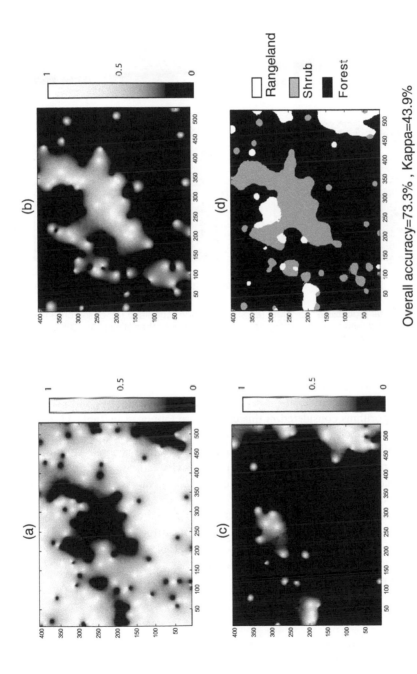

Figure 11.3 Conditional probabilities for forest (a), shrub (b), and rangeland (c), based on ordinary indicator kriging (OIK), and corresponding MAP selection (d).

Overall accuracy=73.3% , Kappa=43.9%

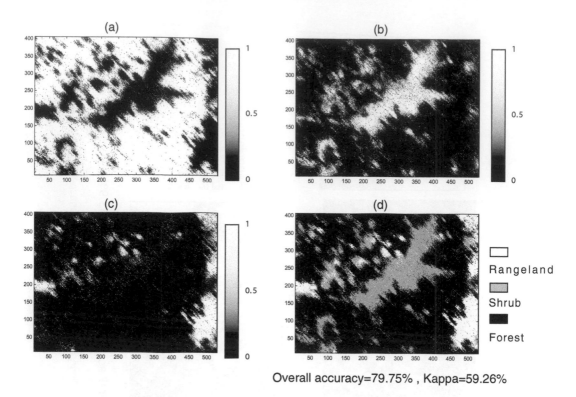

Overall accuracy=79.75% , Kappa=59.26%

Figure 11.4 Conditional probabilities for forest (a), shrub (b), and rangeland (c), based on Bayesian integration of spectrally derived and spacially derived preposterior probabilities (GML/OIK), and corresponding MAP selection (d).

0.59, a 9.6% and 34.1% improvement, respectively, relative to the corresponding accuracy statistics computed from the GML classification.

For comparison, accuracy assessment statistics, including producer's and user's accuracy, for all classification algorithms considered in this chapter are tabulated in Table 11.2. Clearly, classification accuracy using the proposed contextual classification methods was superior to that using only spectral or only spatial information. As stated above, overall accuracy and the Kappa coefficients are significantly higher for the proposed methods. In addition, both producer's and user's accuracy for all three classes are higher than the corresponding values computed from the spectrally derived or the spacially derived classifications.

The reference and classification-derived class proportions are also provided in Table 11.3 for comparison. Clearly, MAP selection from the fused posterior probabilities $p_k[c(\mathbf{u})\,|\,x(\mathbf{u}), \mathbf{c}_g]$ yielded the closest class proportions to the reference ones: 0.69 vs. 0.65 (reference) for forest, 0.21 vs. 0.21 for shrub, and 0.10 vs. 0.14 for rangeland. The other methods performed worse with respect to reproducing the reference class proportions.

11.3.3 Mapping Classification Accuracy

The three spectrally derived preposterior probabilities, $p_1[c(\mathbf{u})\,|\,\mathbf{x}(\mathbf{u})]$, $p_2[c(\mathbf{u})\,|\,\mathbf{x}(\mathbf{u})]$, and $p_3[c(\mathbf{u})\,|\,\mathbf{x}(\mathbf{u})]$ for forest, shrub, and rangeland, respectively, were converted into an accuracy value $a_c(\mathbf{u})$ for the particular class reported at pixel $\mathbf{u}$ (i.e., for the classification of Figure 11.2d), as described in Section 11.2.4. These accuracy values were mapped in Figure 11.5a. The same procedure was repeated using the three fusion-based posterior probabilities $p_1[c(\mathbf{u})\,|\,x(\mathbf{u}), \mathbf{c}_g]$, $p_2[c(\mathbf{u})\,|\,x(\mathbf{u}), \mathbf{c}_g]$, and $p_3[c(\mathbf{u})\,|\,x(\mathbf{u}), \mathbf{c}_g]$, for forest, shrub, and rangeland, respectively, to yield

Table 11.2 Accuracy Statistics for Classification Based on MAP Selection from Conditional Probabilities Computed Using Different Methods: Gaussian Maximum Likelihood (GML), Ordinary Indicator Kriging (OIK), and Bayesian Integration of GML and OIK Probabilities (GML/OIK)

	GML	OIK	GML/OIK
Overall accuracy			
	0.73	0.73	0.80
Kappa			
	0.44	0.44	0.59
Producer's accuracy			
Forest	0.92	0.88	0.91
Shrub	0.44	0.52	0.63
Rangeland	0.30	0.39	0.51
User's accuracy			
Forest	0.82	0.78	0.86
Shrub	0.48	0.61	0.64
Rangeland	0.55	0.63	0.68

Table 11.3 Class Proportions from Reference and Classified Maps Based on MAP Selection from Conditional Probabilities Computed Using Different Methods: Gaussian Maximum Likelihood (GML), Ordinary Indicator Kriging (OIK), and Bayesian Integration of GML and OIK Probabilities (GML/OIK)

	Reference	GML	OIK	GML/OIK
Forest	0.65	0.73	0.73	0.69
Shrub	0.21	0.19	0.18	0.21
Rangeland	0.14	0.08	0.09	0.10

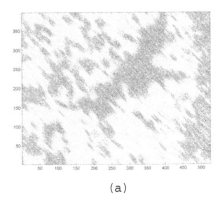
(a)

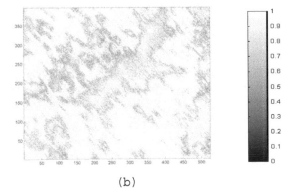
(b)

Figure 11.5 Pixel-specific accuracy values for GML-derived classes (a) and for GML/OIK-derived classes (b).

an accuracy value $a_f(\mathbf{u})$ for the particular class reported at pixel $\mathbf{u}$ (i.e., for the classification of Figure 11.4d). These accuracy values were mapped in Figure 11.5b. The accuracy map of Figure 11.5b exhibited much higher values than the corresponding map of Figure 11.5a, indicating an increased confidence in classification due precisely to the consideration of contextual information. In addition, the low accuracy values (~0.4–0.6) of Figure 11.5b were found near class boundaries, as opposed to the low accuracy values of Figure 11.5a, which just corresponded to pixels classified as shrub and rangeland. This latter characteristic implied that contextual information yielded a more realistic map of classification accuracy, which could be useful for designing additional sampling campaigns.

11.4 DISCUSSION

A geostatistical approach for mapping thematic classification uncertainty was presented in this chapter. The spatial correlation of each class, as inferred from a set of training pixels, along with the actual locations of these pixels, was used via indicator kriging to estimate the location-specific probability that a pixel belongs to a certain class, given the spatial information contained in the training pixels. The proposed approach for estimating the above preposterior probability accounted for texture information via the corresponding indicator covariance model for each class, as well as for the spatial proximity of each pixel to the training pixels after this proximity was discounted for the spatial redundancy (clustering) of the training pixels. Space-derived preposterior probabilities were merged via Bayes' rule with spectrally derived preposterior probabilities, the latter based on the collocated vector of reflectance values at each pixel. The final (fused) posterior probabilities accounted for both spectral and spatial information.

The performance of the proposed methods was evaluated via a case study that used realistically simulated reflectance values. A subset of 0.14% (314) of the image pixels was retained as a training set. The results indicated that the proposed method of context estimation, when coupled with Bayesian integration, yielded more accurate classifications than the conventional maximum likelihood classifier. More specifically, relative improvements of 10% and 34% were found for overall accuracy and the Kappa coefficient. In addition, contextual information yielded more realistic classification accuracy maps, whereby pixels with low accuracy values tended to coincide with class boundaries.

11.5 CONCLUSIONS

The proposed geostatistical methodology constitutes a viable means for introducing contextual information into the mapping of thematic classification uncertainty. Since the results presented in the case study in this chapter appear promising, further research is required to evaluate the performance of the proposed contextual classification and its use for mapping thematic classification uncertainty over a variety of real-world data sets. In particular, issues pertaining to the type and level of spatial correlation, the density of the training pixels, and their effects on the resulting classification uncertainty maps should be investigated in greater detail.

In conclusion, we suggest that the final posterior probabilities of class occurrence be used in a stochastic simulation framework, whereby multiple, alternative, synthetic representations of land cover maps would be generated using various algorithms for simulating categorical variables (Deutsch and Journel, 1998). These alternative representations would reproduce: (1) the observed classes at the training pixels, (2) the class proportions, (3) the spatial correlation of each class inferred from the training pixels, and (4) possible relationships with spectral or other ancillary spatial information. The ensemble of simulated land-cover maps could be then used for error propagation (e.g., Kyriakidis and Dungan [2001]), thus allowing one to go beyond simple map accuracy statistics and address map use (and map value) issues.

11.6 SUMMARY

Thematic classification accuracy constitutes a critical factor in the successful application of remotely sensed products in various disciplines, such as ecology and environmental sciences. Apart from traditional accuracy statistics based on the confusion matrix, maps of posterior probabilities of class occurrence are extremely useful for depicting the spatial variation of classification uncertainty. Conventional classification procedures such as Gaussian maximum likelihood, however, do not account for the plethora of ancillary data that could enhance such a metadata map product.

In this chapter, we propose a geostatistical approach for introducing contextual information into the mapping of classification uncertainty using information provided only by the training pixels. Probabilities of class occurrence that account for context information are first estimated via indicator kriging and are then integrated in a Bayesian framework with probabilities for class occurrence based on conventional classifiers, thus yielding improved maps of thematic classification uncertainty. A case study based on realistically simulated TM imagery illustrates the applicability of the proposed method: (1) regional accuracy scores indicate relative improvements over traditional classification algorithms in the order of 10% for overall accuracy and 34% for the Kappa coefficient and (2) maps of pixel-specific accuracy values tend to pinpoint class boundaries as the most uncertain regions, thus appearing as a promising means for guiding additional sampling campaigns.

REFERENCES

Atkinson, P.M. and P. Lewis, Geostatistical classification for remote sensing: an introduction, *Comput. Geosci.*, 26, 361–371, 2000.

Benediktsson, J.A. and P.H. Swain, Consensus theoretic classification methods, *IEEE Trans. Syst. Man Cybernet.*, 22, 688–704, 1992.

Benediktsson, J.A., P.H. Swain, and O.K. Ersoy, Neural network approaches versus statistical methods in classification of multisource remote sensing data, *IEEE Trans. Geosci. Remote Sens.*, 28, 540–552, 1990.

Bonham-Carter, G.F., *Geographic Information Systems for Geoscientists*, Pergamon, Ontario, 1994.

Congalton, R.G., A review of assessing the accuracy of classifications of remotely sensed data, *Remote Sens. Environ.*, 37, 35–46, 1991.

Congalton, R.G., Using spatial autocorrelation analysis to explore the errors in maps generated from remotely sensed data, *Photogram. Eng. Remote Sens.*, 54, 587–592, 1988.

Congalton, R.G. and K. Green, *Assessing the Accuracy of Remote Sensed Data: Principles and Practices*, Lewis, Boca Raton, FL, 1999.

Cressie, N.A.C., *Statistics for Spatial Data*, John Wiley & Sons, New York, 1993.

De Bruin, S., Predicting the areal extent of land-cover types using classified imagery and geostatistics, *Remote Sens. Environ.*, 74, 387–396, 2000.

Deutsch, C.V., Cleaning categorical variable (lithofacies) realizations with maximum a-posteriori selection, *Comput. Geosci.*, 24, 551–562, 1998.

Deutsch, C.V. and A.G. Journel, *GSLIB: Geostatistical Software Library and User's Guide*, 2nd ed., Oxford University Press, New York, 1998.

Foody, G.M., Status of land-cover classification accuracy assessment, *Remote Sens. Environ.*, 80, 185–201, 2002.

Foody, G.M., N.A. Campbell, N.M. Trood, and T.F. Wood, Derivation and applications of probabilistic measures of class membership from the maximum-likelihood classifier, *Photogram. Eng. Remote Sens.*, 58, 1335–1341, 1992.

Goovaerts, P., Geostatistical incorporation of spatial coordinates into supervised classification of hyperspectral data, *J. Geogr. Syst.*, 4, 99–111, 2002.

Goovaerts, P., *Geostatistics for Natural Resources Evaluation*, Oxford University Press, New York, 1997.

Haralick, R.M. and H. Joo, A context classifier, *IEEE Trans. Geosci. Remote Sens.*, 24, 997–1007, 1986.

Hutchinson, C.F., Techniques for combining Landsat and ancillary data for digital classification improvement, *Photogram. Eng. Remote Sens.*, 48, 123–130, 1982.

Isaaks, E.H. and R.M. Srivastava, *An Introduction to Applied Geostatistics*, Oxford University Press, New York, 1989.

Jensen, J.R., *Introductory Digital Image Processing: A Remote Sensing Perspective*, Prentice Hall, Upper Saddle River, NJ, 1996.

Journel, A.G., Non-parametric estimation of spatial distributions, *Math. Geol.*, 15, 445–468, 1983.

Kyriakidis, P.C. and J.L. Dungan, A geostatistical approach for mapping thematic classification accuracy and evaluating the impact of inaccurate spatial data on ecological model predictions, *Environ. Ecol. Stat.*, 8, 311–330, 2001.

Lee, T., J.A. Richards, and P.H. Swain, Probabilistic and evidential approaches for multisource data analysis, *IEEE Trans. Geosci. Remote Sens.*, 25, 283–293, 1987.

Li, S.Z., *Markov Random Field Modeling in Image Analysis*, Springer-Verlag, Tokyo, 2001.

Solow, A.R., Mapping by simple indicator kriging, *Math. Geol.*, 18, 335–352, 1986.

Steele, B.M., Combing multiple classifiers: an application using spatial and remotely sensed information for land cover type mapping, *Remote Sens. Environ.*, 74, 545–556, 2000.

Steele, B.M. and R.L. Redmond, A method of exploiting spatial information for improving classification rules: application to the construction of polygon-based land cover type maps, *Int. J. Remote Sens.*, 22, 3143–3166, 2001.

Stehman, S.V., Comparing thematic maps based on map value, *Int. J. Remote Sens.*, 20, 2347–2366, 1999.

Stehman, S.V., Selecting and interpreting measures of thematic classification accuracy, *Remote Sens. Environ.*, 62, 77–89, 1997.

Strahler, A.H., Using prior probabilities in maximum likelihood classification of remotely sensed data, *Remote Sens. Environ.*, 47, 215–222, 1980.

Swain, P.H., S.B. Vardeman, and J.C. Tilton, Contextual classification of multispectral image data, *Pattern Recogn.*, 13, 429–441, 1981.

Switzer, P., Extensions of linear discriminant analysis for statistical classification of remotely sensed data, *Math. Geol.*, 12, 367–376, 1980.

Switzer, P., W.S. Kowalik, and R.J.P. Lyon, A prior method for smoothing discriminant analysis classification maps, *Math. Geol.*, 14, 433–444, 1982.

Tso, B. and P.M. Mather, *Classification Methods for Remotely Sensed Data*, Taylor & Francis, London, 2001.

van der Meer, F., Classification of remotely sensed imagery using an indicator kriging approach: application to the problem of calcite-dolomite mineral mapping, *Int. J. Remote Sens.*, 17,1233–1249, 1996.

Vogelmann, J.E., T.L. Sohl, P.V. Campbell, and D.M. Shaw, Regional land cover characterization using Landsat Thematic Mapper data and ancillary data sources, *Environ. Monit. Assess.*, 51, 415–428, 1998.

Woodcock, C.E., A.H. Strahler, and D.L.B. Jupp, The use of variograms in remote sensing. I: scene models and simulated images, *Remote Sens. Environ.*, 25, 323–348, 1988.

Zhang, J. and M. Goodchild, *Uncertainty in Geographic Information*, Taylor & Francis, London, 2002.

CHAPTER 12

An Error Matrix Approach to Fuzzy Accuracy Assessment: The NIMA Geocover Project

Kass Green and Russell G. Congalton

CONTENTS

12.1 INTRODUCTION

As remote sensing applications have grown in complexity, so have the classification schemes associated with these efforts. The classification scheme then becomes a very important factor influencing the accuracy of the entire project. A review of the recent accuracy assessment literature points out some of the limitations of using only an error matrix approach to accuracy assessment with a complex classification scheme. Congalton and Green (1993) recommend the error matrix as a jumping-off point for identifying sources of confusion (i.e., differences between the map created from remotely sensed data and the reference data) and not simply the "error." For example, the variation in human interpretation can have a significant impact on what is considered correct. If photographic interpretation is used as the source of the reference data and that interpretation is not completely correct, then the results of the accuracy assessment could be very misleading. The same holds true even for observations made in the field. As classification schemes become more complex, more variation in human interpretation is introduced (Congalton, 1991; Congalton and Biging, 1992; Gong and Chen, 1992; Lowell, 1992).

Gopal and Woodcock (1994) proposed the use of fuzzy sets to "allow for explicit recognition of the possibility that ambiguity might exist regarding the appropriate map label for some locations on a map. The situation of one category being exactly right and all other categories being equally and exactly wrong often does not exist." They allowed for a variety of responses, such as absolutely right, good answer, acceptable, understandable but wrong, and absolutely wrong. While dealing with the ambiguity, this approach does not allow the accuracy assessment to be reported as an error matrix.

This chapter introduces a technique using fuzzy accuracy assessment that allows for the analyst to incorporate the variation or ambiguity in the map label and also present the results in the form of an error matrix. This approach is applied here to a worldwide mapping effort funded by the National Imagery and Mapping Agency (NIMA) using Landsat Thematic Mapper (TM) imagery. The Earth Satellite Corporation (Earthsat) performed the mapping and Pacific Meridian Resources of Space Imaging conducted the accuracy assessment. The results presented here are for one of the initial prototype test areas (for an undisclosed location of the world) used for developing this fuzzy accuracy assessment process.

12.2 BACKGROUND

The quantitative accuracy assessment of maps produced from remotely sensed data involves the comparison of a map with reference information that is assumed to be correct. The purpose of a quantitative accuracy assessment is the identification and measurement of map errors. The two primary motivations include: (1) providing an overall assessment of the reliability of the map (Gopal and Woodcock, 1994) and (2) understanding the nature of map errors. While more attention is often paid to the first motivation, understanding the errors is arguably the most important aspect of accuracy assessment. For any given map class, it is critical to know the probability of the site's being labeled correctly and what classes are confused with one another. Quantitative accuracy assessment provides map users with a consistent and objective analysis of map quality and error. Quantitative analysis is fundamental to map use; without it, users would make decisions without knowing the reliability of the map as a whole or the sources of confusion.

The error matrix is the most widely accepted format for reporting remotely sensed data classification accuracies (Story and Congalton, 1986; Congalton, 1991). Error matrices simply compare map data to reference data. An error matrix is an array of numbers set out in rows and columns that expresses the number of pixels or polygons assigned to a particular category in one classification relative to those assigned to a particular category in another classification (Table 12.1). One of the classifications is considered to be correct (reference) and may be generated from aerial photography, airborne video, ground observation, or ground measurement, while the other classification is generated from the remotely sensed data (observed).

An error matrix is an effective way to represent accuracy because both the total and the individual accuracies of each category are clearly described and confusion between classes is evident. Also indicated are errors of inclusion (commission errors) and errors of exclusion (omission errors) that may be present in the classification. A commission error occurs when an area is included into a category when it does not belong. An omission error is excluding an area from the category in which it does belong. Every error is an omission from the correct category and a commission to a wrong category. For example, in the error matrix in Table 12.1 four areas were classified as deciduous but the reference data showed that they were actually coniferous. Therefore, four areas were omitted from the correct coniferous category and committed to the incorrect deciduous category. Utilizing this information, users can ascertain the relative strengths and weaknesses of each map class, creating a more solid basis for decision making.

Additionally, the error matrix can be used to compute overall accuracy and producer's and user's accuracies (Story and Congalton, 1986). Overall accuracy is simply the sum of the major diagonal (i.e., the correctly classified sample units) divided by the total number of sample units in

Table 12.1 Example Error Matrix

		Reference Data				Land-Cover Categories
		D	C	AG	SB	row total
	D	63	4	22	24	113
	C	6	79	8	8	101
Classified Data	AG	0	11	85	11	107
	SB	4	7	3	89	103
	column total	73	101	118	132	424

Land-Cover Categories

D = deciduous

C = conifer

AG = agriculture

SB = shrub

OVERALL ACCURACY =
(63 + 79 + 85 + 89)/424 =
316/424 = 75%

PRODUCER'S ACCURACY	USER'S ACCURACY
D = 63/73 = 86%	D = 63/113 = 56%
C = 79/101 = 78%	C = 79/101 = 78%
AG = 85/118 = 72%	AG = 85/107 = 79%
SB = 89/132 = 67%	SB = 89/103 = 86%

the error matrix. This value is the most commonly reported accuracy assessment statistic. User's and producer's accuracies are ways of representing individual category accuracies instead of just the overall classification accuracy.

One of the assumptions of the traditional or deterministic error matrix is that an accuracy assessment sample site can have only one label. However, classification scheme rules often impose discrete boundaries on continuous conditions in nature. In situations where classification scheme breaks represent artificial distinctions along a continuum of land cover (LC), observer variability is often difficult to control and, while unavoidable, it can have profound effects on results (Congalton and Green, 1999). While it is difficult to control observer variation, it is possible to use a fuzzy assessment approach to compensate for differences between reference and map data that are caused not by map error but by variation in interpretation (Gopal and Woodcock, 1994). In this study, both deterministic error matrices and those using the fuzzy assessment approach were compiled.

12.3 METHODS

Accuracy assessment requires the development of a statistically rigorous sampling design of the location (distribution) and type of samples to be taken or collected. Several considerations are critical to the development of a robust design to support an accuracy assessment that is truly representative of the map being assessed. Important design considerations include the following:

- What are the map classes and how are they distributed? How a map is sampled for accuracy will partially be driven by how the categorical information of interest is spatially distributed. These distributions are a function of how the features of interest have been categorized — referred to as the "classification scheme."
- What is the appropriate sample unit? Sampling units are the portions of the landscape that will be sampled for the accuracy assessment.
- How many samples should be taken? Accuracy assessment requires that an adequate number of samples be gathered so that any analysis performed is statistically valid. However, the collection of data at each sample point can be very expensive, requiring that sample size be kept to a minimum to be affordable.

Table 12.2 Classification Labels

Class Number	Class Name
1	Forest, Deciduous
2	Forest, Evergreen
3	Shrub/Scrub
4	Grassland
5	Barren/Sparsely Vegetated
6	Urban/Built-Up
7	Agriculture, Other
8	Agriculture, Rice
9	Wetland, Permanent Herbaceous
10	Wetland, Mangrove
11	Water
12	Ice/Snow
13	Cloud/Cloud Shadow/No Data

- How should the samples be chosen? The choice and distribution of samples, or sampling scheme, is an important part of any inventory design. Selection of the proper scheme is critical to generating results that are representative of the map being assessed. First, the samples must be selected without bias. Second, further data analysis will depend on which sampling scheme is selected. Finally, the sampling scheme will determine the distribution of samples across the landscape, which will significantly affect accuracy assessment costs.

This chapter addresses all of the above considerations relative to the NIMA GeoCover study. Major study elements included (1) the finalization of the NIMA GeoCover classification scheme, (2) accuracy assessment sample design and selection, (3) accuracy assessment site labeling, and (4) the compilation of the deterministic and fuzzy error matrix

12.3.1 Classification Scheme

The first task in this project was to specify the NIMA GeoCover classification system rules. A classification scheme has two critical components: (1) a set of labels (e.g., deciduous forest, urban, shrub/scrub, etc.) and (2) a set of rules or definitions such as a dichotomous key for assigning labels. Without a clear set of rules, the assignment of labels to types can be arbitrary and lack consistency. In addition to having labels and a set of rules, a classification scheme should be mutually exclusive and totally exhaustive. All study partners worked together to develop and finalize a classification scheme with the necessary labels and rules. Table 12.2 presents the labels; the classification rules can be found in Appendix A of this chapter.

12.3.2 Sampling Design

Sample design often requires trade-offs between the need for statistical rigor and the practical constraints of budget and available reference data. To achieve statistically reliable results and keep costs to a minimum, a multistaged, stratified random sample design was employed for this project. Research by Congalton (1988) indicates that random and stratified random samplings are the optimal sampling designs for accuracy assessment.

One of the most important aspects of sample design is that the reference data must be independent from data used to create the map. The need for independence posed a dilemma for the assessment of the NIMA GeoCover prototype because the National Technical Means (NTM) used for reference data development were not available for the entire study area. NTM can be defined as classified intelligence gathering systems and the data they generate.

As a result of this limited NTM availability, a choice needed to be made to either (1) constrain the accuracy assessment sample to the areas with existing NTM data, and thereby risk sampling only some of the mapped area, or (2) allow samples to be chosen randomly, resulting in some samples landing in areas where existing NTM was not immediately available for reference data development. The latter approach was selected because limiting the accuracy assessment area was considered statistically unacceptable. To overcome the NTM data gaps, first-stage samples were chosen prior to receipt of the final map. This provided additional time for the acquisition of new NTM data. Persistent data gaps were supplemented by the interpretation of TM composite images.

First stage sample units were 15-min quadrangle areas. To ensure that an adequate number of accuracy assessment sites per cover class were sampled, quadrangles were selected for inclusion in accuracy assessment based on the diversity and number of cover classes in the quadrangle. A relative diversity index was determined through the screening of TM composite images of the study area. The number and diversity of cover type polygons were summarized for each quadrangle, and the six quadrangles with the greatest cover type diversity and largest number of classes were selected as the first-stage samples.

The second-stage sample units were the polygons of the LC map vector file. Fifty polygons per class were randomly selected across all the six quadrangles. If fewer than 50 polygons of a particular class existed within the six quadrangles, then all the available polygons in that class were selected. Both primary and secondary sample selection was automated using accuracy assessment software developed for this project.

12.3.3 Site Labeling

All accuracy assessment samples had two class labels: a map label and a reference site label. For this project, the "map" label was automatically derived from the LC polygon map label provided by Earthsat and stored for later use in the compilation of the error matrix. An expert analyst, based on image interpretation of NTM data, manually assigned the corresponding "reference" label. Each sample polygon was automatically displayed on the computer screen simultaneously with the assessment data form (Figure 12.1). The analyst entered the label for the site into the form using the imagery and other ancillary data available. To ensure independence, at no time did the image analyst labeling the samples have access to map data.

To account for variation in interpretation, the accuracy assessment analyst also completed a LC-type fuzzy logic matrix for every accuracy assessment site (Figure 12.1). Each polygon was evaluated for the likelihood of being identified as each of the possible cover types. First, the analyst

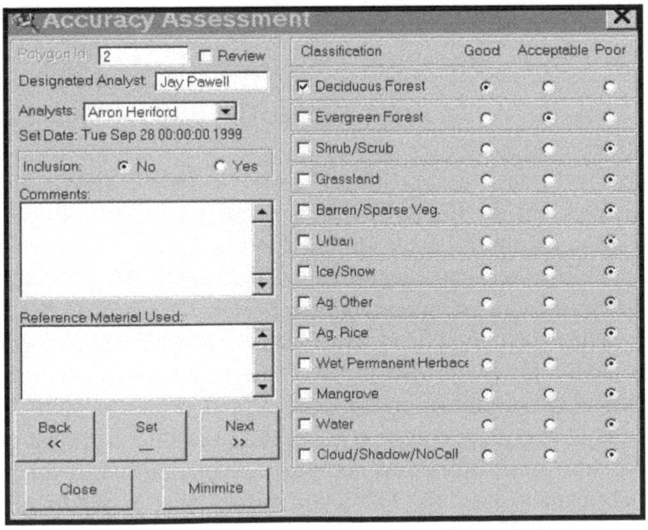

Figure 12.1 Form for labeling accuracy assessment reference sites.

determined the most appropriate label for the site, and the label was entered in the appropriate box under the "classification" column in the form. This label determined in which row of the matrix the site would be tallied and was used for calculation of the deterministic error matrix. After assigning the label for the site, the remaining possible map labels were evaluated as "good," "acceptable," or "poor" candidates for the site's label. For example, a site might fall near the classification scheme margin between forest and shrub/scrub. In this instance, the analyst might rate forest as most appropriate but shrub/scrub as "acceptable." As each site was interpreted, the deterministic and fuzzy assessment reference labels were entered into the accuracy assessment software for creation of the error matrix.

12.3.4 Compilation of the Deterministic and Fuzzy Error Matrix

Following reference site labeling, the error matrix was automatically compiled in the accuracy assessment software. Each accuracy assessment site was tallied in the matrix in the column (based on the map label) and row (based on the most appropriate reference label). The deterministic (i.e., traditional) overall accuracy was calculated by dividing the total of the diagonal by the total number of accuracy assessment sites. The producer's and user's accuracies were calculated by dividing the number of sites in the diagonal by the total number of references (producer's accuracy) or maps (user's accuracy) for each class. That is, from a map producer's viewpoint, given the total number of accuracy assessment sites for a particular class, what was the proportion of sites correctly mapped? Conversely, class accuracy by column represents "user's" class accuracy. For a particular class on the map, user's class accuracy estimates the percentage of times the class was mapped correctly.

Nondiagonal cells in the matrix contain two tallies, which can be used to distinguish class labels that are uncertain or that fall on class margins from class labels that are most probably in error. The first number represents those sites in which the map label matched a "good" or "acceptable" reference label in the fuzzy assessment (Table 12.3). Therefore, even though the label was not considered the most appropriate, it was considered acceptable given the fuzziness of the classification system and the minimal quality of some of the reference data. These sites are considered a "match" for estimating fuzzy assessment accuracy. The second number in the cell represents those sites where the map label was considered poor (i.e., an error).

The fuzzy assessment overall accuracy was estimated as the percentage of sites where the "best," "good," or "acceptable" reference label(s) matched the map label. Individual class accuracy was estimated by summing the number of matches for that class's row or column divided by the row or column total. Class accuracy by row represents "producer's" class accuracy.

12.4 RESULTS

Table 12.3 reports both the deterministic and fuzzy assessment accuracies. The overall and individual class accuracies and the Kappa statistic are displayed. Overall accuracy is estimated in a deterministic way by summing the diagonal and dividing by the total number of sites. For this matrix, overall deterministic accuracy would be estimated at 48.6% (151/311). However, this approach ignores any variation in the interpretation of reference data and the inherent fuzziness at class boundaries. Including the "good" and "acceptable" ratings, overall accuracy is estimated at 74% (230/311). The large difference between these two estimates reflects the difficulty in distinguishing several of the classes, both from TM imagery and from the NTM. For example, a total of 31 sites were labeled as evergreen forest on the map and deciduous forest in the reference data. However, 24 of those sites were labeled as acceptable, meaning they were either at or near the class break or were inseparable from the TM and/or NTM data (Appendix A).

The Kappa statistic was 0.37. The Kappa statistic adjusts the estimate of overall accuracy for the accuracy expected from a purely random assignment of map labels and is useful for comparing

Table 12.3 Error Matrix for the Initial Prototype Area Showing the Computations for the Deterministic and Fuzzy Assessments

Initial Prototype Area

MAP

LABELS	Decid. Forest	EG Forest	Scrub/ Shrub	Grass	Barren/ Sparse	Urban	Ice/ Snow	Ag. Other	Ag. Rice	Wet, Perm. Herb.	Man-grove	Water	Cloud/ Shadow	Deterministic Totals	Percent Deterministic	Fuzzy Totals	% Fuzzy
R E F E R E N C E														Producer's	Accuracies		
Deciduous Forest	48	24.7	0.1	0.3	0.0	0.1	0.0	0.1	0.0	0.0	0.0	0.18	0.0	48/113	42.5%	72/113	63.7%
Evergreen Forest	4.0	17	0.1	0.0	0.0	0.0	0.0	0.1	0.0	0.0	0.0	0.3	0.0	17/26	65.4%	21/26	80.8%
Shrub/Scrub	2.0	0.1	15	8.1	0.0	0.0	0.0	2.2	0.0	0.0	0.0	0.0	0.0	15/31	48.4%	27/31	87.1%
Grassland	0.1	0.0	5.1	14	0.0	0.0	0.0	3.0	0.0	0.0	0.0	0.0	0.0	14/24	58.3%	22/24	91.7%
Barren/Sparse Veg	0.0	0.0	0.2	0.0	0	0.0	0.0	0.1	0.0	0.0	0.0	0.0	0.0	0/3	0.0%	0/3	0.0%
Urban	0.0	0.0	0.0	0.0	0.0	20	0.0	2.0	0.0	0.0	0.0	0.0	0.0	20/22	90.9%	22/22	100.0%
Ice/Snow	0.0	0.0	0.0	0.0	0.0	0.0	0	0.0	0.0	0.0	0.0	0.0	0.0	NA	NA	NA	NA
Agriculture Other	0.1	0.1	7.15	18.6	0.0	2.0	0.0	29	0.0	0.0	0.0	1.2	0.0	29/82	35.4%	57/82	69.5%
Agriculture Rice	0.0	0.0	0.0	0.0	0.0	0.1	0.0	0.0	0	0.0	0.0	1.0	0.0	0/2	0.0%	1/2	50.0%
Wet, Perm.Herb.	0.0	0.0	0.0	0.0	0.0	0.0	0.0	0.0	0.0	0	0.0	0.0	0.0	NA	NA	NA	NA
Mangrove	0.0	0.0	0.0	0.0	0.0	0.0	0.0	0.0	0.0	0.0	0	0.0	0.0	NA	NA	NA	NA
Water	0.0	0.0	0.0	0.0	0.0	0.0	0.0	0.0	0.0	0.0	0.0	8	0.0	8/8	100.0%	8/8	100.0%
Cloud/Shadow	0.0	0.0	0.0	0.0	0.0	0.0	0.0	0.0	0.0	0.0	0.0	0.0	0	NA	NA	NA	NA

User's Accuracies

	Decid. Forest	EG Forest	Scrub/ Shrub	Grass	Barren/ Sparse	Urban	Ice/ Snow	Ag. Other	Ag. Rice	Wet, Perm. Herb.	Man-grove	Water	Cloud/ Shadow
Totals Deterministic	48/56	17/50	15/47	14/50	NA	20/24	NA	29/51	NA	NA	NA	8/33	NA
Percent Det.	85.7%	34.0%	31.9%	28.0%	NA	83.3%	NA	56.9%	NA	NA	NA	24.2%	NA
Fuzzy Totals	54/56	41/50	27/47	40/50	NA	22/24	NA	36/51	NA	NA	NA	10/33	NA
Percent Fuzzy	96.4%	82.0%	57.4%	80.0%	NA	91.7%	NA	70.6%	NA	NA	NA	30.3%	NA

Overall Accuracies

	Deterministic	Fuzzy
	151/311	230/311
	48.6%	74.0%

Kappa: 37.2

different matrices. However, it does not account for fuzzy class membership and variation in interpretation of the reference data. From a map user's perspective, individual fuzzy assessment class accuracies vary from 30% (for water) to 96% (for deciduous forest). Producer's accuracies range from 0% (for barren/sparse vegetation and wet, permanent herbaceous) to 100% (for water and urban). The highest combined user's and producer's accuracies occur in the urban class (100% and 91.7%, respectively).

A useful comparison is the total number of sites for a particular class by row and by column. For example, for deciduous forest there are a total of 113 reference sites and a total of 56 map sites. This indicates that the map underestimates deciduous forest. Another underestimated class is agriculture–other (51 vs. 82). Conversely, for evergreen forest there are a total of 50 map sites and 26 reference sites, indicating that the map overestimates evergreen forest. Other overestimated classes include shrub (47 vs. 31) and grassland (50 vs. 24).

12.5 DISCUSSION AND CONCLUSIONS

The following text discusses and analyzes the major sources of confusion and agreement in the LC map for the initial prototype study. The highest user's accuracy occurs in the deciduous forest class (96.4%). However, producer's accuracy in deciduous forest is low (63.7%), indicating that there is more deciduous forest in the area than is indicated on the map. The highest producer's accuracy is in water and urban (100%). While the urban user's accuracy is also high (91.7%) (indicating that urban is a very reliable class), the user's accuracy for water is low (30.3%), indicating that significant commission errors may exist in the water class. For example, 18 water map sites were determined to be deciduous in the reference data. After the matrix was generated, these sites were reviewed. In each case, the sites were small, scattered polygons in forested areas. Because the water was maintained at full resolution (no filtering was performed), any scattered pixels of water were maintained in the polygon coverage. Many of these polygons came from one or two pixels of water. Because there are many of these small polygons, more than half of the accuracy assessment sites for water came from these polygons.

Confusion also existed in the agriculture–other class, which tends to be confused with shrub/scrub, grassland, or deciduous forest. User's class accuracy for agriculture–other is estimated at 71% (36/51). Eleven sites were labeled as deciduous forest. These sites were also reexamined. In most all cases, the polygons came from small groups of pixels (greater than the minimum mapping unit of 1.4 ha) labeled as agriculture within forested areas. The matrix also identifies confusion between agriculture and shrub and between agriculture and grasslands. For the shrub/scrub map class, 22 sites were labeled as agriculture in the reference data, with 15 sites rated as "poor." Subsequent review of the maps revealed scattered pixels and polygons of shrub within agricultural areas and scattered agriculture within shrub. For grasslands, 24 sites were labeled as agriculture in the reference data, with 18 sites labeled as "acceptable." This reflects the uncertainty with separating grassland from agriculture in many cases. Often, they have identical spectral responses, and unless there are distinct geometric spatial patterns or other contextual features, it is very difficult to distinguish these classes from TM imagery alone.

Map error is often the result of scattered polygons in otherwise homogeneous areas. For example, scattered small polygons of water (particularly in forested areas) accounted for the low estimate of class accuracy for water. Likewise, scattered polygons of agriculture in shrub and grassland and scattered polygons of shrub and grassland in agriculture influenced the accuracies of these classes. This type of error points to the need for increased precision in the image classification algorithms, additional map editing, and/or refinement of the polygon-generating algorithms.

Finally, it should be noted that the first-stage sample units contained no polygons of barren/sparse vegetation, agriculture–rice, ice/snow, mangrove, cloud/shadow or wet, permanent herbaceous. Therefore, these map classes were not sampled for accuracy assessment. Because the first-stage samples are chosen for their diversity, this indicates that the entire map also has no or few polygons with these classes. Considering the location of the prototype, it is reasonable to assume that ice/snow, agriculture–rice, and mangrove do not exist in the area. However, a few reference sites ($n = 5$) were labeled barren/sparse vegetation and wet, permanent herbaceous, indicating that these classes do exist in the area and may be underrepresented in the map.

12.6 SUMMARY

The error matrix or contingency table has become widely accepted as the standard method for reporting the accuracy of GIS data layers derived from remotely sensed data. The matrix provides descriptive statistics including overall, producer's, and user's accuracies as well as sample size information by category and in total. In addition, the matrix is a starting point for a variety of analytical tools, including normalization and Kappa analysis. More recently, the incorporation of fuzzy accuracy assessment has been suggested and adopted by many remote sensing analysts. As proposed, most of these current techniques use a variety of metrics to represent the fuzzy analysis. This chapter introduces the use of a fuzzy error matrix for applying fuzzy accuracy assessment. The fuzzy matrix has the same benefits as a traditional deterministic error matrix, including the computation of all the descriptive statistics. A detailed, practical case study is presented to demonstrate the application of this fuzzy error matrix.

A total of 311 accuracy assessment sites were utilized to estimate the accuracy of the initial prototype area. The traditional estimate of overall accuracy is 48.6%. Accounting for fuzzy class membership and variation in interpretation, overall accuracy is estimated at 74%. The spread between the deterministic and fuzzy assessment estimates is large, but not unusual. Part of this spread is a function of the lack of NTM for several of the reference sites ($n = 84$), resulting in the reference label's being determined from manual interpretation of the TM data. Hopefully, more NTM will be available as the project progresses, which will reduce the spread between deterministic and fuzzy logic estimates. However, some spread will remain because of fuzziness in the boundaries of LC classes. Therefore, acceptable fuzziness between deciduous and evergreen forest (especially in mixed conditions) and deciduous forest and shrub will remain.

REFERENCES

Congalton, R., A comparison of sampling schemes used in generating error matrices for assessing the accuracy of maps generated from remotely sensed data, *Photogram. Eng. Remote Sens.*, 54, 587–592, 1988.

Congalton, R., A review of assessing the accuracy of classifications of remotely sensed data, *Remote Sens. Environ.*, 37, 35–46, 1991.

Congalton, R. and G. Biging, A pilot study evaluating ground reference data collection efforts for use in forest inventory, *Photogram. Eng. Remote Sens.*, 58, 1669–1671, 1992.

Congalton R. and K. Green, A practical look at the sources of confusion in error matrix generation, *Photogram. Eng. Remote Sens.*, 59, 641–644, 1993.

Congalton, R. and K. Green, *Assessing the Accuracy of Remotely Sensed Data: Principles and Practices,* Lewis Publishers, Chelsea, MI, 1999.

Gong, P. and J. Chen, Boundary Uncertainties in Digitized Maps: Some Possible Determination Methods, in Proceedings of GIS/LIS'92, San Jose, CA, 1992, pp. 274–281.

Gopal, S. and C. Woodcock, Theory and methods for accuracy assessment of thematic maps using fuzzy sets, *Photogram. Eng. Remote Sens.*, 60, 181–188, 1994.

Lowell, K., On the Incorporation of Uncertainty into Spatial Data Systems, in Proceedings of GIS/LIS'92, San Jose, CA, 1992, pp. 484–493.

Story, M. and R. Congalton, Accuracy assessment: a user's perspective, *Photogram. Eng. Remote Sens.*, 52, 397–399, 1986.

APPENDIX **A**

Classification Rules

Parcel Appearance	Categorization Call
If pixel appears as water	Water **(Category 11)**
If ≥ 35% man-made impervious material	Urban **(Category 6)**
If cultivated (excluding forest plantations)	Examine for evidence of rice cultivation
If rice	Agriculture, Rice **(Category 8)**
Otherwise	Agriculture, Other **(Category 7)**
If total natural vegetation cover ≥ 10%	Examine for content
If coastal/estuarine AND vegetation cover is mangrove	Wetland, Mangrove **(Category 10)**
If ≥ 35% woody vegetation AND > 3 m in height	Examine for forest type
If woody vegetation deciduous w/ < 25% evergreen intermixture	Forest, Deciduous **(Category 1)**
If woody vegetation deciduous w/≥ 25% evergreen intermixture OR if woody vegetation is 100% evergreen	Forest, Evergreen **(Category 2)**
If woody vegetation ≥ 10% cover AND height < 3 m OR if woody vegetation between 10% and 35% cover at any height	Shrub/Scrub **(Category 3)**
If herbaceous cover ≥ 10% OR mixed shrub and grass AND no evidence of seasonal or permanent saturation (topo position = upland)	Grassland **(Category 4)**
Else	Wetland, Permanent Herbaceous **(Category 9)**
If nonvegetated	Examine for content
If soil intermittently or permanently saturated	Wetland, Permanent Herbaceous **(Category 9)**
If snow or ice cover	Perennial Ice or Snow **(Category 12)**
If view of ground obscured by cloud, shadow, satellite sensor artifact, or lack of TM data	Cloud/Cloud Shadow/No Data **(Category 13)**
Else	Barren/Sparsely Vegetated **(Category 5)**

CHAPTER 13

Mapping Spatial Accuracy and Estimating Landscape Indicators from Thematic Land-Cover Maps Using Fuzzy Set Theory

Liem T. Tran, S. Taylor Jarnagin, C. Gregory Knight, and Latha Baskaran

CONTENTS

13.1 INTRODUCTION

The accuracy of thematic map products is not spatially homogenous, but rather variable across most landscapes. Properly analyzing and representing the spatial distribution (pattern) of thematic map accuracy would provide valuable user information for assessing appropriate applications for land-cover (LC) maps and other derived products (i.e., landscape metrics). However, current thematic map accuracy measures, including the confusion or error matrix (Story and Congalton, 1986) and Kappa coefficient of agreement (Congalton and Green, 1999), are inadequate for analyzing the spatial variation of thematic map accuracy. They are not able to answer several important scientific and application-oriented questions related to thematic map accuracy. For example, are errors distributed randomly across space? Do different cover types have the same spatial accuracy pattern? How do spatial accuracy patterns affect products derived from thematic maps? Within this context, methods for displaying and analyzing the spatial accuracy of thematic maps and bringing the spatial accuracy information into other calculations, such as deriving landscape indicators from thematic maps, are important issues to advance scientifically appropriate applications of remotely sensed image data.

Our study objective was to use the fuzzy set approach to examine and display the spatial accuracy pattern of thematic LC maps and to combine uncertainty with the computation of landscape indicators (metrics) derived from thematic maps. The chapter is organized by (1) current methods for analyzing and mapping thematic map accuracy, (2) presentation of our methodology for constructing fuzzy LC maps, and (3) deriving landscape indicators from fuzzy maps.

There have been several studies analyzing the spatial variation of thematic map accuracy (Campbell, 1981; Congalton, 1988). Campbell (1987) found a tendency for misclassified pixels to form chains along boundaries of homogenous patches. Townshend et al. (2000) explained this tendency by the fact that, in remotely sensed images, the signal coming from a land area represented by a specific pixel can include a considerable proportion of signal from neighboring pixels. Fisher (1994) used animation to visualize the reliability in classified remotely sensed images. Moisen et al. (1996) developed a generalized linear mixed model to analyze misclassification errors in connection with several factors, such as distance to road, slope, and LC heterogeneity. Recently, Smith et al. (2001) found that accuracy decreases as LC heterogeneity increases and patch sizes decrease.

Steele et al. (1998) formulated a concept of misclassification probability by calculating values at training observation locations and then used spatial interpolation (kriging) to create accuracy maps for thematic LC maps. However, this work used the training data employed in the classification process but not the independent reference data usually collected after the thematic map has been constructed for accuracy assessment purposes. Steele et al. (1998) stated that the misclassification probability is not specific to a given cover type. It is a population concept indicating only the probability that the predicted cover type is different from the reference cover type, regardless of the predicted and reference types as well as the observed outcome, and whether correct or incorrect. Although this work brought in a useful approach to constructing accuracy maps, it did not provide information for the relationship between misclassification probabilities and the independent reference data used for accuracy assessment (i.e., the "real" errors). Furthermore, by combining training data of all different cover types together, it produced similar misclassification probabilities for pixels with different cover types that were colocated. This point should be open to discussion, as our analysis described below indicates that the spatial pattern of thematic map accuracy varies from one cover type to another, and pixels with different cover types located in close proximity might have different accuracy levels.

Recently, fuzzy set theory has been applied to thematic map accuracy assessment using two primary approaches. The first was to design a fuzzy matching definition for a crisp classification, which allows for varying levels of set membership for multiple map categories (Gopal and Woodcock, 1994; Muller et al., 1998; Townsend, 2000; Woodcock and Gopal, 2000). The second approach defines a fuzzy classification or fuzzy objects (Zhang and Stuart, 2000; Cheng et al., 2001). Although the fuzzy theory-based methods take into consideration error magnitude and ambiguity in map classes while doing the assessment, like other conventional measures, they do not show spatial variation of thematic map accuracy.

To overcome shortcomings in mapping thematic map accuracy, we have developed a fuzzy set-based method that is capable of analyzing and mapping spatial accuracy patterns of different cover types. We expanded that method further in this study to bring the spatial accuracy information into the calculations of several landscape indicators derived from thematic LC maps. As the method of mapping spatial accuracy was at the core of this study, it will be presented to a reasonable extent in this chapter.

13.2 METHODS

This study used data collected for the accuracy assessment of the National Land Cover Data (NLCD) set. The NLCD is a LC map of the contiguous U.S. derived from classified Landsat

Figure 13.1 The Mid-Atlantic Region; 10 watersheds used in later analysis are highlighted on the map.

Thematic Mapper (TM) images (Vogelmann et al., 1998; Vogelmann et al., 2001). The NLCD was created by the Multi-Resolution Land Characterization (MRLC) consortium (Loveland and Shaw, 1996) to provide a national-scope and consistently classified LC data set for the country. Methodology and results of the accuracy assessment have been described in Stehman et al. (2000), Yang et al. (2000, 2001), and Zhu et al. (1999, 2000). While data for the accuracy assessment were taken by federal region and available for several regions, this study only used data collected for Federal Geographic Region III, the Mid-Atlantic Region (MAR) (Figure 13.1). Table 13.1 shows the number of photographic interpreted "reference" data samples associated with each class in the LC map (Level I) for the MAR. Note that the reference data for Region III did not include alternate reference cover-type labels or information concerning photographic interpretation confidence, unlike data associated with other federal geographic regions.

Table 13.1 Number of Samples by Andersen Level I Classes

Class Name	MRLC Code	No. of Samples
Water	11	79
Developed	20s	222
Barren	30s	127
Forested Upland	40s	338
Shrubland	51	0
Nonnatural Woody	61	0
Herbaceous Upland Natural/Seminatural Vegetation	71	0
Herbaceous Planted/Cultivated	80s	237
Wetlands	90s	101
Total		1104

Major analytical study elements were: (1) to define a multilevel agreement between sampled and mapped pixels, (2) to construct accuracy maps for six LC types, (3) to define cover-type-conversion degrees of membership for mapped pixels, (4) to develop a cover-type-conversion rule set for different conditions of accuracy and LC dominance, (5) to construct fuzzy LC maps, and (6) to develop landscape indicators from fuzzy LC maps.

13.2.1 Multilevel Agreement

In the MRLC accuracy assessment performed by Yang et al. (2001), agreement was defined as a match between the primary or alternate reference cover-type label of the sampled pixel and a majority rule LC label in a 3×3 window surrounding the sample pixel. Here we defined a multilevel agreement at a sampled pixel (Table 13.2) and applied it for all available sampled pixels. It has been demonstrated that the multilevel agreement went beyond the conventional binary agreement and covered a wide range of possible results, ranging from "conservative bias" (Verbyla and Hammond, 1995) to "optimistic bias" (Hammond and Verbyla, 1996). We define a discrete fuzzy set A $(A = \{(a_1, \mu_1),...,(a_6, \mu_6)\})$ representing the multilevel agreement at a mapped pixel regarding a specific cover type as follows:

$$\mu_i = \frac{\sum_{k=1}^{n} \dfrac{\delta_k I_k}{d_k^p} \Big/ \sum_{k=1}^{n} \dfrac{\delta_k}{d_k^p}}{Max_i \left(\sum_{k=1}^{n} \dfrac{\delta_k I_k}{d_k^p} \Big/ \sum_{k=1}^{n} \dfrac{\delta_k}{d_k^p} \right)} = \frac{M_i}{Max_i(M_i)} \tag{13.1}$$

where a_i, $i = 1,...,6$ are six different levels (or categories) of agreement at a mapped pixel; μ_i is fuzzy membership of the agreement level i of the pixel under study; d is the distance from sampled point k to the pixel (k ranges from 1 to n, where n is the number of nearest sampled points taken

Table 13.2 Multilevel Agreement Definitions

Levels	Description
I	A match between the LC label of the sampled pixel and the center pixel's LC type as well as a LC mode of the three-by-three window (662 sampled points)
II	A match between the LC label of the sampled pixel and a LC mode of the three-by-three window (39 sampled points)
III	A match between the LC label of the sampled pixel and the LC type of any pixel in the three-by-three window (199 sampled points)
IV	A match between the LC label of the sampled pixel and the LC type of any pixel in the five-by-five window (84 sampled points)
V	A match between the reference LC label of the sampled pixel and the LC type of any pixel in the seven-by-seven window (31 sampled points)
VI	Failed all of the above (89 sampled points)

into consideration); I_k is a binary function that equals 1 if the sampled point k has the agreement level i and 0 otherwise; p is the exponent of distance used in the calculation; and δ_k is the photographic interpretation confidence score of the sampled pixel k. As information on photographic interpretation confidence was not available for the Region III data set, δ_k was set as constant ($\delta_k =$ 1) in this study. The division by the maximum of A_i was to normalize the fuzzy membership function (Equation 13.1). Verbally, the fuzzy number of multilevel agreement at a mapped pixel defined in Equation 13.1 is a modified inverse distance weighted (IDW) interpolation of the n nearest sample points for each agreement level defined in Table 13.2. But instead of using all n data points together in the interpolation, as in conventional IDW for continuous data, the n sample pixels were divided into six separate groups based on their agreement levels and six iterations of IDW interpolation (one for each agreement level) were run. For each iteration of a particular agreement level, only those samples (among n sample pixels) with that agreement level would be coded as 1, while other reference samples were coded as 0 by the use of the binary function I_k. IDW then returned a value between 0 and 1 for M_i in each iteration. In other words, M_i is an IDW-based weight of sample pixels at the agreement level i among the n closest sample pixels surrounding the pixel under study. With the "winner-takes-all" rule, the agreement level with maximum M_i (i.e., maximum membership value $\mu_i = 1$) will be assigned as the agreement level of the mapped pixel under study.

After the multilevel agreement fuzzy set A was calculated (Equation 13.1), its scalar cardinality was computed as follows (Bárdossy and Duckstein, 1995):

$$car(A) = \sum_{i=1}^{6} \mu_i \qquad (13.2)$$

Thus, the scalar cardinality of the multilevel agreement fuzzy set A is a real number between 1 and 6. This is an indicator of the agreement-level "homogeneity" of sampled pixels surrounding the pixel under study. If $car(A)$ is close to 1, the majority of sampled pixels surrounding the mapped pixel under study have the same agreement level. Conversely, the greater $car(A)$ is, the more heterogeneous in agreement levels the sampled pixels are. Note that there is another way for a mapped pixel to have a near 1 cardinality. That is when the distance between the mapped pixel and a sampled pixel is very close compared to those of other sampled pixels, reflecting the local effect in the IDW interpolation. However, this case occurs only in small areas surrounding each sampled pixel.

13.2.2 Spatial Accuracy Map

Using the above equations, discrete fuzzy sets representing multilevel agreement and their cardinalities were calculated for all mapped pixels associated with a particular cover type. Then, the cardinality values of all pixels were divided into three unequal intervals (1–2, 2–3, and > 3). They were assigned (labeled) to the appropriate category, representing different conditions of agreement-level heterogeneity of neighboring sampled pixels. The three cardinality classes were then combined with six levels of agreement to create 18-category accuracy maps.

13.2.3 Degrees of Fuzzy Membership

This step calculated the possible occurrence of multiple cover types for any given pixel(s) locations expressed in terms of degrees of fuzzy membership. This was done by comparing cover types of mapped pixels and sampled pixels at the same location based on individual pixels and a 3×3 window-based evaluation. To illustrate, assume that the mapped pixel and the sampled pixel had cover types x and y, respectively. In the one-to-one comparison between the mapped and sampled pixels, if x and y are the same, then it is reasonable to state that the mapped pixel was classified correctly. In that case, the degree of membership for cover type x to remain the same is

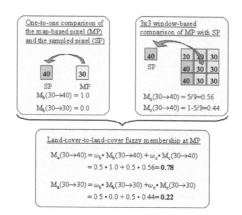

Figure 13.2 Illustration of calculating the cover-type-conversion degrees of membership.

assigned to 1. On the other hand, if x is different from y, then it can be stated that the mapped pixel is wrongly classified, and the degree of membership of x to become y would be 1. The above statements can be summarized as follows:

$$M_a(x \rightarrow x) = 1 \quad \text{if x} = y \tag{13.3}$$

$$M_a(x \rightarrow y) = 1 \text{ and } M_a(x \rightarrow x) = 0 \quad \text{if x} \neq y$$

Using a 3×3 window, if there was a match between x and y, then it is reasonable to state that the cover type of the more dominant pixels (x) in the 3×3 window was probably most representative. However, if the mapped pixels were wrongly classified (e.g., no match between x and y), then the more dominant cover type x is, the higher the possibility that the mapped pixel with cover type x will have cover type y. Within that context, the cover-type-conversion degrees of membership regarding x and y at the mapped pixel were computed as follows:

$$M_b(x \rightarrow x) = n_x/9 \quad \text{if x} = y \tag{13.4}$$

$$M_b(x \rightarrow y) = n_x/9 \text{ and } M_b(x \rightarrow x) = 1 - (n_x/9) \quad \text{if x} \neq y$$

where n_x is the number of pixels in the 3×3 window with cover type x. The ultimate degrees of membership of cover types at the mapped pixel were computed as the weighted-sum average of those from the one-to-one and 3×3-window–based comparisons as follows:

$$M(x \rightarrow y) = \omega_a \bullet M_a(x \rightarrow y) + \omega_b \bullet M_b(x \rightarrow y) \tag{13.5}$$

where ω_a and ω_b were weights for M_a and M_b, respectively, with $\omega_a + \omega_b = 1$ (note that x and y in Equation 13.5 can be different or the same). In this study, we applied equal weights (i.e., $\omega_a = \omega_b = 0.5$) for the two one-to-one and 3×3-window–based comparisons. Figure 13.2 demonstrates how degrees of fuzzy membership of a mapped pixel were computed.

13.2.4 Fuzzy Membership Rules

Here we integrate degrees of membership at individual locations derived from the previous step into a set of fuzzy rules. Theoretically, a fuzzy rule generally consists of a set of fuzzy set(s) as argument(s) $A_{,k}$ and an outcome B also in the form of a fuzzy set such that:

$$\text{If } (A_1 \text{ and } A_2 \text{ and } \dots \text{ and } A_k) \text{ then } B \tag{13.6}$$

where k is the number of arguments. We constructed four fuzzy rules for each cover type for four different combinations of two arguments including (1) accuracy level (i.e., low and high) and (2) majority (i.e., dominant or subordinate). Both of the arguments were available spatially; the first was obtained from the accuracy maps constructed in previous steps and the second was derived directly from the LC thematic map. The four fuzzy rules for cover type x are stated as follows:

- Rule 1: if x is "dominant" and the accuracy is "high," then the degree of membership of x to become y is:

$$\mu_1(x \to y) = \frac{\sum_i A_i^x \cdot n_{x,i} \cdot M_i(x \to y)}{\sum_i A_i^x \cdot n_{x,i}} \qquad (13.7)$$

- Rule 2: if x is "subordinate" and the accuracy is "high," then:

$$\mu_2(x \to y) = \frac{\sum_i A_i^x \cdot (9 - n_{x,i}) \cdot M_i(x \to y)}{\sum_i A_i^x \cdot (9 - n_{x,i})} \qquad (13.8)$$

- Rule 3: if x is "dominant" and the accuracy is "low," then:

$$\mu_3(x \to y) = \frac{\sum_i (1 - A_i^x) \cdot n_{x,i} \cdot M_i(x \to y)}{\sum_i (1 - A_i^x) \cdot n_{x,i}} \qquad (13.9)$$

- Rule 4: if x is "subordinate" and the accuracy is "low," then:

$$\mu_4(x \to y) = \frac{\sum_i (1 - A_i^x) \cdot (9 - n_{x,i}) \cdot M_i(x \to y)}{\sum_i (1 - A_i^x) \cdot (9 - n_{x,i})} \qquad (13.10)$$

where A_i^x is accuracy level for land-cover type x at point i with its values ranging from 0 to 1 and $n_{x,i}$ is the number of pixels labeled x in the 3×3 window surrounding the mapped pixel i. We assigned values of A_i^x based on the multilevel agreement for cover type x at that point. A_i^x is equal to 1 if the agreement level is I and is equal to 0.8, 0.6, 0.4, 0.2, and 0 for agreement levels II, III, IV, V, and VI, respectively. While Equations 13.7–13.10 are based on fuzzy set theory and the error or confusion matrix is associated with probability theory, outcomes of Equations 13.7–13.10 are somewhat similar to information in a row of the error matrix. Note that while one sampled point is used only once in computing the error matrix, it is employed four times at different degrees in constructing the four fuzzy rules. For example, a sampled point in a high accuracy area dominated by cover type x will contribute more to rule1 than to rules 2–4. In contrast, a sampled point in a low accuracy area and subordinate cover type x will have a more significant contribution to rule 4 above than to the other rules. Consequently, each rule represents the degrees of membership of cover type conversion for specific conditions of accuracy and dominance that vary spatially on the map. In contrast, a row in the error matrix is a global summary of a cover type for the whole map and does not provide any localized information.

13.2.5 Fuzzy Land-Cover Maps

The fuzzy rule set derived in the previous step was used to construct various LC conversion maps representing the degrees of fuzzy membership (or possibility) from x to y of all mapped pixels associated with cover type x. For example, to construct the "barren-to-forested upland" map, the four fuzzy rules were applied to all pixels mapped as barren (Table 13.3a through Table 13.3d). In contrast to ordinary rules, where only one rule is activated at a time, the four fuzzy rules were activated simultaneously at different degrees depending on levels of accuracy and LC dominance at that particular location. Consequently, four outcomes resulted from the four fuzzy rules. There are different methods for combining fuzzy rule outcomes (Bárdossy and Duckstein, 1995). Here we applied the weighted sum combination method whose details and application can be found in Bárdossy and Duckstein (1995) and Tran (2002).

A fuzzy LC map for a given cover type was constructed by combining six cover-type-conversion maps. For example, to develop the fuzzy forested upland map, six maps were merged: (1) forested upland-to-forested upland, (2) water-to-forested upland and developed-to-forested upland, (3) barren-to-forested upland, (4) herbaceous planted/cultivated-to-forested upland, and (5) wetlands-to-forested upland. The final fuzzy forested upland map represented the degrees of membership of forested upland for all pixels on the map. The degree of membership at a pixel on the fuzzy LC map was a result of several factors, including the thematic mapped cover type at that pixel and the dominance and accuracy of that LC type in the area surrounding the pixel under study. To illustrate, in a forest-dominated upland area with high accuracy, the degrees of membership of forested upland will be high (i.e., close to 1). Conversely, in a barren-dominated area with high accuracy, the degrees of membership of forested upland will be very low (i.e., close to 0) for barren-labeled pixels. In contrast, in a barren-dominated area with low accuracy, the degrees of membership of forested upland increases to some extent (i.e., approximately 0.3 to 0.4) for barren-labeled pixels. Focusing on forest-related landscape indicators, we used only the fuzzy forested upland map in the next section.

13.2.6 Deriving Landscape Indicators

First, several α-cut maps were created from the fuzzy forested upland map. Each α-cut map was a binary map of forested upland with the degrees of membership $< \alpha$. For example, a 0.5-cut forested upland map is a binary map with two lumped categories: forest for pixels with degrees of membership for forested upland < 0.5 and non-forest otherwise. Then, landscape indicators of interest were derived from these α-cut maps in a similar way to those from an ordinary LC map. The difference was that instead of having a single number for the indicator under study (as with an ordinary LC map) there were several values of the indicator in accordance to various α-cut maps. Generally, the more variable those values were, the more uncertain the indicator was for that particular watershed.

13.3 RESULTS AND DISCUSSION

Plate 13.1 presents accuracy maps for six cover types. All maps were created with the values of 10 for the number of sampled pixels n and 2 for the exponent of distance p (Equation 13.1). The smaller the number of n and/or the larger the value of p, the more the local effects of sampled points on the accuracy maps are taken into account. One important point illustrated by these maps is that the spatial accuracy patterns were different from one cover type to another. For example, while forested upland was understandably more accurate in highly forested areas, herbaceous

Table 13.3 The Fuzzy Cover-Type-Conversion Rule Set

Land-Cover Types		Rules	Low Accuracy						Rules	High Accuracy					
			11	20s	30s	40s	80s	90s		11	20s	30s	40s	80s	90s
Water (11)	Dominant	1-a	0.20	0.09	0.43	0.06	0.00	0.21	1-c	0.98	0.01	0.00	0.00	0.00	0.01
	Subordinate	1-b	0.35	0.07	0.17	0.23	0.00	0.28	1-d	0.53	0.16	0.00	0.09	0.00	0.22
Developed (20s)	Dominant	2-a	0.03	0.08	0.17	0.33	0.35	0.03	2-c	0.00	0.91	0.00	0.03	0.05	0.00
	Subordinate	2-b	0.00	0.032	0.08	0.32	0.27	0.00	2-d	0.00	0.72	0.00	0.11	0.16	0.00
Barren (30s)	Dominant	3-a	0.01	0.21	0.06	0.47	0.24	0.01	3-c	0.00	0.05	0.67	0.21	0.06	0.00
	Subordinate	3-b	0.01	0.36	0.17	0.33	0.10	0.03	3-d	0.00	0.08	0.48	0.36	0.07	0.01
Natural forested upland (40s)	Dominant	4-a	0.04	0.04	0.36	0.08	0.35	0.13	4-c	0.00	0.01	0.01	0.91	0.06	0.01
	Subordinate	4-b	0.02	0.16	0.09	0.36	0.34	0.04	4-d	0.01	0.14	0.03	0.58	0.20	0.04
Herbaceous planted/ cultivated (80s)	Dominant	5-a	0.01	0.18	0.37	0.27	0.12	0.04	5-c	0.00	0.05	0.01	0.06	0.88	0.00
	Subordinate	5-b	0.04	0.20	0.19	0.16	0.42	0.00	5-d	0.02	0.12	0.05	0.14	0.67	0.00
Wetlands (90s)	Dominant	6-a	0.07	0.06	0.07	0.69	0.06	0.06	6-c	0.02	0.01	0.01	0.13	0.02	0.82
	Subordinate	6-b	0.05	0.11	0.16	0.34	0.13	0.21	6-d	0.04	0.06	0.03	0.24	0.08	0.55

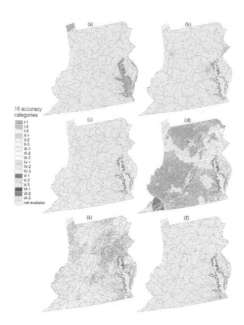

Plate 13.1   (See color insert following page 114.) Fuzzy accuracy maps of (a) water, (b) developed, (c) barren, (d) forested upland, (e) herbaceous planted/cultivated, and (f) wetlands.

planted/cultivated tended to be more accurate in populated areas. On the other hand, developed areas around Richmond and Roanoke had lower accuracy levels compared with other urbanized areas, such as Baltimore, Washington, DC, Philadelphia, and Pittsburgh.

For the forested upland accuracy map, some areas had abnormally low accuracy levels, such as those in central and southern Pennsylvania. The southwestern corner of Virginia had a very low level of accuracy (agreement level 6), indicating that there was almost no match at all between sampled pixels and mapped pixels in this area. This raised questions about both the thematic map classification process and the quality of the reference data. Thus, the fuzzy accuracy maps indicated irregularities or accumulated errors associated with both the thematic map and reference data set. This information is not illustrated using conventional accuracy measure; however, it is very beneficial for designing sampling schemes to support reference data cross-examination.

Table 13.3 presents the fuzzy cover-type-conversion rule set that is, as mentioned above, somewhat similar to a combination of four error matrices in one. The possibilities derived from each fuzzy rule should be interpreted relatively. For example, for a low accuracy, barren-dominant area, the possibility for a barren-labeled pixel to be forested upland (i.e., rule 3-a) was the highest compared with other cover types, including barren, and it was double the second highest possibility of barren-to-herbaceous planted/cultivated (i.e., 0.47 vs. 0.24). Note that the outcomes of each fuzzy rule were not normalized (i.e., to have the highest possibility equal 1) for the purpose of global rule-to-rule comparison. For instance, the wetlands-to-forested upland possibility of a wetlands-labeled pixel in a low-accuracy, wetlands-dominant area (rule 6-a) was double (0.69 vs. 0.33) the developed-to-forested upland possibility of a developed-labeled pixel in a low-accuracy, developed-dominant area (rule 2-a). Unlike an error matrix, the fuzzy rule set table provided significant insights into spatial accuracy variation of the thematic map under study. As the size of the referenced data set was relatively small compared with the area it covered, we used only two arguments (inputs): the accuracy levels and cover type dominance. If there are more sampled data in future analyses, additional arguments (factors) that might affect the classification process (e.g., slope, altitude, sun angle, and fragmentation) can be included in the fuzzy rules, and potentially more insights into the thematic map spatial accuracy patterns can be revealed.

Figure 13.3 presents six fuzzy cover-type-conversion maps of water-to-forested upland, developed-to-forested upland, barren-to-forested upland, forested upland-to-forested upland, herbaceous planted/cultivated-to-forested upland, and wetlands-to-forested upland. These maps resulted from spatially applying the fuzzy rule set to six LC types on the thematic map. Each map had a distinct

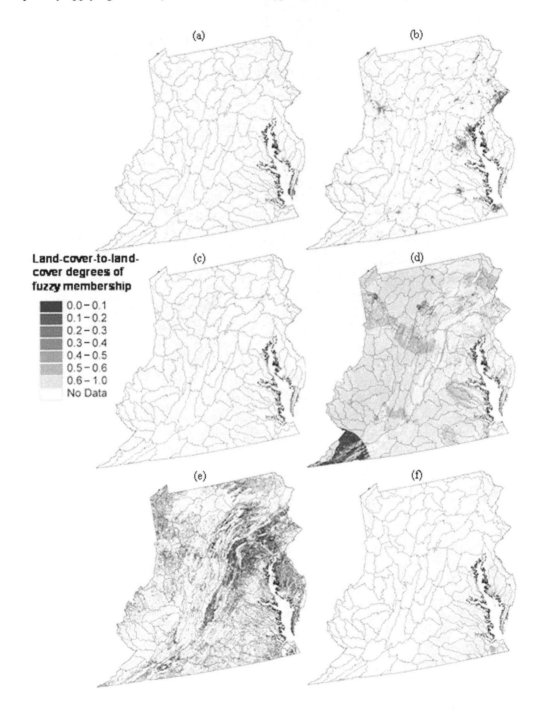

Figure 13.3 Fuzzy cover-type-conversion maps of: (a) water-to-forested upland, (b) developed-to-forested upland, (c) barren-to-forested upland, (d) forested upland-to-forested upland, (e) herbaceous planted/cultivated-to-forested upland, and (f) wetlands-to-forested upland.

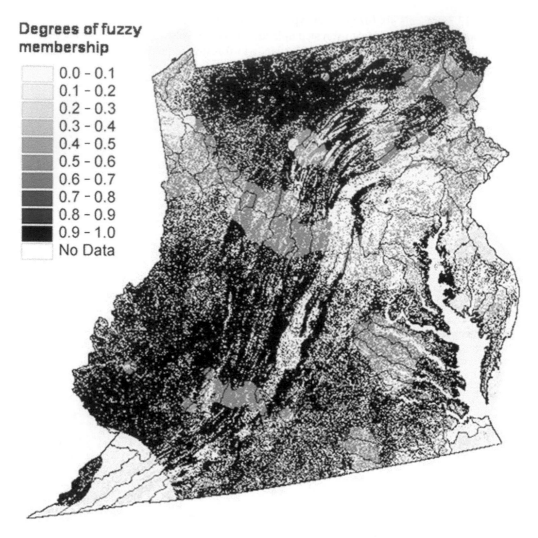

Figure 13.4 Fuzzy forested upland map.

pattern as the degree of membership of a cover type reclassified as forested upland at each location on a map was decided by the dominance and accuracy of that cover type at that spot. Figure 13.4 shows the fuzzy forested upland map that was a combination of the six cover type conversion maps (Figure 13.3). An abnormality in the southwestern corner of Virginia apparently resulted from a very low level of accuracy for most of the forest upland sampled pixels in the vicinity. This made the forested upland degrees of membership for this area very low, although the area was dominated by forest. This irregularity can be verified only through the additional reference data. For other forested areas with low accuracy levels, like southern Pennsylvania, the degrees of membership were greater (around 0.5 to 0.6). This value implies that a forested upland-labeled pixel in such an area has a low probability (0.1 to 0.2) of being another cover type (i.e., herbaceous planted/cultivated or developed).

Figure 13.5a–d presents the crisp binary map and three α-cut maps of the fuzzy forested upland map at the levels of 0.1, 0.25, and 0.5. One can see that the 0.1-cut forested upland map (b) had more forest than the crisp binary map (a) in all areas other than southwestern Virginia. This result is because the 0.1-cut map included pixels that were labeled to other cover types but had possibilities > 0.1 of being forested upland. This was somewhat similar to the result if a rule to include only

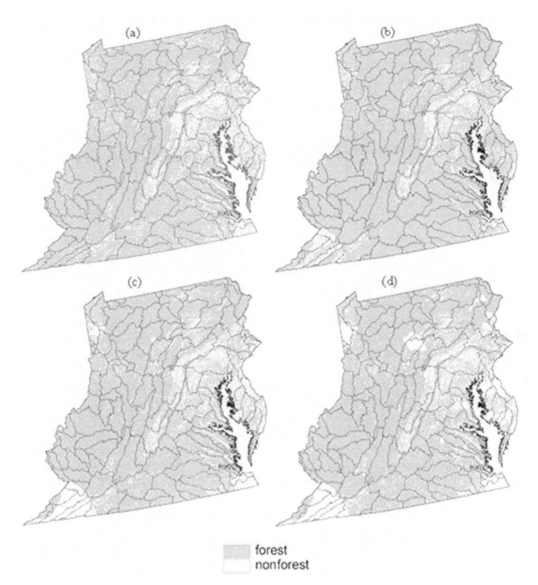

forest
nonforest

Figure 13.5 Crisp binary forested upland map (a) and three α-cut maps derived from the fuzzy forested upland map: (b) 0.1-cut, (c) 0.25-cut, and (d) 0.5-cut.

the forested upland omission errors into the forested upland category had been used. Conversely, the 0.25-cut forested upland map (c) appeared to be similar to the crisp binary map in terms of forest coverage. This can be explained by the fact that only pixels with moderate forested upland degrees of membership (> 0.25) were included in the 0.25-cut map. This excluded the forested upland omission errors but maintained the commission errors. For the 0.50-cut map (d), forest coverage was proportionately less than on the binary map and areas with low forest accuracy were excluded from the map. By exploring various α-cut maps of forested upland, the different forested upland map outcomes can be explored including and/or excluding omission and commission errors.

Table 13.4 presents two forested landscape indicators (FOR% and INT20) for 10 watersheds in MAR (Figure 13.5). FOR% was computed to extract the number of pixels with forested upland cover on a watershed basis divided by the total number of pixels for each watershed to yield the watershed-based index value. INT20 was used to calculate the proportion of forested upland cover within each

Table 13.4 Values of FOR% and INT20 for 10 Watersheds in the Mid-Atlantic Region

Watershed	FOR%				INT20			
	Crisp	0.10-cut	0.25-cut	0.50-cut	Crisp	0.10-cut	0.25-cut	0.50-cut
Schuylkill	47.5	55.4	47.7	45.4	23.6	31.1	24.0	22.8
Lower West Branch Susquehanna	68.8	73.0	69.0	68.8	54.9	60.3	55.3	54.9
Lower Susquehanna	29.0	36.3	29.2	28.1	10.8	16.2	10.9	10.7
Nanticoke	30.1	57.5	31.2	21.9	6.8	37.6	8.0	4.5
Cacapon-Town	84.9	96.0	84.9	84.0	72.0	92.3	72.6	71.1
Pamunkey	64.2	78.4	65.2	60.1	39.1	61.9	40.5	36.4
Upper James	86.9	95.3	87.1	86.9	77.4	91.4	77.8	77.3
Hampton Roads	16.2	35.0	7.3	4.4	2.4	14.0	1.6	1.1
Connoquenessing	55.4	65.2	54.1	50.3	25.0	39.4	25.0	23.3
Little Kanawha	86.2	90.4	86.4	86.2	71.8	80.5	72.4	71.8

window using a threshold of 90% to determine interior habitat suitability (i.e., suitable if $\geq$ 90% forest coverage). Then, the proportion of watershed with suitable interior habitat was determined as INT20 (based on a 450- $\times$ 450-m window). Various values of FOR% and INT20 at three α-cut maps provided possible values of these landscape indicators for the watersheds under study.

For the Schuylkill watershed (2040203) located in an urbanized area with moderate accuracy for forested upland pixels, FOR% ranged from 55.4 to 45.4 with a 10% change from 0.1- to 0.5-cut. Also, the FOR% value at 0.25-cut was very close to those for the crisp binary forested upland map (i.e., 47.7 vs. 47.5) and INT20 values at this watershed changed about 8.3% from 0.1- to 0.5-cut. The Lower Susquehanna watershed (2050306), also located in an urbanized area, had a relatively higher accuracy level; 0.10- and 0.25-cut variations of FOR% and INT20 were only 8.2% and 5.5%, respectively. Conversely, for the Little Kanawha watershed (5030203), located in a forested area with a high accuracy level, FOR% changed only 4.2% from 0.1- to 0.5-cut (from 90.4 to 86.2%). However, the INT20 0.10- to 0.25-cut variation increased to 8.7%. These analyses can be applied to other watersheds, providing valuable insights into the accuracy of the landscape indicators across the region. These two landscape indicators serve as an example of how landscape indicators derived from thematic LC maps can be analyzed to reveal their spatial accuracy and possible value in the study area.

13.4 CONCLUSIONS

We have developed a fuzzy set-based method to map the spatial accuracy of thematic maps and compute landscape indicators while taking into account the spatial variation of accuracy associated with different LC types. This method provides valuable information not only on the spatial patterns of accuracy associated with various cover types but also on the possible values of landscape indicators across the study area. Such insights have not previously been incorporated into any of the existing thematic map-related accuracy assessment methods. We believe that including a spatial assessment in the accuracy assessment process would greatly enhance the user's capability to evaluate map suitability for numerous environmental applications.

13.5 SUMMARY

This chapter presented a fuzzy set-based method of mapping the spatial accuracy of thematic maps and computing landscape indicators while taking into account the spatial variation of accuracy associated with different LC types. First, a multilevel agreement was defined, providing a framework to accommodate different levels of matching between sampled pixels and mapped pixels. Then, the multilevel agreement data at the sampled pixel locations were used to construct spatial accuracy maps for six cover types approximating an Anderson Level I classification for the Mid-Atlantic region. A set of fuzzy rules was developed that determined degrees of fuzzy membership for cover type conversion under different conditions of accuracy and cover type dominance. Operations of the fuzzy rule set created a set of fuzzy cover-type-conversion maps. Fuzzy LC maps were then created from a combination of six fuzzy cover-type-conversion maps from all cover types. Then, the LC maps were used to derive several α-cut maps that were binary maps for representative cover types in accordance with different degrees of fuzzy membership. Finally, landscape indicators were derived from those binary α-cut LC maps. Variations in the value of indicator values derived from different α-cut maps illustrated the level of accuracy (uncertainty) associated with watershed-specific indicators.

ACKNOWLEDGMENTS

The authors would like to thank James Wickham, U.S. EPA Technical Director of the Multi-Resolution Land Characterization (MRLC) consortium, for his valuable remarks. In addition, comments from Elizabeth R. Smith and Robert O'Neill were greatly appreciated. The first author gratefully acknowledges partial financial support from the National Science Foundation and National Oceanic and Atmospheric Administration (Grant SBE-9978052, Brent Yarnal, principal investigator) and from the U.S. Environmental Protection Agency via cooperative agreement number R-82880301 with Pennsylvania State University. Any opinions, findings, and conclusions or recommendations expressed in this material are those of the authors and do not necessarily reflect those of the National Science Foundation or the U.S. Environmental Protection Agency.

REFERENCES

Bárdossy, A. and L. Duckstein, *Fuzzy Rule-Based Modeling with Applications to Geophysical, Biological and Engineering Systems,* CRC Press, Boca Raton, FL, 1995.

Campbell, J.B., Spatial autocorrelation effects upon the accuracy of supervised classification of land cover, *Photogram. Eng. Remote Sens.,* 47, 355–363, 1981.

Campbell, J.B., *Introduction to Remote Sensing,* Guilford Press, New York, 1987.

Cheng, T., M. Molenaar, and H. Lin, Formalizing fuzzy objects from uncertain classification results, *Int. J. Geogr. Info. Sci.,* 15, 27–42, 2001.

Congalton, R.G., Using spatial autocorrelation analysis to explore the errors in maps generated from remotely sensed data, *Photogram. Eng. Remote Sens.,* 54, 587–592, 1988.

Congalton, R.G. and K. Green, *Assessing the Accuracy of Remote Sensed Data: Principles and Practices,* Lewis Publishers, Boca Raton, FL, 1999.

Fisher, P.F., Visualization of the reliability in classified remotely sensed images, *Photogram. Eng. Remote Sens.,* 60, 905–910, 1994.

Gopal, S. and C. Woodcock, Theory and methods of accuracy assessment of thematic maps using fuzzy sets, *Photogram. Eng. Remote Sens.,* 60, 181–189, 1994.

Hammond, T.O. and D.L. Verbyla, Optimistic bias in classification accuracy assessment, *Int. J. Remote Sens.,* 17, 1261–1266, 1996.

Loveland, T.R. and D.M. Shaw, Multiresolution land characterization: building collaborative partnerships, in *Proceedings of the ASPRS/GAP Symposium on Gap Analysis: A Landscape Approach to Biodiversity Planning,* Scott, J.M., T. Tear, and F. Davis, Eds., Charlotte, NC, National Biological Service, Moscow, ID, pp. 83–89, 1996.

Moisen, G.C., D.R. Cutler, and T.C. Edwards, Jr., Generalized linear mixed models for analyzing error in a satellite-based vegetation map of Utah, in *Spatial Accuracy Assessment in Natural Resources and Environmental Sciences: Second International Symposium,* Mowrer, H.T., R.L. Czaplewski, and R.H. Hamre, Eds., General Technical Report RM-GTR-277, USDA Forest Service, Fort Collins, CO, pp. 459–466, 1996.

Muller, S.V., D.A. Walker, F.E. Nelson, N.A. Auerbach, J.G. Bockheim, S. Guyer, and D. Sherba, Accuracy assessment of a land-cover map of the Kuparuk River Basin, Alaska: considerations for remote regions, *Photogram. Eng. Remote Sens.*, 64, 619–628, 1998.

Smith, J.H., J.D. Wickham, S.V. Stehman, and L. Yang, Impacts of patch size and land-cover heterogeneity on thematic image classification accuracy, *Photogram. Eng. Remote Sens.*, 68, 65–70, 2001.

Steele, B.M., J.C. Winne, and R.L. Redmond, Estimation and mapping of misclassification probabilities for thematic land cover maps, *Remote Sens. Environ.*, 66, 192–202, 1998.

Stehman, S.V., J.D. Wickham, L. Yang, and J.H. Smith, Assessing the accuracy of large-area land cover maps: experiences from the Multi-Resolution Land-Cover Characteristics (MRLC) project, in Proceedings of the Fourth International Symposium on Spatial Accuracy Assessment in Natural Resources and Environmental Sciences, Heuvelink, G.B.M. and M.J.P.M. Lemmens, Eds., Amsterdam, July 12–14, 2000, pp. 601–608.

Story, M. and R.G. Congalton, Accuracy assessment: a user's perspective, *Photogram. Eng. Remote Sens.*, 52, 397–399, 1986.

Townsend, P., A quantitative fuzzy approach to assess mapped vegetation classifications for landscape applications, *Remote Sens. Environ.*, 72, 253–267, 2000.

Townshend, J.R.G., C. Huang, S.N.V. Kalluri, R.S. DeFries, and S. Liang, Beware of per-pixel characterization of land cover, *Int. J. Remote Sens.*, 21, 839–843, 2000.

Tran, L.T., M.A. Ridgley, L. Duckstein, and R. Sutherland, Application of fuzzy logic-based modeling to improve the performance of the Revised Universal Soil Loss Equation, *Catena* 47(ER3), 203–226, 2002.

Verbyla, D.L. and T.O Hammond, Conservative bias in classification accuracy assessment due to pixel-by-pixel comparison of classified images with reference grids, *Int. J. Remote Sens.*, 16, 581–587, 1995.

Vogelmann, J.E., S.M. Howard, L. Yang, C.R. Larson, B.K. Wylie, and N. Van Driel, Completion of the 1990s National Land Cover Data Set for the conterminous United States from Landsat Thematic Mapper data and ancillary data source, *Int. J. Remote Sens.*, 67, 650–662, 2001.

Vogelmann, J.E., T. Sohl, and S.M. Howard, Regional characterization of land cover using multiple sources of data, *Photogram. Eng. Remote Sens.*, 64, 45–57, 1998.

Woodcock, C. and S. Gobal, Fuzzy set theory and thematic maps: accuracy assessment and area estimation, *Int. J. Geogr. Info. Sci.*, 14, 153–172, 2000.

Yang, L., S.V. Stehman, J.D. Wickham, J.H. Smith, and N.J. Van Driel, Thematic validation of land cover data for the eastern United Stated using aerial photography: feasibility and challenges, in Proceedings of the Fourth International Symposium on Spatial Accuracy Assessment in Natural Resources and Environmental Sciences, Heuvelink, G.B.M. and M.J.P.M. Lemmens, Eds., Amsterdam, July 12–14, 2000, pp. 747–754.

Yang, L., S.V. Stehman, J.H. Smith, and J.D. Wickham, Thematic accuracy of MRLC land cover for the eastern United States, *Remote Sens. Environ.*, 76, 418–422, 2001.

Zhang, J. and N. Stuart, Fuzzy methods for categorical mapping with image-based land cover data, *Int. J. Geogr. Info. Sci.*, 15, 175–195, 2000.

Zhu, Z., L. Yang, S.V. Stehman, and R.L. Czaplewski, Accuracy assessment for the U.S. Geological Survey regional land cover mapping program: New York and New Jersey region, *Photogram. Eng. Remote Sens.*, 66, 1425–1435, 2000.

Zhu, Z., L. Yang, S.V. Stehman, and R.L. Czaplewski, Designing an accuracy assessment for a USGS regional land cover mapping program, in *Spatial Accuracy Assessment: Land Information Uncertainty in Natural Resources*, Lowell, K. and A. Jaton, Eds., Sleeping Bear Press/Ann Arbor Press, Chelsea, MI, 1999, pp. 393–398.

CHAPTER 14

Fuzzy Set and Spatial Analysis Techniques for Evaluating Thematic Accuracy of a Land-Cover Map

Sarah R. Falzarano and Kathryn A. Thomas

CONTENTS

14.1 INTRODUCTION

14.1.1 Accuracy Assessment

Accuracy assessments of thematic maps have often been overlooked. With the increasing popularity and availability of geographic information systems (GIS), maps can readily be produced with minimal regard for accuracy. Frequently, a map that looks good is assumed to be 100% accurate.

Understanding the accuracy of meso-scale (1:100,000 to 1:500,000 scale) digital maps produced by government agencies is especially important because of the potential for broad dissemination and use. Meso-scale maps encompass large areas, and thus the information may affect significantly large populations. Additionally, digital information can be shared much more easily than hard-copy maps in the rapidly growing technological world. Finally, information produced by public agencies is freely available and sometimes actively disseminated. These combined factors highlight that a thorough understanding of the thematic accuracy of a map is essential for proper use.

A rigorous assessment of a map allows users to determine the suitability of the map for particular applications. For example, estimates of thematic accuracy are needed to assist land managers in providing a defensible basis for use of the map in conservation decisions (Edwards et al., 1998).

Errors can occur and accumulate throughout a land-cover (LC) mapping project (Lunetta et al., 1991). The final map can have spatial (positional) and/or thematic (classification) errors. Spatial errors may occur during the registration of the spatial data to ground coordinates or during sequential analytical processing steps, while thematic errors occur as a result of cover-type misclassifications. Thematic errors may include variation in human interpretation of a complex classification scheme or an inappropriate classification system for the data used (e.g., understory classification when satellite imagery can only visualize the overstory).

This chapter focuses on analysis and estimation of thematic accuracy of a LC map containing 105 cover types. Using a single reference data set, three methods of analysis were conducted to illustrate the increase in accuracy information portrayed by fuzzy set theory and spatial visualization. This added information allows a user to better evaluate use of the map for any given application.

14.1.2 Analysis of Reference Data

14.1.2.1 Binary Analysis

The analysis and estimation of thematic accuracy of meso-scale LC maps has traditionally been limited to a binary analysis (i.e., right/wrong) (Congalton, 1996; Congalton and Green, 1999). This type of assessment provides information about agreement between cover types as mapped (classified data) and corresponding cover types as determined by an independent data source (reference data). The binary assessment is summarized in an error matrix (Congalton and Green, 1999), also referred to as a confusion or contingency table. In the matrix, the cover type predicted by the classified data (map) is assigned to rows and the observed cover type (reference data) is displayed in columns. The values in each cell represent the count of sample points matching the combination of classified and reference data (Congalton, 1996). Errors of inclusion (commission errors) and errors of exclusion (omission errors) for each cover type and overall map accuracy can be calculated using the error matrix. "User's accuracy" corresponds to the area on the map that actually represents that LC type on the ground. "Producer's accuracy" represents the percentage of sampling points that were correctly classified for each cover type.

A binary analysis of accuracy data using an error matrix omits information in two ways: (1) it does not take into account the degree of agreement between reference and map data and (2) it ignores spatial information from the reference data. The error matrix forces each map label at each reference point into a correct or incorrect classification. However, a LC classification is often not discrete (i.e., one type is exclusive of all others). Instead, types grade from one to another and may be related, justifying one or more map labels for the same geographic area. The binary assessment does not take into account that the reference data may be incorrect. In addition, the error matrix does not use the locations of the reference points directly, and accuracy is assumed to be spatially constant within each LC type. Instead, accuracy may vary spatially across the landscape in a manner partially or totally unrelated to LC type (Steele et al., 1998). This has led to the utilization of two additional analysis techniques, fuzzy set analysis and spatial analysis, to describe the thematic accuracy of a LC map.

14.1.2.2 Fuzzy Set Analysis

An alternative method of analysis of thematic accuracy uses fuzzy set theory (Zadeh, 1965). Adapted from its original application to describe the ability of the human brain to understand vague relationships, Gopal and Woodcock (1994) developed fuzzy set theory for thematic accuracy assessment of digital maps. A fuzzy set analysis provides more information about the degree of agreement between the reference and mapped cover types. Instead of a right or wrong analysis, map labels are considered partially right or partially wrong, generally on a five-category scale. This is more useful for assessing vegetation types that may grade into one another yet must be classified into discrete types by a human observer (Gopal and Woodcock, 1994). The fuzzy set analysis provides a number of measures with which to judge the accuracy of a LC map.

Fuzzy set theory aids in the assessment of maps produced from remotely sensed data by analyzing and quantifying vague, indistinct, or overlapping class memberships (Gopal and Woodcock, 1994). Distinct boundaries between LC types seldom exist in nature. Instead, there are often gradations from one cover (vegetation) type to another. Confusion results when a location can legitimately be labeled as more than one cover type (i.e., vegetation transition zones). Unlike a binary assessment, fuzzy set analysis allows partial agreement between different LC types. Additionally, the fuzzy set analysis provides insight into the types of errors that are being made. For example, the misclassification of ponderosa pine woodland as juniper woodland may be a more acceptable error than classifying it as a desert shrubland. In the first instance, the misclassification may not be important if the map user wishes to know where all coniferous woodlands exist in an area.

14.1.2.3 Spatial Analysis

Advanced techniques in assessing the thematic accuracy of maps are continually evolving. A new technique proposed in this chapter uses the spatial locations of the reference data to interpolate accuracy between sampling sites to create a continuous spatial view of accuracy. This technique is termed a *thematic spatial analysis*; however, it should not be confused with assessing the *spatial error of the map*. The thematic spatial analysis portrays thematic accuracy in a spatial context.

Reference data inherently contain spatial information that is usually ignored in both binary and fuzzy set analyses. For both analyses, the spatial locations of the reference data are not utilized in the summary statistics, and results are given in tabular, rather than spatial, format. The most fundamental drawback of the confusion matrix is its inability to provide information on the spatial distribution of the uncertainty in a classified scene (Canters, 1997). A thematic spatial analysis addresses this spatial issue by using the geographic locations gathered using a global positioning system (GPS) with the reference data. These locations are used in an interpolation process to assign accuracy to locations that were not directly sampled. Accuracy is not tied to cover type, but rather to the location of the reference sites. Therefore, accuracy can be displayed for specific locations on the LC map.

Data that are close together in space are often more alike than those that are far apart. This spatial autocorrelation of the reference data is accounted for in spatial models. In fact, spatial models are more general than classic, nonspatial models (Cressie, 1993) and have less-strict assumptions, specifically about independence of the samples. Therefore, randomly located reference data will be accounted for in a spatial model.

Literature on the spatial variability of thematic map accuracy is limited. Congalton (1988) proposed a method of displaying accuracy by producing a binary difference image to represent agreement or disagreement between the classified and reference images. Fisher (1994) proposed a dynamic portrayal of a variety of accuracy measures. Steele et al. (1998) developed a map of accuracy illustrating the magnitude and distribution of classification errors. The latter used kriging to interpolate misclassification estimates (produced from a bootstrapping method) at each reference point. The interpolated estimates were then used to construct a contour map showing accuracy estimates over the map extent. This work provided a starting point for this study. The fuzzy set analysis described earlier was used in conjunction with kriging to produce a fuzzy spatial view of accuracy.

14.2 BACKGROUND

A LC map, or map of the natural vegetation communities, water, and human alterations that represent the landscape (e.g., agriculture, urban, etc.), provides basic information for a multitude of applications by federal, state, tribal, and local agencies. Several public (i.e., the USDA Forest Service and USDI Fish & Wildlife Service) and private (i.e., The Nature Conservancy) agencies use meso-scale LC maps for local and regional conservation planning. LC maps can be used in land-use planning, fire modeling, inventory, and other applications. Because of their potential for utilization in a variety of applications by different users, it is important to determine the thematic map accuracies.

A thematic accuracy assessment was conducted on the northern half of a preliminary Arizona Gap Analysis Program (AZ-GAP) LC map (Graham, 1995). The map (Plate 14.1) was derived primarily from Landsat Thematic Mapper (TM) satellite imagery from 1990. Aerial video and ground measurements were used to facilitate classification of spectral classes into 105 discrete cover types for Arizona using a modification of the classification system by Brown et al. (1979). This system attempted to model natural hierarchies in the southwestern U.S. However, Graham's procedures were not well described or documented.

The preliminary LC map consists of polygons labeled with cover types contained in a GIS with a 40-ha minimum mapping unit (MMU); MMUs were smaller in riparian locations. This resolution is best suited for interpretation at the 1:100,000 scale (meso-scale).

14.3 METHODOLOGY

14.3.1 Reference Data

A random sampling design, stratified according to cover type, was used to determine the set of polygons to be sampled in the accuracy assessment. A total of 930 sampling sites representing 59 different cover types in northern Arizona were visited during the summer of 1997. Field technicians identified dominant, codominant, and associate plant species and ancillary data for a 1-ha area. The field data at each site were assigned to one of the 105 cover classes by the project plant ecologist using the incomplete definitions provided by Graham. Each reference site was tied to the GPS-measured point location at the center of the 1-ha field plots. The resulting reference data set, therefore, consisted of 930 points with a field assigned cover type and associated point location.

14.3.2 Binary Analysis

Traditional measures of map accuracy were calculated by comparing the cover label at each reference site to the map. Matches between the two were coded as either agreed (1) or disagreed (0). These statistics were incorporated into an error matrix from which user's and producer's accuracies for each cover type were calculated, as well as overall accuracy of the LC map.

14.3.3 Fuzzy Set Analysis

The Gopal–Woodcock (1994) fuzzy set ranking system was refined for application to the reference data for the northern AZ-GAP LC map (Table 14.1). The fuzzy set ranks reflected a hierarchical approach to LC classification. While Gopal and Woodcock (1994) suggested that fuzzy

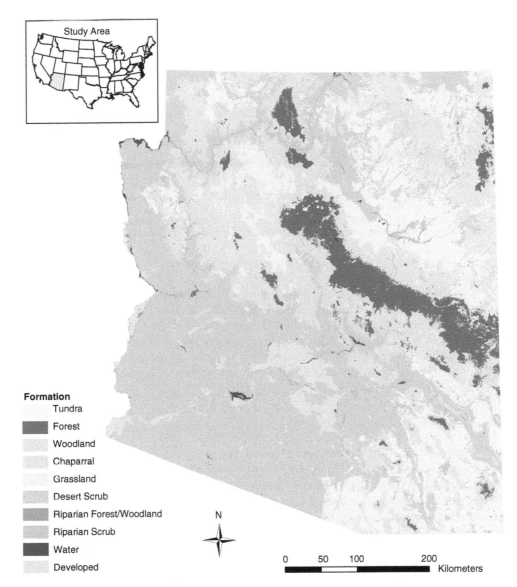

Plate 14.1   (See color insert following page 114.) Preliminary AZ-GAP land-cover map to formation level classification. See Appendix A for a complete list of all cover classes. The preliminary map contained 58,170 polygons describing 105 vegetation types (Appendix A).

set ranks for each cover type be assigned at each sampling point, this method would have been impractical in the field. Instead, the fuzzy set ratings were predefined rather than assessed at each sampling site. A matrix of the 105 cover classes (reference vs. map) assigned a fuzzy set rank to each reference site by comparing its reference data assignment to the map assignment.

Using the fuzzy set rank for each reference site, the functions that described the thematic accuracy of the classification were calculated (Gopal and Woodcock 1994). For this study, we calculated the following functions:

Max (M) = number of sites with an absolutely right answer (accuracy rank of 5)

Right (R) = number of sites with a reasonable, good, or absolutely right answer (accuracy ranks of 3, 4, and 5)

Table 14.1 Accuracy Ranks Assigned to the Reference Data of the AZ-GAP Land-Cover Map

Rank	Answer	Description
1	Wrong	The reference and map types did not correspond, and there was no ecological reason for the noncorrespondence.
2	Understandable but Wrong	The reference and map types did not correspond, but the reason for non-correspondence was understood[a].
3	Reasonable or Acceptable	The reference and map types were all the same life form (i.e., formation types[b]).
4	Good	The reference and map types were characterized by the same species at the dominant species level.
5	Absolutely Right	The reference and map types were exactly the same.

[a] These reasons include vegetation types that are ecotonal and/or vegetation types that can occur as inclusions within other vegetation types.
[b] Tundra, Coniferous Forest, Evergreen Woodland, Chaparral, Grasslands, Desert Scrub, Riparian Broadleaf Woodland/Forest, Riparian Leguminous Woodland/Forest, Riparian Scrub, Wetlands, Water, and Developed.

Increase (R – M) = difference between the Right and Max functions

The Max (M) function calculated the same information as user's accuracy in a binary assessment. The Right (R) function allowed reasonable and better answers to be counted. For this study, the R function calculated the accuracy of the LC map to the life form level or better. The Increase (R – M) function reflected the improvement in accuracy associated with using the R function instead of the M function. Since the Gopal–Woodcock (1994) fuzzy set assessment was altered to save time in the field, certain data for calculating membership, difference, ambiguity, and confusion statistics were not collected.

14.3.4 Spatial Analysis

The nature of the accuracy ranks were explored by calculating the mean, median, and mode, and a histogram was plotted. The points were mapped to display the accuracy rank and location of the data. Interpolating the accuracy ranks produces a continuous map of thematic accuracy. Kriging was data driven and exploited the spatial autocorrelation exhibited by the data. An ordinary kriging regression technique for estimating the best linear unbiased estimate of variables at an unsampled location was applied to reduce the local variability by calculating a moving spatial average.

The kriging interpolation produces continuous values even though the accuracy ranks are ordinal. However, a value between two of the ranks is meaningful, and this suggests that the kriged results are also meaningful. For example, a value between "reasonable or acceptable" and "good" can be characterized as "reasonably good."

The first step in the kriging process was to calculate the empirical variogram, or an analogous measure of the spatial autocorrelation present in the data. The variogram is one of the most common measures of spatial autocorrelation used in geostatistics. It is calculated as 0.5 the average difference squared of all data values separated by a specified distance (lag):

$$\gamma(h) = \frac{1}{2\,|\,N(h)\,|} \sum_{N(h)} (z_i - z_j)^2 \qquad (14.1)$$

where h = distance measure with magnitude only, $N(h)$ = set of all pair-wise Euclidean distances $i - j = h$, $|N(h)|$ = number of distinct pairs in $N(h)$, and z_i and z_j = fuzzy set ranks at spatial locations i and j.

For the accuracy ranks in this study, we chose to use a modified version of the variogram to calculate the empirical variogram, as follows (Cressie and Hawkins, 1980):

$$\hat{\gamma}(h) = \frac{\left(\frac{1}{2|N(h)|} \sum_{N(h)} \sqrt{|z_i - z_j|} \right)^4}{0.457 + \dfrac{0.494}{|N(h)|}} \tag{14.2}$$

This modified form of the variogram has the advantage of reducing the effect of outliers in the data without removing specific data points. The estimation is based on the fourth power of the square root of the absolute differences in z-values.

Once an appropriate empirical variogram is calculated, a model is fit to the data (Figure 14.1). The model variogram has known mathematical properties (such as positive definiteness) and is used in kriging equations to determine the estimator weights. Possible valid models include exponential, spherical, gaussian, linear, and power (Goovaerts, 1997).

The nugget effect (C_0) represents the random variability present in a data set at small distances. By definition, the value of the variogram at a distance of zero is zero; however, data values can display a discontinuity at very small distances. This apparent discontinuity at the origin could reflect the unaccounted-for spatial variability at distances smaller than the sampling distance or could be an artifact of the error associated with measurement.

The range (A_0) is the distance over which the samples are spatially correlated. The sill ($C_0 + C$) is the point of maximum variance and is the sum of the structural variance (C, variance attributed purely to the process) and the nugget effect (Royle, 1980). It is the plateau that the model variogram reaches at the range, and it is estimated by the sample variance only in the case of a model showing a pure nugget effect. The model is fit to the empirical variogram visually and is optimized by calculating the residual sum of squares (RSS). The values of the three main parameters are changed iteratively to reduce the RSS value and fit the model.

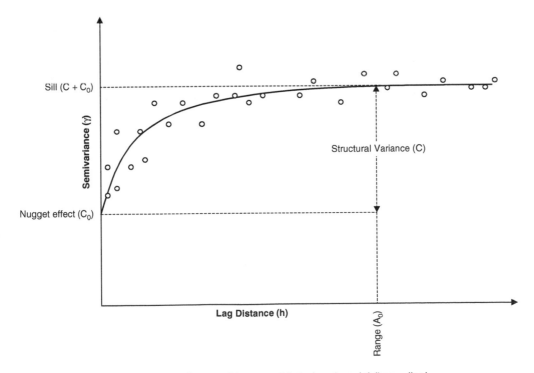

Figure 14.1 Generic variogram including empirical data (circles) and model (heavy line).

Ordinary kriging was performed on the fuzzy set reference data. The model and parameters were selected to produce a regularly spaced lattice of points representing accuracy ranks. Kriging predicted continuous (rather than ordinal) accuracy ranks ranging from one to five. The resulting tabular file of coordinate locations and predicted accuracy ranks was converted to a grid format, with predicted accuracy rank as the value of each 1-km^2 cell. The result is a fuzzy spatial view of accuracy, a map of predicted accuracy ranks for northern Arizona. The continuous accuracy rank estimates were rounded into ordinal ranks for ease of interpretation and display. A frequency histogram was produced from the predicted accuracy ranks.

14.4 RESULTS

14.4.1 Binary Analysis

User's and producer's accuracies for each cover type and overall accuracy were low (Table 14.2). The highest producer's accuracies were for anthropogenically defined cover types industrial (60%) and mixed agriculture/urban/industrial (80%). Producer's accuracies for natural cover types ranged between zero and 50%; the best performers were Encinal mixed oak/mixed chaparral/semi-desert grassland – mixed scrub (50%) and Mohave blackbush – Yucca scrub (50%). Likewise, the highest user's accuracies were also for anthropogenically defined cover types urban (91%) and industrial (86%). Natural cover types ranged between 0 and 48.3%; the best performer was Engelmann spruce – mixed conifer (48.3%). The standard error was < 5% for almost all sampled vegetation types, and overall map accuracy was 14.8%.

14.4.2 Fuzzy Set Analysis

The Max statistic for the fuzzy set reference data yields the same information as user's accuracy for the binary accuracy assessment (Table 14.3). However, the R function provided a different view. Accuracy improves across the table for all cover types because the R function was more inclusive than the M function. For example, in cover class 18 (ponderosa pine – pinyon – juniper), the M statistic indicates this type has very low accuracy (5%). The R statistic indicated that when assessed at the life-form level it was 74% correct. The range for R statistics was large, between 0 and 100%. However, the cover types were more often correct to the life form (mean 52.7% ± 33.4%) compared to the M statistic (mean 13.8% ± 18.8%). The mean increase in accuracy when viewed at the life form level was 38.8% ± 31.5%.

14.4.3 Spatial Analysis

The accuracy ranks had a mean and median near 3.0 with a large standard deviation; however, the mode did not correspond to the mean and median (Figure 14.2). The distribution had a fairly broad shape but is mostly symmetrical. The fuzzy set reference data (Figure 14.3) illustrated classic signs of being positively spatially autocorrelated at shorter distance separations (Figure 14.4 and Figure 14.5). This was substantiated by the lower variance values at shorter lag distances. Also, the variance values seem to reach a plateau at a lag distance where they become uncorrelated. The empirical variogram was best fit with a spherical model (Figure 14.4). The parameters were iteratively changed to achieve a low residual sum of squares and resulted in a nugget of 0.6638, sill of 1.4081, and range of 22.6 km.

The spherical model and parameters were used to determine the weights in the kriging equations. The predicted accuracy ranks produced from kriging do not reach the extremes of "wrong" and

Table 14.2 Producer's and User's Accuracies by Land-Cover Type

Code	Cover Type	No. of Sites	Producer's Accuracy (%)	Standard Error	User's Accuracy (%)	Standard Error
3	Engelmann Spruce-Mixed Conifer	29	41.2	7.0	48.3	7.2
4	Rocky Mountain Lichen-Moss	1	0.0	0.0	0.0	0.0
5	Rocky Mountain Bristlecone-Limber Pine	2	0.0	0.0	0.0	0.0
6	Pinyon-Juniper-Shrub/Ponderosa Pine-Gambel Oak-Juniper	21	0.0	0.0	0.0	0.0
7	Pinyon-Juniper/Sagebrush/Mixed-Grass-Scrub	34	18.2	6.5	11.8	5.5
8	Pinyon-Juniper-Shrub Live Oak-Mixed Scrub	13	8.0	7.3	15.4	9.6
9	Pinyon-Juniper (Mixed)/Chaparral-Scrub	33	8.3	4.6	3.0	2.9
10	Pinyon-Juniper-Mixed Shrub	18	0.0	0.0	0.0	0.0
11	Pinyon-Juniper-Mixed Grass-Scrub	34	5.3	3.7	2.9	2.9
12	Pinyon-Juniper (Mixed)	41	6.7	3.9	2.4	2.1
13	Douglas Fir-Mixed Conifer	35	38.5	7.2	28.6	6.7
14	Arizona Cypress	8	25.0	12.8	12.5	9.9
15	Ponderosa Pine	45	12.5	4.8	13.3	4.8
16	Ponderosa Pine-Mixed Conifer	23	11.5	5.4	13.0	5.6
17	Ponderosa Pine-Gambel Oak-Juniper/Pinyon-Juniper Complex	36	11.8	5.1	16.7	5.9
18	Ponderosa Pine-Pinyon-Juniper	39	16.7	5.8	5.1	3.3
20	Ponderosa Pine-Mixed Oak-Juniper	3	10.0	18.2	33.3	28.6
21	Encinal Mixed Oak	1	0.0	0.0	0.0	0.0
22	Encinal Mixed Oak-Pinyon-Juniper	5	16.7	18.1	40.0	23.7
23	Encinal Mixed Oak-Mexican Pine-Juniper	2	0.0	0.0	0.0	0.0
24	Encinal Mixed Oak-Mexican Mixed Pine	1	0.0	0.0	0.0	0.0
25	Encinal Mixed Oak-Mesquite	1	0.0	0.0	0.0	0.0
26	Encinal Mixed Oak/Mixed Chaparral/Semidesert Grassland-Mixed Scrub	10	50.0	15.0	10.0	9.0
27	Great Basin Juniper	2	0.0	0.0	0.0	0.0
28	Interior Chaparral Shrub Live Oak-Pointleaf Manzanita	14	20.0	10.7	35.7	12.9
29	Interior Chaparral Mixed Evergreen Schlerophyll	18	33.3	11.0	27.8	10.5
30	Interior Chaparral (Mixed)/Sonoran-Paloverde-Mixed Cacti	1	0.0	0.0	0.0	0.0
31	Interior Chaparral (Mixed)/Mixed Grass-Mixed Scrub Complex	10	0.0	0.0	0.0	0.0
32	Rocky Mountain/Great Basin Dry Meadow	18	20.0	6.4	27.8	7.2
33	Madrean Dry Meadow	22	0.0	0.0	0.0	0.0
34	Great Basin (or Plains) Mixed Grass	20	9.5	6.1	10.0	6.1
35	Great Basin (or Plains) Mixed Grass-Mixed Scrub	40	8.5	4.2	15.0	5.5
36	Great Basin (or Plains) Mixed Grass-Sagebrush	4	11.1	16.4	25.0	22.7
37	Great Basin (or Plains) Mixed Grass-Saltbush	24	35.7	8.1	20.8	6.9
38	Great Basin (or Plains) Mixed Grass-Mormon Tea	20	11.1	6.8	5.0	4.8
42	Semidesert Mixed Grass-Mixed Scrub	2	0.0	0.0	0.0	0.0
43	Great Basin Sagebrush Scrub	12	0.0	0.0	0.0	0.0

Table 14.2 Producer's and User's Accuracies by Land-Cover Type (*Continued*)

Code	Cover Type	No. of Sites	Producer's Accuracy (%)	Standard Error	User's Accuracy (%)	Standard Error
44	Great Basin Big Sagebrush-Juniper-Pinyon	30	20.0	6.5	13.3	5.5
45	Great Basin Sagebrush-Mixed Grass-Mixed Scrub	27	20.0	7.0	22.2	7.3
46	Great Basin Shadscale-Mixed Grass-Mixed Scrub	24	0.0	0.0	0.0	0.0
47	Great Basin Greasewood Scrub	11	37.5	14.8	27.3	13.6
48	Great Basin Saltbush Scrub	7	6.7	9.5	14.3	12.9
49	Great Basin Blackbrush-Mixed Scrub	36	16.0	5.8	11.1	5.0
50	Great Basin Mormon Tea-Mixed Scrub	18	19.4	9.1	33.3	10.9
51	Great Basin Winterfat-Mixed Scrub	11	0.0	0.0	0.0	0.0
52	Great Basin Mixed Scrub	26	9.1	5.6	11.5	6.3
53	Great Basin Mormon Tea/Pinyon-Juniper	16	0.0	0.0	0.0	0.0
55	Mohave Creosotebush-Bursage Mixed Scrub	7	28.6	17.9	28.6	17.9
58	Mohave Blackbush-Yucca Scrub	13	50.0	10.4	23.1	8.7
59	Mohave Saltbush Yucca Scrub	5	0.0	0.0	0.0	0.0
61	Mohave Creosotebush-Brittlebush Mohave Globemallow Scrub	5	0.0	0.0	0.0	0.0
63	Mohave Joshua Tree	1	0.0	0.0	0.0	0.0
64	Mohave Mixed Scrub	9	9.1	9.8	11.1	10.7
75	Sonoran Paloverde-Mixed Cacti-Mixed Scrub	1	0.0	0.0	0.0	0.0
82	Agriculture	1	0.0	0.0	0.0	0.0
83	Urban	11	41.7	14.9	90.9	8.6
84	Industrial	7	60.0	18.9	85.7	13.4
85	Mixed Agriculture/Urban/Industrial	20	80.0	7.7	20.0	7.7
87	Water	2	0.0	0.0	0.0	0.0

"absolutely right." Instead, they range from a minimum of 1.039 to a maximum of 4.934, and mean and median are very close to 3.0 (Figure 14.5).

The fuzzy spatial view of accuracy displays the predicted accuracy ranks reclassified as an ordinal variable (Figure 14.6). High accuracy is lighter in color than low accuracy. The frequency histogram of accuracy ranks shows that approximately 85% of the fuzzy spatial view of accuracy had a rank of 3, 4, or 5 (Figure 14.5). In ecological terms, the LC map was accurate to the life form level or better for a majority of the study area.

14.5 DISCUSSION

A binary analysis using an error matrix provides limited information about thematic accuracy of a LC map. In fact, an overall accuracy of 14.8% for the map was dismal and discourages use of the map for any application. However, this was not unexpected given the preliminary nature of the map, high number of cover types, small reference data sample size (*n*) compared to the number of cover types and lack of documentation of the Graham vegetation types. In fact, a binary analysis is conservatively biased against a classification system that is poorly defined and numerous in classes (Verbyla and Hammond, 1995). The lack of descriptions in the Graham classification system made labeling the cover type of each reference point difficult. In addition, division of the cover types of Arizona into 105 classes made distinguishing between types problematic. Therefore, a binary analysis likely assigned a wrong answer to locations with partially correct LC classification.

Table 14.3 Fuzzy Set Accuracy by Land-Cover Type

Code	Cover Type	No. of Sites	Max (M) Best Answer		Right (R) Correct		Increase (R - M)	
			#	%	#	%	#	%
3	Engelmann Spruce-Mixed Conifer	29	14	48.3	25	86.2	11	37.9
4	Rocky Mountain Lichen-Moss	1	0	0.0	0	0.0	0	0.0
5	Rocky Mountain Bristlecone-Limber Pine	2	0	0.0	2	100.0	2	100.0
6	Pinyon-Juniper-Shrub/Ponderosa Pine-Gambel Oak-Juniper	21	0	0.0	10	47.6	10	47.6
7	Pinyon-Juniper/Sagebrush/Mixed-Grass-Scrub	34	4	11.8	19	55.9	15	44.1
8	Pinyon-Juniper-Shrub Live Oak-Mixed Scrub	13	2	15.4	11	84.6	9	69.2
9	Pinyon-Juniper (Mixed)/Chaparral-Scrub	33	1	3.0	12	36.4	11	33.3
10	Pinyon-Juniper-Mixed Shrub	18	0	0.0	7	38.9	7	38.9
11	Pinyon-Juniper-Mixed Grass-Scrub	34	1	2.9	18	52.9	17	50.0
12	Pinyon-Juniper (Mixed)	41	1	2.4	21	51.2	20	48.8
13	Douglas Fir-Mixed Conifer	35	10	28.6	28	80.0	18	51.4
14	Arizona Cypress	8	1	12.5	1	12.5	0	0.0
15	Ponderosa Pine	45	6	13.3	28	62.2	22	48.9
16	Ponderosa Pine-Mixed Conifer	23	3	13.0	16	69.6	13	56.5
17	Ponderosa Pine-Gambel Oak-Juniper/Pinyon-Juniper Complex	36	6	16.7	20	55.6	14	38.9
18	Ponderosa Pine-Pinyon-Juniper	39	2	5.1	29	74.4	27	69.2
20	Ponderosa Pine-Mixed Oak-Juniper	3	1	33.3	2	66.7	1	33.3
21	Encinal Mixed Oak	1	0	0.0	0	0.0	0	0.0
22	Encinal Mixed Oak-Pinyon-Juniper	5	2	40.0	3	60.0	1	20.0
23	Encinal Mixed Oak-Mexican Pine-Juniper	2	0	0.0	0	0.0	0	0.0
24	Encinal Mixed Oak-Mexican Mixed Pine	1	0	0.0	0	0.0	0	0.0
25	Encinal Mixed Oak-Mesquite	1	0	0.0	1	100.0	1	100.0
26	Encinal Mixed Oak/Mixed Chaparral/Semidesert Grassland-Mixed Scrub	10	1	10.0	2	20.0	1	10.0
27	Great Basin Juniper	2	0	0.0	0	0.0	0	0.0
28	Interior Chaparral Shrub Live Oak-Pointleaf Manzanita	14	5	35.7	6	42.9	1	7.1
29	Interior Chaparral Mixed Evergreen Schlerophyll	18	5	27.8	7	38.9	2	11.1
30	Interior Chaparral (Mixed)/Sonoran-Paloverde-Mixed Cacti	1	0	0.0	1	100.0	1	100.0
31	Interior Chaparral (Mixed)/Mixed Grass-Mixed Scrub Complex	10	0	0.0	0	0.0	0	0.0
32	Rocky Mountain/Great Basin Dry Meadow	18	5	27.8	5	27.8	0	0.0
33	Madrean Dry Meadow	22	0	0.0	6	27.3	6	27.3
34	Great Basin (or Plains) Mixed Grass	20	2	10.0	3	15.0	1	5.0
35	Great Basin (or Plains) Mixed Grass-Mixed Scrub	40	6	15.0	15	37.5	9	22.5
36	Great Basin (or Plains) Mixed Grass-Sagebrush	4	1	25.0	2	50.0	1	25.0
37	Great Basin (or Plains) Mixed Grass-Saltbush	24	5	20.8	10	41.7	5	20.8
38	Great Basin (or Plains) Mixed Grass-Mormon Tea	20	1	5.0	9	45.0	8	40.0
42	Semidesert Mixed Grass-Mixed Scrub	2	0	0.0	0	0.0	0	0.0
43	Great Basin Sagebrush Scrub	12	0	0.0	7	58.3	7	58.3
44	Great Basin Big Sagebrush-Juniper-Pinyon	30	4	13.3	13	43.3	9	30.0
45	Great Basin Sagebrush-Mixed Grass-Mixed Scrub	27	6	22.2	17	63.0	11	40.7

Table 14.3 Fuzzy Set Accuracy by Land-Cover Type (*Continued*)

Code	Cover Type	No. of Sites	Max (M) Best Answer		Right (R) Correct		Increase (R - M)	
			#	%	#	%	#	%
46	Great Basin Shadscale-Mixed Grass-Mixed Scrub	24	0	0.0	13	54.2	13	54.2
47	Great Basin Greasewood Scrub	11	3	27.3	10	90.9	7	63.6
48	Great Basin Saltbush Scrub	7	1	14.3	4	57.1	3	42.9
49	Great Basin Blackbrush-Mixed Scrub	36	4	11.1	24	66.7	20	55.6
50	Great Basin Mormon Tea-Mixed Scrub	18	6	33.3	9	50.0	3	16.7
51	Great Basin Winterfat-Mixed Scrub	11	0	0.0	5	45.5	5	45.5
52	Great Basin Mixed Scrub	26	3	11.5	18	69.2	15	57.7
53	Great Basin Mormon Tea/Pinyon-Juniper	16	0	0.0	10	62.5	10	62.5
55	Mohave Creosotebush-Bursage Mixed Scrub	7	2	28.6	6	85.7	4	57.1
58	Mohave Blackbush-Yucca Scrub	13	3	23.1	11	84.6	8	61.5
59	Mohave Saltbush Yucca Scrub	5	0	0.0	5	100.0	5	100.0
61	Mohave Creosotebush-Brittlebush Mohave Globemallow Scrub	5	0	0.0	5	100.0	5	100.0
63	Mohave Joshua Tree	1	0	0.0	1	100.0	1	100.0
64	Mohave Mixed Scrub	9	1	11.1	9	100.0	8	88.9
75	Sonoran Paloverde-Mixed Cacti-Mixed Scrub	1	0	0.0	0	0.0	0	0.0
82	Agriculture	1	0	0.0	0	0.0	0	0.0
83	Urban	11	10	90.9	11	100.0	1	9.1
84	Industrial	7	6	85.7	7	100.0	1	14.3
85	Mixed Agriculture/Urban/Industrial	20	4	20.0	19	95.0	15	75.0
87	Water	2	0	0.0	0	0.0	0	0.0
	Sum	930	138		523		385	
	Accuracy of the whole map			14.8		56.2		41.4

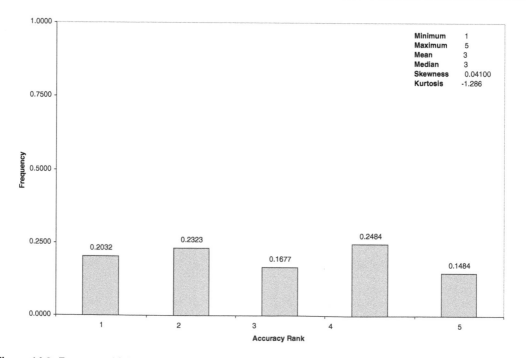

Figure 14.2 Frequency histogram of accuracy ranks.

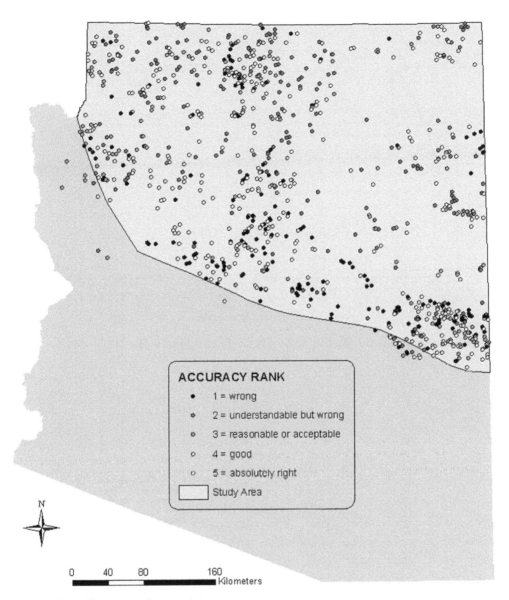

Figure 14.3 Map of fuzzy set reference data.

A fuzzy set analysis provided more information about the agreement between the reference data and the map and was less biased against a small sample size compared to number of cover types. The M statistics were disturbing, but less so when the R statistics were considered. The R function indicates that many cover types were more accurately classified to the life form level. Yet, even for this statistic, accuracies did not reach the targeted 80% in most instances. This added information allows the user and producer to judge the value of the LC map for different applications. For example, for certain cover types, the map performed adequately to the life form level and could be used in applications where this determination is all that is required. Fuzzy set theory was particularly appropriate for LC classification systems that must be discrete but represent a continuum.

Adding the spatial location of accuracy to the accuracy ranks contributed additional accuracy information to the LC map. Thematic map accuracy may vary spatially across a landscape in a manner partially or totally unrelated to cover type. In other words, a cover type may be misclassified more often when it occurs in certain contexts, such as on steep slopes. Also, cover types that were

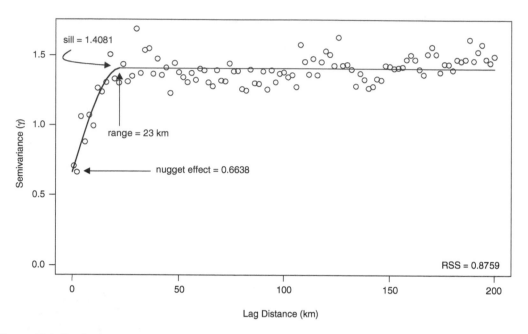

Figure 14.4 Semivariogram and spherical model of the fuzzy set reference data (930 points).

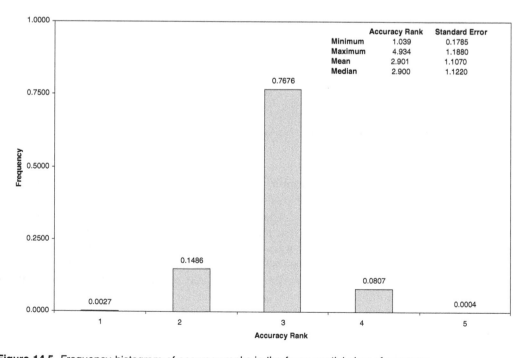

Figure 14.5 Frequency histogram of accuracy ranks in the fuzzy spatial view of accuracy.

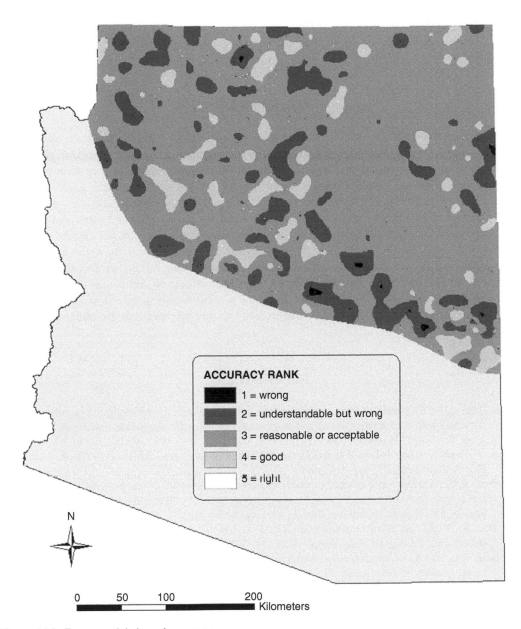

Figure 14.6 Fuzzy spatial view of accuracy.

located near ground control data used in the map development tended to be more correct than remote areas for which only imagery was used to develop the map.

The fuzzy spatial view of accuracy built upon the information produced by the fuzzy set analysis and created a map of accuracy of the preliminary AZ-GAP LC map. Not only was accuracy displayed as it varied across the northern Arizona landscape, but the degree of accuracy was conveyed by accuracy ranks. Overall, the fuzzy spatial view of accuracy indicated that the LC map was accurate to the life-form level, with locations of higher and lower accuracy. The histogram of accuracy ranks for northern Arizona indicated that the interpolated accuracy was 85% at the life-form level for all cover types. However, where classification required identification of the dominant and, in some cases, associate, species, accuracy remained low (8%).

The fuzzy spatial view of accuracy facilitated the identification of areas with low accuracy that needed focused attention to refine the map and allowed users to assess the accuracy of the map for their specific area of interest.

14.6 SUMMARY

Using the same reference data and LC map, three methods of thematic accuracy assessments were conducted. First, a traditional thematic accuracy assessment using a binary rule (right/wrong) was used to compare mapped and reference data. Results were summarized in an error matrix and presented in tabular form by thematic class. Second, a fuzzy set assessment was used to rank and express the degree of agreement between the mapped and reference data. This allowed for the expression of accuracy to reflect the fuzzy nature of the classes. Results were also displayed in tabular form by class but included several estimates of accuracy based on the degree of agreement defined. Lastly, a spatial analysis using the accuracy rank of the reference data was interpolated across the study area and displayed in map form. Fuzzy set theory and spatial visualization help portray the accuracy of the LC map more effectively to the user than a traditional binary accuracy assessment. The approach provided a substantially greater level of information about map accuracy, which allows the map users to thoroughly evaluate its utility for specific project applications.

REFERENCES

Brown, D.E., C.H. Lowe, and C.P. Pase, A digitized classification system for the biotic communities of North America, with community (series) and association examples for the Southwest, *J. Arizona-Nevada Acad. Sci.,* 14, 1–16, 1979.

Canters, F., Evaluating uncertainty of area estimates derived from fuzzy land-cover classification, *Photogram. Eng. Remote Sens.,* 63, 403–414, 1997.

Congalton, R.G., Accuracy assessment: a critical component of land cover mapping, in *Gap Analysis: A Landscape Approach to Biodiversity Planning,* Scott, J.M., T.H. Tear, and F.W. Davis, Eds., American Society for Photogrammetry and Remote Sensing, Bethesda, MD, 1996, pp. 119–131.

Congalton, R., Using spatial autocorrelation analysis to explore the errors in maps generated from remotely sensed data, *Photogram. Eng. Remote Sens.*, 54, 587–592, 1988.

Congalton, R.G. and K. Green, *Assessing the Accuracy of Remotely Sensed Data: Principles and Practices,* Lewis Publishers, Boca Raton, FL, 1999.

Cressie, N. *Statistics for Spatial Data,* revised ed., Wiley & Sons, New York, 1993.

Cressie, N. and D.M. Hawkins, Robust estimation of the variogram: I, *Math. Geol.,* 12, 115–125, 1980.

Edwards, T.C. Jr., G.G. Moisen, and D.R. Cutler, Assessing map accuracy in a remotely sensed, ecoregion-scale cover map, *Remote Sens. Environ.,* 63, 73–83, 1998.

Fisher, P.F., Visualization of the reliability in classified remotely sensed images, *Photogram. Eng. Remote Sens.,* 60, 905–910, 1994.

Goovaerts, P., *Geostatistics for Natural Resources Evaluation,* Oxford University Press, New York, 1997.

Gopal, S. and C. Woodcock, Theory and methods for accuracy assessment of thematic maps using fuzzy sets, *Photogram. Eng. Remote Sens.,* 60, 181–188, 1994.

Graham, L.A., Preliminary Arizona Gap Analysis Program Land Cover Map, University of Arizona, Tucson, 1995.

Lunetta, R.S., R.G. Congalton, L.F. Fenstermaker, J.R. Jensen, K.C. McGwire, and L.R. Tinney, Remote sensing and geographic information system data integration: error sources and research issues, *Photogram. En. Remote Sens.,* 57, 677–687, 1991.

Royle, A.G., Why geostatistics? in *Geostatistics,* Royle, A.G., I. Clark, P.I. Brooker, H. Parker, A. Journel, J.M. Endu, R. Sandefur, D.C. Grant, and P. Mousset-Jones, Eds., McGraw Hill, New York, 1980.

Steele, B.M., J.C. Winne, and R.L. Redmond, Estimation and mapping of local misclassification probabilities for thematic land cover maps, *Remote Sens. Environ.*, 66, 192–202, 1998.

Verbyla, D.L. and T.O. Hammond, Conservative bias in classification accuracy assessment due to pixel-by-pixel comparison of classified images with reference grids, *Int. J. Remote Sens.*, 16, 581–587, 1995.

Zadeh, L.A., Fuzzy sets, *Info. Control*, 8, 338–353, 1965.

APPENDIX A

Arizona Gap Analysis Classification System

Formation	Land-Cover Class
Tundra	Rocky Mountain Lichen-Moss
Forest	Engelmann Spruce-Mixed Conifer
Forest	Rocky Mountain Bristlecone-Limber Pine
Forest	Douglas Fir-Mixed Conifer
Forest	Arizona Cypress
Forest	Ponderosa Pine
Forest	Ponderosa Pine-Mixed Conifer
Forest	Ponderosa Pine-Gambel Oak-Juniper/Pinyon-Juniper Complex
Forest	Ponderosa Pine/Pinyon-Juniper
Forest	Ponderosa Pine-Aspen
Forest	Ponderosa Pine-Mixed Oak-Juniper
Forest	Douglas Fir-Mixed Conifer (Madrean)
Forest	Ponderosa Pine (Madrean)
Woodland	Pinyon-Juniper-Shrub/Ponderosa Pine-Gambel Oak-Juniper
Woodland	Pinyon-Juniper/Sagebrush/Mixed Grass-Scrub
Woodland	Pinyon-Juniper-Shrub Live Oak-Mixed Shrub
Woodland	Pinyon-Juniper (Mixed)/Mixed Chaparral-Scrub
Woodland	Pinyon-Juniper-Mixed Shrub
Woodland	Pinyon-Juniper-Mixed Grass-Scrub
Woodland	Pinyon-Juniper (Mixed)
Woodland	Encinal Mixed Oak
Woodland	Encinal Mixed Oak-Pinyon-Juniper
Woodland	Encinal Mixed Oak-Mexican Pine-Juniper
Woodland	Encinal Mixed Oak-Mexican Mixed Pine
Woodland	Encinal Mixed Oak-Mesquite
Woodland	Encinal Mixed Oak/Mixed Chaparral/Semidesert Grassland-Mixed Scrub
Woodland	Great Basin Juniper
Chaparral	Interior Chaparral-Shrub Live Oak-Pointleaf Manzanita
Chaparral	Interior Chaparral-Mixed Evergreen Sclerophyll
Chaparral	Interior Chaparral (Mixed)/Son. Paloverde-Mixed Cacti
Chaparral	Interior Chaparral (Mixed)/Mixed Grass-Scrub Complex
Grassland	Rocky Mountain/Great Basin Dry Meadow
Grassland	Madrean Dry Meadow
Grassland	Great Basin Mixed Grass
Grassland	Great Basin Mixed Grass-Mixed Scrub
Grassland	Great Basin Mixed Grass-Sagebrush
Grassland	Great Basin Mixed Grass-Saltbush
Grassland	Great Basin Mixed Grass-Mormon Tea
Grassland	Semidesert Tobosa Grass-Scrub
Grassland	Semidesert Mixed Grass-Yucca-Agave
Grassland	Semidesert Mixed Grass-Mesquite
Grassland	Semidesert Mixed Grass-Mixed Scrub
Desert Scrub	Great Basin Sagebrush
Desert Scrub	Great Basin Big Sagebrush-Juniper-Pinyon
Desert Scrub	Great Basin Sagebrush-Mixed Grass-Mixed Scrub
Desert Scrub	Great Basin Shadscale-Mixed Grass-Mixed Scrub
Desert Scrub	Great Basin Greasewood Scrub
Desert Scrub	Great Basin Saltbush Scrub
Desert Scrub	Great Basin Blackbrush-Mixed Scrub

Formation	Land-Cover Class
Desert Scrub	Great Basin Mormon Tea-Mixed Scrub
Desert Scrub	Great Basin Winterfat-Mixed Scrub
Desert Scrub	Great Basin Mixed Scrub
Desert Scrub	Great Basin Mormon Tea/Pinyon-Juniper
Desert Scrub	Mohave Creosotebush Scrub
Desert Scrub	Mohave Creosotebush-Bursage-Mixed Scrub
Desert Scrub	Mohave Creosotebush-Yucca spp. (incl. Joshuatree)
Desert Scrub	Mohave Blackbrush-Mixed Scrub
Desert Scrub	Mohave Blackbrush-Yucca spp. (incl. Joshuatree)
Desert Scrub	Mohave Saltbush-Mixed Scrub
Desert Scrub	Mohave Brittlebush-Creosotebush Scrub
Desert Scrub	Mohave Creosotebush-Brittlebush/Mohave Globemallow Scrub
Desert Scrub	Mohave Catclaw Acacia-Mixed Scrub
Desert Scrub	Mohave Joshuatree
Desert Scrub	Mohave Mixed Scrub
Desert Scrub	Chihuahuan Creosotebush-Tarbush Scrub
Desert Scrub	Chihuahuan Mesquite Shrub Hummock
Desert Scrub	Chihuahuan Whitethorn Scrub
Desert Scrub	Chihuahuan Mixed Scrub
Desert Scrub	Sonoran Creosotebush Scrub
Desert Scrub	Sonoran Creosotebush-Bursage Scrub
Desert Scrub	Sonoran Creosotebush-Mesquite Scrub
Desert Scrub	Sonoran Creosotebush-Bursage-Paloverde-Mixed Cacti (wash)
Desert Scrub	Sonoran Brittlebush-Mixed Scrub
Desert Scrub	Sonoran Saltbush-Creosote Bursage Scrub
Desert Scrub	Sonoran Paloverde-Mixed Cacti-Mixed Scrub
Desert Scrub	Sonoran Paloverde-Mixed Cacti/Sonoran Creosote-Bursage
Desert Scrub	Sonoran Paloverde-Mixed Cacti/Semidesert Grassland-Mixed Scrub
Desert Scrub	Sonoran Crucifixion Thorn
Desert Scrub	Sonoran Smoketree
Desert Scrub	Sonoran Catclaw Acacia
Riparian Forest/Woodland	Great Basin Riparian/Cottonwood-Willow Forest
Riparian Forest/Woodland	Interior Riparian/Cottonwood-Willow Forest
Riparian Forest/Woodland	Interior Riparian/Mixed Broadleaf Forest
Riparian Forest/Woodland	Interior Riparian/Mesquite Forest
Riparian Forest/Woodland	Sonoran Riparian/Cottonwood-Willow Forest
Riparian Forest/Woodland	Sonoran Riparian/Cottonwood-Mesquite Forest
Riparian Forest/Woodland	Sonoran Riparian/Mixed Broadleaf Forest
Riparian Forest/Woodland	Sonoran Riparian/Mesquite Forest
Riparian Scrub	Madrean Riparian/Wet Meadow
Riparian Scrub	Playa/Semipermanent Water
Riparian Scrub	Great Basin Riparian Forest/Mixed Riparian Scrub
Riparian Scrub	Great Basin Riparian/Sacaton Grass Scrub
Riparian Scrub	Great Basin Riparian/Reed-Cattail Marsh
Riparian Scrub	Great Basin Riparian/Wet Mountain Meadow
Riparian Scrub	Interior Riparian/Mixed Riparian Scrub
Riparian Scrub	Sonoran Riparian/Leguminous Short-Tree Forest/Scrub
Riparian Scrub	Sonoran Riparian/Mixed Riparian Scrub
Riparian Scrub	Sonoran Riparian/Sacaton Grass Scrub
Riparian Scrub	Sonoran Riparian/Low-lying Riparian Scrub
Riparian Scrub	Sonoran/Chihuahuan Riparian/Reed-Cattail Marsh
Riparian Scrub	Riparian/Flood-damaged 1993
Water	Water
Developed	Agriculture
Developed	Urban
Developed	Industrial
Developed	Mixed Agriculture/Urban/Industrial

CHAPTER 15

The Effects of Classification Accuracy on Landscape Indices

Guofan Shao and Wenchun Wu

CONTENTS

15.1 INTRODUCTION

Remote sensing technology has advanced markedly during the past decades. Accordingly, remote sensor data formats have evolved from image (pre-1970s) to digital formats subsequent to the launch of Landsat (1972), resulting in a proliferation of derivative map products. The accuracy of these products has become an integral analysis step essential to evaluate appropriate applications (Congalton and Green, 1999). During the past three decades, accuracy assessment has become widely applied and accepted. Although methodologies have improved, little attention has been given to the effects of classification accuracy on the development of landscape metrics or indices.

Thematic maps derived from image classification are not always the final product from the user's perspective (Stehman and Czaplewski, 1998). Because all image processing or classification inevitably introduces errors into the resultant thematic maps, any subsequent quantitative analyses will reflect these errors (Lunetta et al., 1991). Landscape metrics are commonly derived from remote sensing-derived LC maps (O'Neill et al., 1988; McGarigal and Marks, 1994; Frohn, 1998). Metrics are commonly used to compare landscape configurations through time or across space, or as independent variables in modeling linking spatial pattern and process (Gustafson, 1998). Therefore, conclusions drawn directly or indirectly from analyzing landscape metrics contain uncertainties. The relationships between the accuracy of LC maps and specific derived landscape metrics are quite variable (i.e., metric dependent), which complicates assessment efforts (Hess, 1994; Shao et al., 2001).

A major obstacle to assessing the accuracy of LC maps is the high cost of generating reference data or multiple thematic maps for subsequent comparative analysis. Commonly employed solutions include (1) selecting subsectional maps from a region (Riitters et al., 1995), (2) subdividing regional maps into smaller maps (Cain et al., 1997), or (3) creating multiple maps using computer simulations (Wickham et al., 1997; Yuan, 1997). Maps created using the first or second method are spatially incompatible or incomparable, while maps created using the third method contain errors that do not necessarily represent those found in actual LC maps. Therefore, it is necessary to create multiple maps for a specific geographic area using different analysts or different classification methods (Shao et al., 2001). The approach presented here represents an actual image data analysis and, therefore, conclusions drawn from the analysis should be broadly applicable.

Past studies have focused on only a few indices. Hess and Bay (1997) made a breakthrough in quantifying the uncertainties of adjusted diversity indices. Various statistical models have also been developed to assess the accuracy of total area (%LAND) for individual cover types (Bauer et al., 1978; Card, 1982; Hay, 1988; Czaplewski, 1992; Dymond, 1992; Woodcock, 1996). However, few have used modeling to perform area calibrations (Congalton and Green, 1999). Shao et al. (2003) derived the Relative Area Error (REA) index, which has causal relationships with area estimates of LC categories. This study employed multiple classifications and reference maps to demonstrate how classification accuracy affects landscape metrics. Here the overall accuracy and REA were compared and a simple method was demonstrated to revise %LAND values using corresponding REA index values.

15.2 METHODS

Multiple thematic maps were derived from subscenes of Landsat Thematic Mapper (TM) data for two sites (A and B) located in central Indiana and the temperate forest zone on the eastern Eurasian continent (at the border of China and North Korea). LC mapping was performed to approximate a Level I classification product (Anderson et al., 1976). Site A thematic maps included the following classes: (1) agriculture (including grassland), (2) forest (including shrubs), (3) urban, and (4) water. The second site included only forest and nonforest (clear cuts and other open areas) cover types. A total of 23 independent thematic maps were developed for site A. Analysts ($n = 23$) were allowed to use any method to classify the TM imagery acquired on October 5, 1992. LC maps were evaluated based on the overall accuracy. All the accuracies were comparable because all assessments were performed using the same reference data set. Students performed the image analysis, thus representing work performed by nonprofessionals (Shao et al., 2001).

Eighteen thematic maps were created for site B using a single TM data set acquired on September 4, 1993, and a stack data set combining the 1993 data with other TM data acquired on September 21, 1987. Training samples were acquired using three methods, including (1) computer image interpretation, (2) field observations, and (3) and a combination of the two. Three classification algorithms were used, including (1) the minimum distance (MD), (2) maximum likelihood (ML), and (3) extraction and classification of homogeneous objects (ECHO). Our goal was to make the classification process repeatable, and therefore to represent a professional work process (Wu and Shao, 2002). Two additional maps with 94.0% and 94.5% overall accuracy that were created with alternative approaches were also incorporated into this study. The overall accuracy of these maps ranged from 82.6% to 94.5% (Wu and Shao, 2002). More importantly, a reference map was manually digitized for site B. The errors of landscape metrics of each map were computed as:

$$E_{index} = (I_{map} - I_{ref}) / I_{ref} \times 100$$

(15.1)

where E_{index} = relative errors (in percentage) of a given landscape index for a given thematic map, I_{map} = landscape index value derived from a thematic map, and I_{ref} = landscape index value derived from a reference map.

Thematic maps were assigned to three accuracy groups based on the overall accuracy maps at site A ($n = 23$). Landscape metrics were computed for each map with the FRAGSTATS for site A (McGarigal and Marks, 1994) and with patch analyst (PA) for site B (Elkie et al., 1999). Nine landscape indices were used for site A: largest patch index (LPI), patch density (PD), mean patch size (MPS), edge density (ED), area-weighted mean shape index (AWMSI), mean nearest neighbor distance (MNN), Shannon's diversity index (SHDI), Simpson's diversity index (SDI), and contagion index (CONTAG). Thirteen landscape indices were used for site B: PD, MPS, patch size coefficient of variance (PSCOV), patch site standard deviation (PSSD), ED, mean shape index (MSI), AWMSI, mean patch fractal dimension (MPFD), area-weighted mean patch fractal dimension (AWMPFD), MNN, mean proximity index (MPI), SDI, and %LAND. These landscape indices had broad representation within the different cover categories (McGarigal and Marks, 1994).

15.2.1 Relative Errors of Area (REA)

If a thematic map contains n classes or types, its accuracy can be assessed with an error matrix (Table 15.1).

For a given patch type k ($1 \leq k \leq n$), the reference value of %LAND (LR_k) is computed as:

$$LR_k = \frac{f_{+k}}{N} = \frac{\sum\limits_{i=1}^{n} f_{ik}}{N} = \frac{\sum\limits_{\substack{i=1 \\ i\neq k}}^{n} f_{ik} + f_{kk}}{N} \qquad (15.2)$$

The classification value of %LAND (LC_k) is derived as:

$$LC_k = \frac{f_{k+}}{N} = \frac{\sum\limits_{j=1}^{n} f_{kj}}{N} = \frac{\sum\limits_{\substack{j=1 \\ j\neq k}}^{n} f_{kj} + f_{kk}}{N} \qquad (15.3)$$

Table 15.1 A General Presentation of an Error Matrix Adapted from Congalton and Green (1999)

Classified Cover Type	Reference Data					
	1	$\cdots$	j	$\cdots$	n	Total
1	f_{11}	$\cdots$	f_{1j}	$\cdots$	f_{1n}	f_{1+}
$\vdots$	$\vdots$	$\cdots$	$\cdots$	$\cdots$	$\vdots$	$\vdots$
i	f_{i1}	$\cdots$	f_{ij}	$\cdots$	f_{in}	f_{i+}
$\vdots$	$\vdots$	$\cdots$	$\cdots$	$\cdots$	$\vdots$	$\vdots$
n	f_{n1}	$\cdots$	f_{nj}	$\cdots$	f_{nn}	f_{n+}
Total	f_{+1}	$\cdots$	f_{+j}	$\cdots$	f_{+n}	N

Note: n = the total number of land cover types; N = the total number of sampling points; f_{ij} (i and j = 1, 2, ..., n) = the joint frequency of observations assigned to type i by classification and to type j by reference data; f_{i+} = the total frequency of type i as derived from the classification; and f_{+j} = the total frequency of type j as derived from the reference data.

Thus, the difference between LC_k and LR_k is:

$$LC_k - LR_k = \frac{f_{k+} - f_{+k}}{N} = \frac{\sum_{j=1}^{n} f_{kj} - \sum_{i=1}^{n} f_{ik}}{N} = \frac{\sum_{\substack{j=1 \\ j \neq k}}^{n} f_{kj} - \sum_{\substack{i=1 \\ i \neq k}}^{n} f_{ik}}{N} \tag{15.4}$$

If $LC_k - LR_k = 0$, there are two possibilities: classification errors are zero, or commission errors (CE) and omission errors (OE) are the same for patch type k. The first possibility is normally untrue in reality. In many situations, the second possibility is also untrue. If $CE_k > OE_k$, $LC_k - LR_k > 0$, the value of %LAND of type k is overestimated; if $CE_k < OE_k$, $LC_k - LR_k < 0$, the value of %LAND of type k is underestimated. Therefore, the components of CE_k and OE_k in Equation 15. 4 determine the accuracy of %LAND for patch type k.

Mathematically, CE_k is just as follows:

$$CE_k = \sum_{\substack{j=1 \\ j \neq k}}^{n} f_{kj} \tag{15.5}$$

OE_k is just expressed as:

$$OE_k = \sum_{\substack{i=1 \\ i \neq k}}^{n} f_{ik} \tag{6}$$

The balance between CE_k and OE_k indicates the absolute errors of area estimate for patch type k. The relative errors of area (REA) are then defined as:

$$REA_k = \frac{\sum_{\substack{j=1 \\ j \neq k}}^{n} f_{kj} - \sum_{\substack{i=1 \\ i \neq k}}^{n} f_{ik}}{f_{kk}} \times 100 \tag{15.7}$$

where f_{kk} is an element of the k-th row and k-th column in an error matrix. It represents the frequency of sample points that are correctly classified.

According to Congalton and Green (1999), user's accuracy of type k (UA_k) can be expressed as:

$$UA_k = \frac{f_{kk}}{f_{k+}} = \frac{f_{kk}}{\sum_{j=1}^{n} f_{kj}} = \frac{f_{kk}}{f_{kk} + \sum_{\substack{j=1 \\ j \neq k}}^{n} f_{kj}} \tag{15.8}$$

and producer's accuracy of type k (PA_k) can be expressed as:

$$PA_k = \frac{f_{kk}}{f_{+k}} = \frac{f_{kk}}{\displaystyle\sum_{i=1}^{n} f_{ik}} = \frac{f_{kk}}{f_{kk} + \displaystyle\sum_{\substack{i=1 \\ i \neq k}}^{n} f_{ik}} \qquad (15.9)$$

By substituting Equation 15.8 and Equation 15.9 into Equation 15.7, it is easily derived that:

$$REA_k = \left(\frac{1}{UA_k} - \frac{1}{PA_k} \right) \times 100 \qquad (15.10)$$

Thus, REA can be obtained using information on the error matrix or the user's and producer's accuracy.

Under the assumption that the distribution of errors in the error matrix is representative of the types of misclassification made in the entire area classified, it is easy to calibrate area estimates with REA or UA and PA as follows:

$$A_{c,k} = A_{pc,k} - \frac{f_{kk}}{N} \times REA_k = A_{pc,k} - \frac{f_{kk}}{N} \times \left(\frac{1}{UA_k} - \frac{1}{PA_k} \right) \times 100 \qquad (15.11)$$

where $A_{c,k}$ = calibrated area in percentage for a given land cover type k and $A_{pc,k}$ = precalibrated area in percentage for a given land cover type k.

15.3 RESULTS

Figure 15.1 shows the means and standard deviations of nine landscape indices for three accuracy groups. Except for PD and MPS, landscape indices had < 10% differences in their means among three accuracy groups. The standard deviations of the landscape indices in the lowest accuracy group are much higher than those in the higher accuracy groups. The differences in standard deviations between the lowest accuracy group and other two accuracy groups exceeded 100%, indicating that the uncertainties were higher when classification accuracy was lower.

The statistics of classification accuracy, including the overall accuracy, producer's accuracy, and user's accuracy, all have differences of < 20% among the three accuracy groups (Figure 15.2a). The standard deviation values for overall accuracy are also about the same among the three accuracy groups but are clearly different for producer's accuracy and user's accuracy (Figure 15.2b). Maps in the lowest accuracy group have much higher variations in producer's accuracy and user's accuracy than those in the other two accuracy groups.

For a few indices, such as MPDF, AWMPFD, and SDI at the landscape level, no matter what the classification accuracy was, the errors of landscape indices were within a range of 10% (Figure 15.3). If classification accuracy was poor, the errors of some other landscape indices exceeded 100%. They include PD, PSCOV, ED, AWMSI, and MPI for entire landscapes or forest patches (Figure 15.3 and Figure 15.4). Although no constant relationships were found between the overall accuracy and landscape indices, maps with higher classification accuracy resulted in lower errors for most landscape indices (Figure 15.3 and Figure 15.4). However, overall accuracy did not have good control over the variations of landscape index errors and therefore was not a reliable predictor for the errors of landscape indices. This was particularly true when the overall accuracy was relatively low.

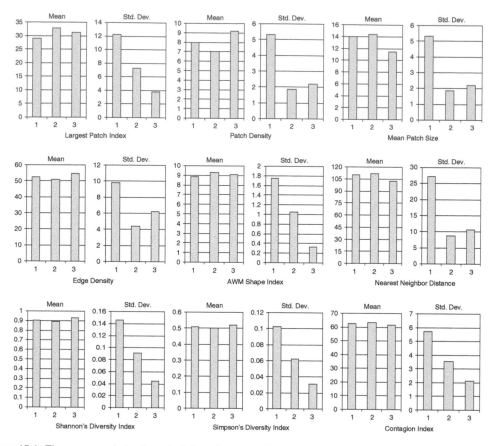

Figure 15.1 The mean and standard deviations for nine selected landscape indices for three accuracy groups; 1 = lowest accuracy, 2 = intermediate accuracy, 3 = highest accuracy.

The errors of %LAND have a perfect linear relationship with REA (R^2 = 0.98), but the errors of all other indices did not show a simple relationship with REA (Figure 15.5). The REA seemed to have a better control over landscape indices errors than did overall accuracy; the variations of landscape index errors corresponding to REA were smaller than those corresponding to overall accuracy (Figure 15.4 and Figure 15.5). Also, the lowest errors of landscape indices normally occurred when REA reached zero (Figure 15.5). Both overall accuracy and REA were not reliable indicators for explaining variations of spatially sophisticated landscape indices, such as MNN and MPI.

The relative errors of %LAND for the forest from the 20 maps ranged from 12 to 25% before calibration (Figure 15.6a). Based on Equation 15.11, the values of %LAND for the forest were calibrated and resulting errors of %LAND for the forest were between 2 and 5% (Figure 15.6b), much lower than the errors before calibration.

15.4 DISCUSSION

Methods used for image classification determine thematic maps' classification content and quality. Although different statistics are used for assessing the accuracy of image data classifications, most are derived directly or indirectly from error matrices. Indices of thematic map accuracy indicate how well image data are classified but do not tell how thematic maps correspond to a landscape's structure and function. This is partly because there is no effective approach to quantify classification

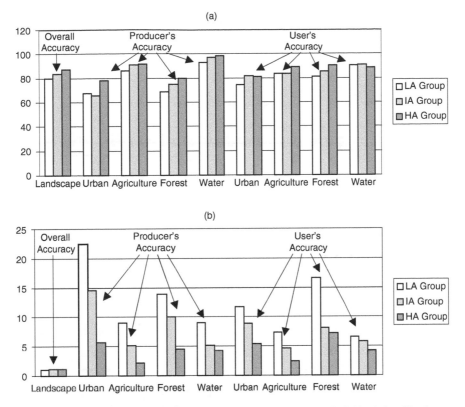

Figure 15.2 The mean (a) and standard deviation (b) values for overall and individual classification accuracies; LA = lowest accuracy, IA = intermediate accuracy, HA – highest accuracy.

errors that have causal relationships with landscape function. Overall accuracy is the most frequently used accuracy statistics, but it has limited control over the errors of landscape indices. In practice, greater overall accuracy resulted in more controllable errors associated with landscape indices. Only an unrealistic, 100% accurate map represents perfect source data for computing landscape indices. For example, the overall accuracy of LC and LU maps derived from TM data for the eastern U.S. was 81% for Anderson Level I (i.e., water, urban, barren land, forest, agricultural land, wetland, and rangeland) and was 60% for Anderson Level II (Vogelmann et al., 2001). Such classification accuracies are not high enough for ensuring reliable landscape index calculations.

Overall accuracy did not have a causal control over the variability of index accuracies. When overall accuracy was relatively low, it also lost control over the difference between user's and producer's accuracies. It also appeared that the uncertainties of landscape indices were more sensitive to the variations in user's and producer's accuracies than to overall accuracy values alone. REA values reflected the differences between user's and producer's accuracies and therefore had a better control over the errors of landscape indices than did overall accuracy, particularly when overall accuracy was relatively low.

Because REA is derived for assessing the accuracy of %LAND, this index alone can be used to predict the errors of %LAND. The linear relationship with REA and the area of forested land verifies the reliability of such predictions with REA. While the overall accuracy is approximately the average of user's and producer's accuracy, REA reveals the differences between user's and producer's accuracy. Therefore, the overall accuracy and REA explained different aspects of classification accuracy. Although the lowest errors of landscape indices often occur when REA is near zero, variations in the errors of landscape indices still existed. When REA and the overall accuracy were used together, the errors of landscape indices were better predicted

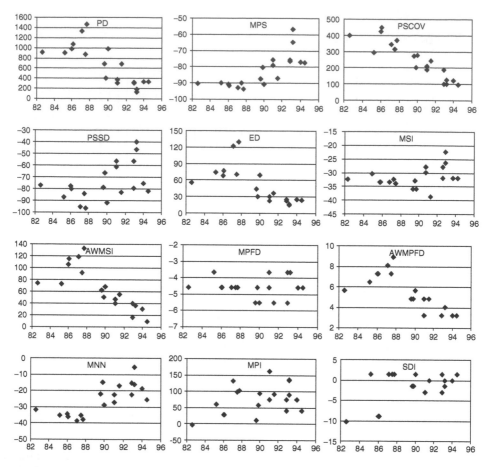

Figure 15.3 The relative errors of 12 selected landscape indices for the landscape (y-axis) against the overall accuracy (x-axis).

(the greater overall accuracy, the smaller REA). However, overall accuracy and REA explained some aspects of classification errors but did not explain other possible sources of classification errors (e.g., the spatial distributions of misclassifications). Therefore, these accuracy measures alone were not adequate to assess the accuracy of the MNN and MPI, which have particularly strong spatial features.

The variations of landscape index errors were different among different landscape indices. For example, the errors of MPDF, AWMPFD, and SDI at the landscape level were within a range of 10%, whereas the errors of PD, PSCOV, ED, AWMSI, and MPI for entire landscapes or forest patches exceeded 100%. The former group of landscape indices was not as sensitive to image data classification and the errors of these landscape indices were not controlled by classification accuracy measures. Landscape indices in this group were unreliable despite the image classification accuracy values. The latter group of landscape indices was sensitive to image data classifications, and therefore a small difference in classification accuracy resulted in a large difference in landscape index values. In this case, classification accuracy was always superior when accuracy-sensitive landscape indices were used. Intermediate indices exhibited intermediate sensitivity to image data classifications. The rule of higher overall accuracy and smaller absolute values of REA was particularly applicable to this intermediate group. Further systematic studies are needed to determine which landscape index belongs to these sensitive groups.

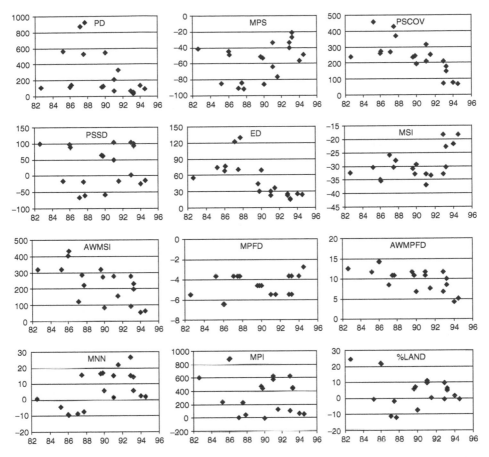

Figure 15.4 The relative errors of 12 selected landscape indices for forest class (y-axis) against the overall accuracy (x-axis).

15.5 CONCLUSIONS

The uncertainties or errors associated with landscape indices vary in their responses to image data classifications. Also, the existing statistical methods for assessing classification accuracy have different controls relative to the uncertainties or errors of landscape indices. Assessing accuracy of landscape indices requires combined knowledge of the overall accuracy (means of user's accuracy and producer's accuracy) and the REA (differences between user's accuracy and producer's accuracy). To reliably characterize landscape conditions using landscape indices, our results indicate it is necessary to use maps with high overall accuracy and low absolute REA. The selections of landscape indices are also important because different landscape indices have different sensitivities to image data classifications. Based on commonly achievable levels of classification accuracy, the magnitudes of errors associated with landscape indices can be higher than the values of landscape indices. Comparisons between different thematic maps should consider these errors. Assuming that the distribution of errors identified by the error matrix is representative of the misclassifications across the area of interest, the total land area of different class categories can be revised with REA and the errors of this landscape index can be lowered. Revised values of %LAND should be used when quantifying landscape conditions.

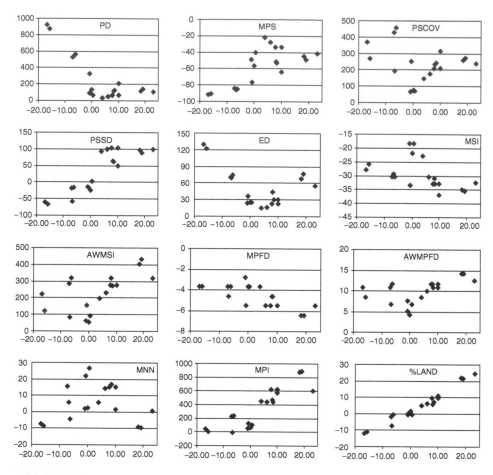

Figure 15.5 The relative errors of 12 selected landscape indices for forest class (y-axis) against the REA (x-axis).

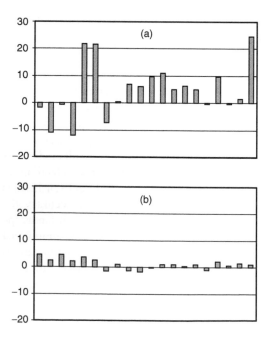

Figure 15.6 A comparison of %LAND errors for forest class among thematic maps ($n = 20$) before calibrations (a) and after calibrations (b).

15.6 SUMMARY

A total of 43 LC maps from two study sites were used to demonstrate the effects of classification accuracy on the uncertainties or errors of 15 selected landscape indices. The measures of classification accuracy used in this study were the overall accuracy and REA. The REA was defined as the difference between the reciprocals of user's accuracy and producer's accuracy. Under variable levels of classification accuracy, different landscape indices had different uncertainties or errors. These variations or errors were explained by both the overall accuracy and REA. Thematic maps with relatively high overall accuracy and low absolute REA ensured lower uncertainties or errors of at least several landscape indices. For landscape indices that were sensitive to classification accuracy, a small increase in classification accuracy resulted in a large increase in their accuracy. Assuming that the error matrix truly represents misclassification errors, the total areas of different class categories can be calibrated using the REA index and the accuracy of quantifying or comparing relative landscape characteristics can be increased.

ACKNOWLEDGMENTS

Thematic LC maps used in this study were partially provided by 23 students from a remote sensing class offered at Purdue University in 1999. The Cooperative Ecological Research Program in cooperation with the China Academy of Sciences and the German Department of Science and Technology provided the TM data used in this study. The authors would like to thank the book editors and anonymous reviewers for their insightful comments and suggestions on the manuscript.

REFERENCES

Anderson, J.R., E.E. Hardy, J.T. Toach, and R.E. Witmer, A Land Use and Land Cover Classification System for Use with Remote Sensor Data, U.S. Geological Survey professional paper 964, U.S. Government Printing Office, Washington, DC, 1976.

Bauer, M.E., M.M. Hixson, B.J. Davis, and J.B. Etheridge, Area estimation of crops by digital analysis of Landsat data, *Photogram. Eng. Remote Sens.*, 44, 1033–1043, 1978.

Cain, D.H., K. Riitters, and K. Orvis, A multi-scale analysis of landscape statistics, *Landsc. Ecol.*, 12, 199–212, 1997.

Card, D.H., Using map categorical marginal frequencies to improve estimates of thematic map accuracy, *Photogram. Eng. Remote Sens.*, 48, 431–439, 1982.

Congalton, R.G. and K. Green, *Assessing the Accuracy of Remotely Sensed Data: Principles and Practices*, Lewis, Boca Raton, FL, 1999.

Czaplewski, R.L., Misclassification bias in areal estimates, *Photogram. Eng. Remote Sens.*, 58, 189–192, 1992.

Dymond, J.R., How accurately do image classifier estimate area? *Int. J. Remote Sens.*, 13, 1735–1342, 1992.

Elkie, P., R. Rempel, and A. Carr, *Patch Analyst User's Manual, Ontario Ministry of Natural Resources, Northwest Science & Technology,* Thunder Bay, Ontario, Tm-002, 1999.

Frohn, R.C., *Remote Sensing for Landscape Ecology: New Metric Indicators for Monitoring, Modeling, and Assessment of Ecosystems*, Lewis, Boca Raton, FL, 1998.

Gustafson, E.J., Quantifying landscape spatial pattern: what is the state of the art? *Ecosystems*, 1, 143–156, 1998.

Hay, A.M., The derivation of global estimates from a confusion matrix, *Int. J. Remote Sens.*, 9, 1395–1398, 1988.

Hess, G.R., Pattern and error in landscape ecology: a commentary, *Landsc. Ecol.*, 9, 3–5, 1994.

Hess, G.R. and J.M. Bay, Generating confidence intervals for composition-based landscape indexes, *Landsc. Ecol.*, 12, 309–320, 1997.

Lunetta, R.S., R.G. Congalton, L.F. Fenstermaker, J.R. Jensen, K.C. McGwire, and L.R. Tinney, Remote sensing and geographic information system data integration: error sources and research issues, *Photogram. Eng. Remote Sens.*, 57, 677–687, 1991.

McGarigal, K. and B.J. Marks, FRAGSTATS: Spatial Patterns Analysis Program for Quantifying Landscape Structure, unpublished software, USDA Forest Service, Oregon State University, 1994.

O'Neill, R.V., J.R. Krummel, R.H. Gardner, G. Sugihara, B. Jackson, D.L. DeAngelis, B.T. Milne, M.G. Turner, B. Zygnut, S.W. Christensen, V.H. Dale, and R.L. Graham, Indices of landscape pattern, *Landsc. Ecol.,* 1, 152–162, 1988.

Riitters, K.H., R.V. O'Neill, C.T. Hunsaker, J.D. Wickham, D.H. Yankee, S.P. Timmins, K.B. Jones, and B.L. Jackson, A factor analysis of landscape pattern and structure metrics, *Landsc. Ecol.,* 10, 23–39, 1995.

Shao, G., D. Liu, and G. Zhao, Relationships of image classification accuracy and variation of landscape statistics, *Can. J. Remote Sens.,* 27, 33–43, 2001.

Shao, G., W. Wu, G. Wu, X. Zhou, and J. Wu, An explicit index for assessing the accurace of cover class areas, *Photogram. Eng. Remote Sens.*, 69(8), 907–913, 2003.

Stehman, S.V. and R.L. Czaplewski, Design and analysis for thematic map accuracy assessment: fundamental principles, *Remote Sens. Environ.,* 64, 331–344, 1998.

Vogelmann, J.E., M.H. Stephen, L. Yang, C.R. Clarson, B.K. Wylie, and N. Van Driel, Completion of the 1990s national land cover data set for the conterminous United States from Landsat Thematic Mapper data and ancillary data sources, *Photogram. Eng. Remote Sens.,* 67, 650–662, 2001.

Wickham, J.D., R.V. O'Neill, K.H. Riitters, T.G. Wade, and K.B. Jones, Sensitivity of selected landscape pattern metrics to landcover misclassification and differences in landcover composition, *Photogram. Eng. Remote Sens.,* 63, 397–402, 1997.

Wookcock, C.E., On the roles and goals for map accuracy assessment: a remote sensing perspective, in Proceedings of the 2nd International Symposium on Spatial Accuracy Assessment in Natural Resources and Environmental Sciences, Fort Collins, CO, 1996, USDA Forest Service Rocky Mountain Forest and Range Experiment Station, Technical Report RM-GTR-277, 1996, pp. 535–540.

Wu, W. and G. Shao, Optimal combinations of data, classifiers, and sampling methods for accurate characterizations of deforestation, *Can. J. Remote Sens.,* 28, 601–609, 2002.

Yuan, D., A simulation of three marginal area estimates for image classification, *Photogram. Eng. Remote Sens.,* 63, 385–392, 1997.

CHAPTER 16

Assessing Uncertainty in Spatial Landscape Metrics Derived from Remote Sensing Data

Daniel G. Brown, Elisabeth A. Addink, Jiunn-Der Duh, and Mark A. Bowersox

CONTENTS

16.1 INTRODUCTION

Recent advances in the field of landscape ecology have included the development and application of quantitative approaches to characterize landscape condition and processes based on landscape patterns (Turner et al., 2001). Central to these approaches is the increasing availability of spatial data characterizing landscape constituents and patterns, which are commonly derived using various remote sensor data (i.e., aerial photography or multispectral imagery). Spatial pattern metrics provide quantitative descriptions of the spatial composition and configurations of habitat or land-cover (LC) types that can be applied to provide useful indicators of the habitat quality, ecosystem function, and the flow of energy and materials within a landscape. Landscape metrics have been used to compare ecological quality across landscapes (Riitters et al., 1995) and across scales (Frohn, 1997) and to track changes in landscape patterns through time (Henebry and Goodin, 2002). These comparisons can often provide quantitative statements of the relative quality of landscapes with respect to some spatial pattern concept (e.g., habitat fragmentation).

Uncertainty associated with landscape metrics has several components, including (1) accuracy (how well the calculated values match the actual values), (2) precision (how closely repeated measurements get to the same value), and (3) meaning (how comparisons between metric values should be interpreted). In practical terms, accuracy, precision, and the meaning of metric values are affected by several factors that include the definitions of categories on the landscape map, map accuracy, and validity and uniqueness of the metric of interest. Standard methods for assessing LC map accuracy provide useful information but are inadequate as indicators of the spatial metric accuracy because they lack information concerning spatial patterns of uncertainty and the correspondence between the map category definitions and landscape concepts of interest. Further, direct estimation of the accuracy of landscape metric values is problematic. Unlike LC maps, standard procedures are currently not available to support landscape metric accuracy assessment. Also, the scale dependence of landscape metric values complicates comparisons between field observations and map-based calculations.

As a transformation process, in which mapped landscape classes are transformed into landscape measurements describing the composition and configuration of that landscape, landscape metrics can be evaluated using precision and meaning diagnostics (Figure 16.1). The primary objective is to acquire a metric with a known and relatively high degree of accuracy and precision that is interpretable with respect to the landscape characteristic(s) of interest. The research presented in this chapter addresses the following issues: (1) precision estimates associated with various landscape metrics derived from satellite images, (2) sensitivity of landscape metrics relative to differences in landscape class definitions, and (3) sensitivities of landscape metrics to landscape pattern concepts of interest (e.g., ecotone abruptness or forest fragmentation) vs. potential confounding concepts (e.g., patchiness or amount of forest).

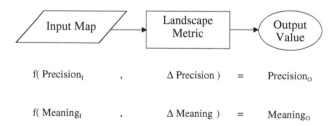

Figure 16.1 Illustration of the issues affecting the quality and utility affecting landscape pattern metric values derived from landscape class maps. The precision and meaning of output values from landscape metrics are functions of the precision and meaning of the input landscape maps and the effect of the metric transformation.

This chapter presents results from recent research that seeks to evaluate uncertainty in landscape metrics, as defined above. To calculate the precision of landscape metrics, repeated estimates of metric values are used to observe the variation in the estimates. Because measures of precision are based on multiple calculations, they are more practical for landscape metric applications than are measures of accuracy. Here we discuss two different approaches to performing multiple calculations of landscape metric values. First, redundant mapping of landscapes was used to calculate the variation in metric values resulting from the redundant maps. Second, spatial simulation was used to evaluate the response of landscape metric values to repeated landscape mapping under a neutral model (Gustafson and Parker, 1992).

Following a general discussion of alternative types of landscape metrics, we compare past research and our results to illustrate how landscape metric values vary using redundant mapping and simulation methods. First, the precision of estimates of change in metric values between two images was investigated using redundant mapping of sample areas that were defined by the overlap of adjacent satellite scenes (Brown et al., 2000a,b). Next, the variations in metrics were calculated using landscape maps derived from the same remote sensing source but classified using different definitions and class. Comparisons illustrating the effects of alternative definitions of "forest" and the application of LC vs. land-use (LU) classes for calculating metrics are presented. Finally, we evaluate the use of simulation to investigate the interpretability of the construct being measured, the degree of similarity among several landscape metrics, and the concept of ecotone abruptness and present simulations to illustrate the problem of interpreting the degree of fragmentation from landscape metrics (Bowersox and Brown, 2001).

16.2 BACKGROUND

Several approaches to characterizing landscape pattern are available, each with its own implications for the accuracy, precision, and meaning of a landscape pattern analysis. With the goal of quantitatively describing the landscape structure, landscape metrics provide information about both landscape composition and configuration (McGarigal and Marks, 1995). The most common approach to quantifying these characteristics has been to map defined landscape classes (e.g., habitat types) and delineate patches of representative landscape classes. Patches are then defined as contiguous areas of homogenous landscape condition. Landscape composition metrics describe the presence, relative abundance, and diversity of various cover types. Landscape configuration refers to the "physical distribution or spatial character of patches within the landscape" (McGarigal and Marks, 1995). Summaries of pattern can be made at the level of the individual patch (e.g., size, shape, and relative location), averaged across individual landscape classes (e.g., average size, shape, and location), or averaged across all patches in the landscape (e.g., average size, shape, and location of all patches).

An alternative to patch-based metrics are metrics focused on identifying transition zone boundaries that are present in continuous data. This approach has not been used as extensively as the patch approach in landscape ecology (Johnston and Bonde, 1989; Fortin and Drapeau, 1995). One approach to using boundaries is to define "boundary elements," defined as cells that exhibit the most rapid spatial rates of change, and "subgraphs," which are strings of connected boundary elements that share a common orientation (direction) of change (Jacquez et al., 2000). The landscape metrics characterize the numbers of boundary elements and subgraphs and the length of sub-graphs, which is defined by the number of boundary elements in a subgraph. An important advantage is that boundary-based statistics can be calculated from images directly, skipping the classification step through which errors can propagate. Throughout this chapter, we refer to patch-based metrics, which were calculated using Fragstats (McGarigal and Marks, 1995), and boundary-based metrics calculated using the methods described by Jacquez et al. (2000).

16.3 METHODS

16.3.1 Precision of Landscape Change Metrics

To measure imprecision in metric values, overlapping Landsat Multi-Spectral Scanner (MSS) path/row images were redundantly processed for two different study areas in the Upper Midwest to create classifications representing forest, nonforest, water, and other and maps of the normalized difference vegetation index (NDVI). Images on row 28 and paths 24–25 overlapped in the northern Lower Peninsula of Michigan and on row 29 and paths 21–22 overlapped on the border between northern Wisconsin and the western edge of Michigan's Upper Peninsula (Brown et al., 2000a).

The georeferenced MSS images at 60-m resolution were acquired from the North American Landscape Characterization (NALC) project during the growing seasons corresponding to three periods: 1973–1975, 1985–1986, and 1990–1991 (Lunetta et al., 1998). Subsequent LC classifications of the four images resulted in accuracies ranging from 72.5% to 91.2% (average 80.5%), based on comparison with aerial photograph interpretations.

For landscape pattern analysis, the two study areas were partitioned into 5- × 5-km² cells. A total of 325 cells in the Michigan site and 250 in the Wisconsin-Michigan site were used in the analysis. The partitions were treated as separate landscapes for calculating the landscape metric values. The values of eight pattern metrics, four patch-based and four boundary-based, were calculated for each partition using each of two overlapping images at each of three time periods in both sites.

The precision of landscape metric values was calculated using the difference between metric values calculated for the same landscape partition within the same time period. For each metric, these differences were summarized across all landscape partitions using the root mean squared difference (RMSD). To standardize the measure of error for comparison between landscape metrics, the relative difference (RD) was calculated as the RMSD divided by the mean of the metric values obtained in both images of a pair.

16.3.2 Comparing Class Definitions

16.3.2.1 Landsat Classifications

To evaluate the sensitivity of maps to differences in class definitions we calculated landscape metric values from two independent LC classifications derived from Landsat Thematic Mapper (TM) imagery of for the Huron River watershed located in southeastern Michigan. The only significant difference between the two LC maps was the class definitions. Accuracy assessments were not performed for either map. Therefore, the analysis serves only as an illustration for evaluating the importance of class definitions.

For the first map, Level I LU/LC classes were mapped for the early 1990s using the National Land Cover Data (NLCD) classification for the region. We developed the second data set using TM imagery from July 24, 1988. It was classified to identify all areas of forest, defined as pixels with > 40% canopy cover, vs. nonforest. Spectral clusters, derived through unsupervised classification (using the ISODATA technique), were labeled through visual interpretation of the image and reclassified. Landscape metrics were computed using Fragstats applied to the forest class from both data sets across the entire watershed. Also, the two data sets were overlaid to evaluate their spatial correspondence.

16.3.2.2 Aerial Photography Interpretations

We also compared two classifications of aerial photography over a portion of Livingston County in southeastern Michigan. The first data set consisted of a manual interpretation of LU and LC using color infrared (CIR) aerial photographs (1:24,000 scale) collected in 1995 (SEMCOG, 1995). The classes were based on a modified Anderson et al. (1976) system, which we reclassified to high-density residential, low-density residential, other urban, and other. The second was a LC classification created through unsupervised clustering and subsequent cluster labeling of scanned color-infrared photography (1:58,000-scale) collected in 1998. The LC classes were forest, herbaceous, impervious, bare soil, wetland, and open water. The two maps were overlaid to identify the correspondence between the LC classes and the urban LU classes. The percentages of forest and impervious cover were calculated within each of the urban LU types.

16.3.3 Landscape Simulations

16.3.3.1 Ecotone Abruptness

An experiment was designed in which 25 different landscape types were defined, each representing a combination of among five different levels of abruptness and five levels of patchiness (Bowersox and Brown, 2001). Ecotone abruptness (i.e., how quickly an ecotone transitioned from forest to nonforest) was controlled by altering the parameters of a mathematical function to model the change from high to low values along the gradient representing forested cover. Patchiness was introduced by combining the mathematical surface with a randomized surface that was smoothed to introduce varying degrees of spatial autocorrelation. Once the combined gradient was created, all cells with a value above a set threshold were classified as forest, and those below were classified as nonforest. The threshold was set so that each simulated landscape was 50% forested and 50% nonforested.

For each type of landscape, 50 different simulations were conducted. The ability of each landscape metric to detect abruptness was then tested by comparing the values of the 50 simulations among the different cover types. The landscape metric values were compared among the abruptness and patchiness levels using analysis of variance (ANOVA). The ANOVA results were analyzed to identify the most suitable metrics for measuring abruptness (i.e., those exhibiting a high degree of variation between landscape types with variable abruptness levels but a low degree of variation between landscape types with variable patchiness).

In addition to several patch-based metrics (including area-weight patch fractal dimension, area-weighted mean shape index, contagion, and total edge), boundary-based metrics were used, including (1) number of boundary elements, (2) number of subgraphs, and (1) maximum subgraph length. The analysis compared the ability of two new boundary-based metrics designed specifically to measure ecotone abruptness and distinguish different levels of abruptness. These new metrics characterize the dispersion of boundary elements around an "average ecotone position," calculated as the centroid of all boundary elements, and the area under the curve of the number of boundary elements vs. the slope threshold level.

16.3.3.2 Fragmentation

The sensitivity of several potential measures of forest fragmentation to the amount of forest was also investigated through simulation. The simulation included: (1) generating a random map for 100- × 100-grid cells with pixel values randomly drawn from a normal distribution (mean = 0, standard deviation = 1), (2) smoothing with a five-by-five averaging filter to introduce spatial autocorrelation, and (3) creating maps ($n = 10$) by classifying cells as forest or nonforest based on different threshold levels. The threshold levels were defined so that the different maps had a uniformly increasing amount of forest from about 9% to about 91% (Figure 16.2). By extracting the maps with different proportions of forest from the same simulated surface, patterns were controlled and the dominant difference among maps was the amount forested. The simulation process was repeated 10 times to produce a range of output values at each landscape proportion level.

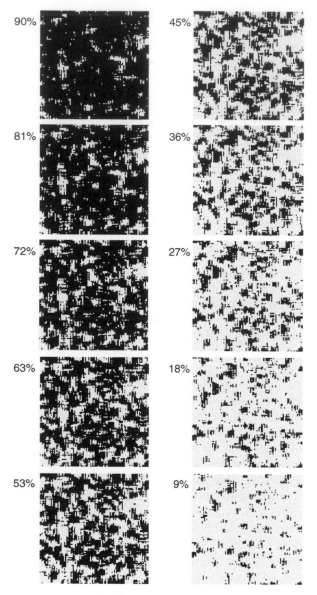

Figure 16.2 One of 10 realizations of landscape simulations created to illustrate the influence of the proportion of the landscape covered by a class on the values of landscape pattern metrics. The number indicates the percentage of the landscape in forest (shown in black).

16.4 RESULTS

16.4.1 Precision of Landscape Metrics

Comparison among the patch-based metrics indicated that the number and size of patches were much less precise than the area of forest and the edge density (Table 16.1). A likely explanation is that the number of patches and mean patch size metrics required that the pixel classification and patch aggregation processes be consistent. Both of these can be sensitive to spatially patterned classification error, thus suggesting that there are differences among metrics in the Δ precision described in Figure 16.1.

Table 16.1 The Average Relative Error for Eight
Different Landscape Metrics, Four
Based on Identifying Landscape
Patches and Four Based on Description
of Boundaries in a Continuous Image

Metric	Average Relative Error
Percentage forest	0.23
Edge density	0.35
No. of patches	0.75
Mean patch size	1.52
No. of boundary elements	0.02
No. of subgraphs	0.11
No. of singletons	0.24
Max subgraph length	0.40

Comparing the patch- vs. boundary-based metrics indicated that the majority of boundary metrics had greater precision than the patch-based statistics (Table 16.1). This can best be explained by the way in which changes in precision were affected by the procedures used to calculate the metric values. All of the patch-based metrics involved an image classification step, and two of them added a patch identification step. Both of these steps are sensitive to spatial variations in image quality and to the specific procedures used. Because the boundary-based metrics were calculated directly from the NDVI images, there was less opportunity for propagation of the spatial pattern of error. Further, the boundary-based metrics used only local information to characterize pattern, but the patch-based metrics used global information (i.e., spectral signatures from throughout the image). This use of global information introduced more opportunities for error in metric calculation.

Additionally, we evaluated the effects of various processing choices on the precision of metrics (Brown et al., 2000a). The results of this work suggest that haze in the images and differences in seasonal timing were important determinants of metric variability. Specifically, less precision resulted from hazier images and image pairs that were separated by more Julian days, irrespective of the year. Also, summarizing landscape metrics over larger areas (i.e., using larger landscape partitions) increased the precision of the estimates, although it reduced the spatial resolution. Further, postclassification processing, such as sieving and filtering, did not consistently increase the precision, and can actually reduce the precision.

The obvious cost associated with obtaining precise estimates through the empirical approach of redundant mapping is that the areas need to be mapped twice. However, the costs may be lower than the costs of obtaining reference data for accuracy assessment, and redundant mapping can provide reasonable estimates of precision in a pattern analysis context, where comparison with a reference data set is much more problematic. Guindon et al. (2003) used a similar approach to dealing with the precision of LC maps.

16.4.2 Comparing Class Definitions

16.4.2.1 Comparing TM Classifications

Across all landscape metrics tested, our forest cover classification of the Huron River watershed suggested that the landscape was much less fragmented than did the NLCD forest class (i.e., that there was more forest, in fewer but larger patches, with less forested/nonforested edge and more core area) (Table 16.2). Comparisons of forested cells indicated that forest cover occurred in several of the nonforest NLCD classes. The definitions of NLCD classes allowed for substantial amounts of forest cover in nonforest classes. For example, in the low-density residential class "vegetation" could account for 20 to 70% of the cover (USGS, 2001). Also, the NLCD forest classes were not 100% forested. Although 65% of the forested cover in the region (by our definition) was contained

Table 16.2 Patch-Based Landscape Metrics Describing Forest in
          Michigan's Huron River Watershed as Mapped in the NLCD
          Data Compared with a Separate Classification of Forested
          Cover Derived from Landsat TM; All Metrics Are Summarized
          at the Class Level for the Forest Patches

Metric	NLCD Forest Data	Forest Cover Classification
Percentage forest	28.1	31.1
No. of patches	28857	19137
Mean patch size (ha)	2.17	3.62
Edge density (m/ha)	106.05	85.33
Mean shape index	1.37	1.37
Total core area index	37.33	50.53

Table 16.3 Percentage of Generalized NLCD Forest Classes Based on
          the Classification of Landsat TM data and the Percentage
          of the Total Forested Cover within each NLCD Class

Generalized NLCD Class	% Forest Cover	% Forest Cover Total
Urban	5.6	5.6
Forest	57.1	65.1
Agriculture and Herbaceous	13.2	25.8
Other	17.5	3.5
Total		100.0

Note: The first column indicates how much forested cover was contained
       within each NLCD class. The second indicates the amount of the
       forested cover within each class.

within forest classes as defined by NLCD, 25% was located in agricultural areas and < 6% in urban areas (Table 16.3).

These findings indicate that landscape metrics are sensitive to the definitions of the input classes. This sensitivity is a result of differences in the meaning of the classes themselves rather than the lack of classification detail or because of inaccuracy in the classification. For some landscape analysis purposes (e.g., habitat of a wildlife species), accounting for forested urban areas may be important. Therefore, some LC classifications, while not necessarily inaccurate, may be inadequate for some purposes.

16.4.2.2 Comparing Photographic Classifications

Urban LU classes, as identified from aerial photographs, all had some amount of forest and impervious cover (Table 16.4). This comparison again illustrated the importance of class definitions but raised the additional issue of class definitions based on LC vs. LU. In the case of LU, the diversity of cover types that made up residential areas was lumped together to map the LU type termed "residential." Cover types contained within urban LU regions included impervious surfaces, forest, and others (e.g., grasslands).

Table 16.4 The Percentage of Impervious Surface
          and Forested Cover within Three Urban
          Land-Use Classes

Land-Use Class	% Impervious	% Forest
High-density residential	36.1	15.4
Low-density residential	23.1	16.8
Other urban	45.4	19.1

16.4.3 Landscape Simulations

16.4.3.1 Ecotone Abruptness

The results of the analysis of metric sensitivity to the abruptness of ecotones suggested that existing landscape metrics were not as useful for quantifying ecotone abruptness as were the new metrics specifically designed for that purpose (Bowersox and Brown, 2001). Some metrics (e.g., total edge, maximum subgraph length) were not sensitive to abruptness. Those sensitive to abruptness were also sensitive to landscape patchiness, which confounded their interpretation (i.e., numbers of boundary elements and subgraphs, area-weight mean shape index). The new metrics, dispersion of boundary elements and cumulative boundary elements, were most consistently related to abruptness while not exhibiting the confounding effects of sensitivity to patchiness. There was not a clear indication that patch- or boundary-based metrics were more or less sensitive to abruptness.

16.4.3.2 Forest Fragmentation

Using simulated landscapes, each of several patch-based metrics exhibited a significant degree of variation when calculated at different levels of percent forested (Figure 16.3, top panel). Edge density was clearly highest when the landscape was 50% forest and lowest when the landscape was either 100% or 0% forest. The largest patch index and the total core area index both increased with increasing percentage of forest. The number of patches decreased with increasing forest percentage, after an initial increase.

The number of patches exhibited the highest degree of variation across different simulation runs (Figure 16.3, top panel). The coefficient of variation across simulation runs varied at different levels of percentage forest, depending on the average value of the metric and its variance (Figure

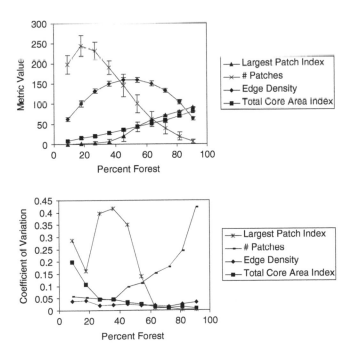

Figure 16.3 Top panel: The relationships between mean landscape pattern metric values across 10 simulations and the proportion of the landscape covered by forest. The error bars show the two times the standard deviation across the 10 runs. Bottom panel: The coefficient of variation of the metric values across simulations, indicating their relative errors.

16.3, bottom panel). The largest patch index and the number of patches exhibit the highest coefficient of variation across the runs, indicating a higher degree of relative error and lower precision. This finding was consistent with the redundant mapping work described above, and it highlights the relative instability of metrics that require patch delineation. Both the empirical and simulation work show that slight changes in the maps of a landscape, as the result of remote sensing image quality issues or just random perturbations, can result in relatively large variations in the number of patches identified and, as a result, in the mean patch size.

16.5 DISCUSSION

The results indicated the difficulty involved in distinguishing the effects of changes in the amount of forest from changes in the pattern of forest. The question is relevant in attempts to understand the effects of landscape structure on ecological processes. Some have argued that the concept of fragmentation is meant to include both the amount of forest and its spatial configuration (Forman, 1997). Others define fragmentation to mean a spatial pattern characteristic of the forest, independent of the effects of how much forest there is (Trzcinski et al., 1999). If the latter definition is used, then a measure of forest fragmentation that is not sensitive to the amount of forest is required. For example, do changes in the pattern of forest have impacts on ecological processes beyond the effects resulting from changes in the amount of forest? Trzcinski et al. (1999) dealt with this question by first evaluating the correlation between bird populations and forest amount, then correlating bird populations with the residuals that resulted from the regression of forest amount vs. forest pattern. The results indicated that there was little effect of forest pattern on bird populations independent of forest amount. However, more work is needed to understand the interactions of land-cover amount and pattern from both the perspective of how to measure pattern independently and how to understand its independent effects.

16.6 CONCLUSIONS

This chapter summarizes work on the precision and meaning of landscape pattern metrics derived from remote sensing. The transformations involved in calculating landscape metrics are complex, and analytical approaches to estimating their uncertainty are likely not to be practical. For that reason, this study has focused on two approaches to evaluating this propagation. First, we used redundant mapping of areas and evaluation of the variation in metric values derived from different imagery acquired near to each other in time. Second, simulation was used to explore the sensitivity of various metrics to differences in landscapes by controlling certain landscape characteristics.

We determined that uncertainty in input data propagates throughout the calculations and ultimately affects landscape metric precision. The precision of landscape metric values calculated to measure forest fragmentation is affected by the similarity in seasonal date of the imagery, atmospheric disturbances in the imagery (clouds and haze), and the amount of forest in the landscape. Metrics calculated for larger landscapes tend to exhibit less variation, but postprocessing of imagery (e.g., through seiving to remove small patches) did not result in increased precision. Landscape metrics whose calculation required more steps (e.g., image classification and patch delineation) were more likely to be susceptible to slight variations in the input data. Therefore, patch-based metrics (e.g., number of patches and mean patch size) tend to be less precise than boundary-based metrics.

Landscape class definitions, whether intentionally different or different because of the mapping method used, are important determinants of landscape pattern. It is possible to achieve significantly different landscape pattern metric values based on different class definitions. This suggests extreme caution should be used when attempting to compare pattern metric values for landscape maps derived from different sources and methods. For example, urban areas typically contain significant forest cover that is not represented in LU class definitions. Also, landscape metric values describing fragmentation of forest calculated from a LU/LC map were found to be different from those calculated from a classification specifically designed to map forest and nonforest. Applications that target the habitat quality for specific animal species, especially small animals, may not be well served by aggregated LU classes.

Landscape pattern metrics transform spatial data in complex ways and users need to exercise caution when interpreting the calculated values. Spatial simulation was a valuable tool for evaluating the behavior of landscape metrics and their sensitivity to various inputs. Ecotone abruptness can be detected using existing landscape metrics, but simulation illustrates that new metrics that measure the variation in boundary element locations are more sensitive to abruptness than existing metrics. Most measures of spatial pattern are also sensitive to both the composition and configuration of the landscape. More work is needed to evaluate the various influences of landscape configuration and composition on metric values.

The results presented here raise several issues for both the users and producers of remote sensing-based LU and LC products for landscape ecology investigation. First, in addition to the issue of data accuracy, the user is well advised to consider the appropriateness of class definitions for a specific application. For example, a general LU/LC classification may not be appropriate for calculation of landscape metrics in a study of habitat quality for a specific species. For this type of application, landscape maps may need to be developed specifically for the intended application. Second, the nature of the spatial transformations taken to compute pattern metrics can have dramatic implications for the precision of the estimated values. Metrics that require image classification and patch delineation are subject to greater imprecision than those based on local characterizations of pattern. Third, the meaning of metric values can be confounded and difficult to interpret. Applications of landscape metrics that seek empirical relationships between metric values and ecosystem characteristics may be able to bypass concerns about meaning and instead focus on correlations with ecosystem outcomes of interest (e.g., based on independent measurements of ecosystem characteristics). However, when directed toward spatial land management goals (e.g., a less fragmented forest), understanding the meaning of metrics is important to improve the probability of achieving the desired objectives.

16.7 SUMMARY

Landscape pattern metrics have been increasingly applied in support of environmental and ecological assessment for characterizing the spatial composition and configuration of landscapes to relate and evaluate ecological function. This chapter summarizes a combination of previously published and new work that investigates the precision and meaning of spatial landscape pattern metrics. The work was conducted on landscapes of the upper midwestern U.S. using satellite images, aerial photographs, and simulated landscapes. By applying a redundant mapping approach, we assessed and compared the degree of precision in the values of landscape metrics calculated over landscape subsets. While increasing landscape size had the effect of increasing precision in the landscape metric estimates, by giving up spatial resolution, postprocessing methods such as filtering and sieving did not have a consistent effect. Comparing multiple classifications of the same area that use different class definitions, we demonstrate that conclusions about landscape composition and configuration are affected by how the landscape classes are defined. Finally, using landscape simulation experiments, we demonstrate that metric sensitivity to a pattern characteristic of interest (e.g., ecotone abruptness of forest fragmentation) can be confounded by sensitivity to other landscape characteristics (e.g., landscape patchiness or amount of forest), making direct measurement of the desired characteristic difficult.

ACKNOWLEDGMENTS

NASA's Land Cover and Land Use Change Program, the National Science Foundation, the USGS's Global Change Program, and the USDA's Forest Service North Central Forest Experiment Station funded the research described in this chapter.

REFERENCES

Anderson, J.R., E.E. Hardy, J.T. Roach, and R.E. Witmer, A Land Use and Land Cover Classification System for Use with Remote Sensing Data, U.S. Geological Survey, professional paper 964, U.S. Government Printing Office, Washington, DC, 1976.

Bowersox, M.A. and D.G. Brown, Measuring the abruptness of patchy ecotones: a simulation-based comparison of patch and edge metrics, *Plant Ecol.*, 156, 89–103, 2001.

Brown, D.G., J.-D. Duh, and S. Drzyzga, Estimating error in an analysis of forest fragmentation change using North American Landscape Characterization (NALC) data, *Remote Sens. Environ.*, 71, 106–117, 2000a.

Brown, D.G., G.M. Jacquez, J.-D. Duh, and S. Maruca, Accuracy of remotely sensed estimates of landscape change using patch- and edge-based pattern statistics, in *Spatial Accuracy 2000*, Lemmens, M.J. et al., Eds., Delft University Press, Amsterdam, 2000b, pp. 75–82.

Forman, R.T.T., *Land Mosaics: The Ecology of Landscapes and Regions,* Cambridge University Press, New York, 1997.

Fortin, M.J. and P. Drapeau, Delineation of ecological boundaries: comparison of approaches and significance tests, *Oikos*, 72, 323–332, 1995.

Frohn, R.C., *Remote Sensing for Landscape Ecology: New Metric Indicators for Monitoring, Modeling, and Assessment of Ecosystems*, Lewis, Boca Raton, FL, 1997.

Guindon, B. and C.M. Edmonds, Using Classification Consistency in Inter-scene Overlap Regions to Model Spatial Variation in Land-Cover Accuracy over Large Geographic Regions, in Geospatial Data Accuracy Assessment, Lunetta, R.S. and J.G. Lyon, Eds., U.S. Environmental Protection Agency, report no. EPA/600/R-03/064, 2003.

Gustafson, E.J. and G.R. Parker, Relationships between land cover proportion and indices of landscape spatial pattern, *Landsc. Ecol.*, 7, 101–110, 1992.

Henebry, G.M. and D.G. Goodin, Landscape trajectory analysis: toward spatio-temporal models of biogeophysical fields for ecological forecasting, Workshop on Spatio-temporal Data Models for Biogeophysical Fields, La Jolla, CA, April 8–10, 2002, available at http://www.calmit.unl.edu/BDEI/papers/henebry_goodin_position.pdf.

Jacquez, G.M., S.L. Maruca, and M.J. Fortin, From fields to objects: a review of geographic boundary analysis, *J. Geogr. Syst.*, 2, 221–241, 2000.

Johnston, C.A. and J. Bonde, Quantitative analysis of ecotones using a geographic information system, *Photogram. Eng. Remote Sens.*, 55, 1643–1647, 1989.

Lunetta, R.S., J.G. Lyon, B. Guindon, and C.D. Elvidge, North American Landscape Characterization dataset development and data fusion issues, *Photogram. Eng. Remote Sens.*, 64, 821–829, 1998.

McGarigal, K. and B.J. Marks, FRAGSTATS: Spatial Pattern Analysis Program for Quantifying Landscape Structure, technical report PNW-GTR-351, USDA Forest Service, Pacific Northwest Research Station, Portland, OR, 1995.

Riitters, K.H., R.V. O'Neill, C.T. Hunsaker, J.D. Wickhan, D.H. Yankee, S.P. Timmins, K.B. Jones, and B.L. Jackson, A factor analysis of landscape pattern and structure metrics, *Landsc. Ecol.*, 10, 23–39, 1995.

SEMCOG (Southeastern Michigan Council of Governments), Land Use/Land Cover, Southeast Michigan, digital data product from SEMCOG, Detroit, MI, 1995.

Trzcinski, M.K., L. Fahrig, and G. Merriam, Independent effects of forest cover and fragmentation on the distribution of forest breeding birds, *Ecol. Appl.*, 9, 586–593, 1999.

Turner, M. G., R. H. Gardner, and R. V. O'Neill, *Landscape Ecology in Theory and Practice*, Springer-Verlag, New York, 2001.

Turner, M.G., R.V. O'Neill, R.H. Gardner, and B.T. Milne, Effects of changing spatial scale on the analysis of landscape pattern, *Landsc. Ecol.*, 3, 153–162, 1989.

USGS (United States Geological Survey). National Land Cover Data, 2001, product description available at http://landcover.usgs.gov/prodescription.html.

CHAPTER 17

Components of Agreement between Categorical Maps at Multiple Resolutions

R. Gil Pontius, Jr. and Beth Suedmeyer

CONTENTS

17.1 INTRODUCTION

17.1.1 Map Comparison

Map comparisons are fundamental in remote sensing and geospatial data analysis for a wide range of applications, including accuracy assessment, change detection, and simulation modeling. Common applications include the comparison of a reference map to one derived from a satellite image or a map of a real landscape to simulation model outputs. In either case, the map that is considered to have the highest accuracy is used to evaluate the map of questionable accuracy. Throughout this chapter, the term *reference map* refers to the map that is considered to have the highest accuracy and the term *comparison map* refers to the map that is compared to the reference map. Typically, one wants to identify similarities and differences between the reference map and the comparison map.

There are a variety of levels of sophistication by which to compare maps when they share a common categorical variable (Congalton, 1991; Congalton and Green, 1999). The simplest method is to compute the proportion of the landscape classified correctly. This method is an obvious first step; however, the proportion correct fails to inform the scientist of the most important ways in which the maps differ, and hence it fails to give the scientist information necessary to improve the comparison map. Thus, it would be helpful to have an analytical technique that budgets the sources of agreement and disagreement to know in what respects the comparison map is strong and weak. This chapter introduces map comparison techniques to determine agreement and disagreement between any two categorical maps based on the quantity and location of the cells in each category; these techniques apply to both hard and soft (i.e., fuzzy) classifications (Foody, 2002).

This chapter builds on recently published methods of map comparison and extends the concept to multiple resolutions (Pontius, 2000, 2002). A substantial additional contribution beyond previous methods is that the methods described in this chapter support stratified analysis. In general, these new techniques serve to facilitate the computation of several types of useful information from a generalized confusion matrix (Lewis and Brown, 2001). The following puzzle example illustrates the fundamental concepts of comparison of quantity and location.

17.1.2 Puzzle Example

Figure 17.1 shows a pair of maps containing two categories (i.e., light and dark). At the simplest level of analysis, we compute the proportion of cells that agree between the two maps. The agreement is 12/16 and the disagreement is 4/16. At a more sophisticated level, we can compute the disagreement in terms of two components: (1) disagreement due to quantity and (2) disagreement due to location. A disagreement of quantity is defined as a disagreement between the maps in terms of the quantity of a category. For example, the proportion of cells in the dark category in the comparison map is 10/16 and in the reference map is 12/16; therefore, there is a disagreement of 2/16. A disagreement of location is defined as a disagreement such that a swap of the location of a pair of cells within the comparison map increases overall agreement with the reference map. The disagreement of location is determined by the amount of spatial rearrangement possible in the comparison map, so that its agreement with the reference map is maximized. In this example, it would be possible to swap the #9 cell with the #3, #10, or #13 cell within the comparison map to increase its agreement with the reference map (Figure 17.1). Either of these is the only swap we

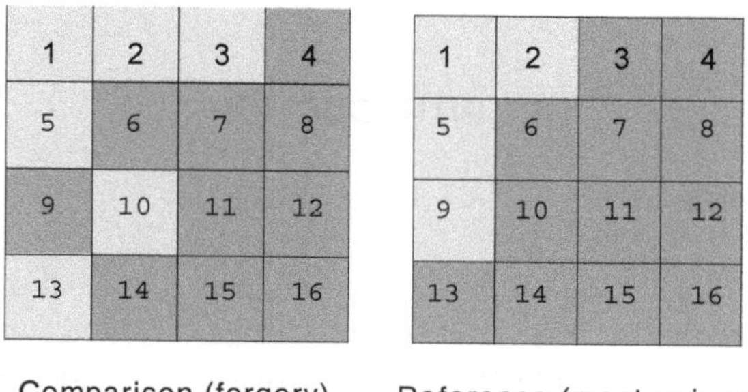

Comparison (forgery)   Reference (masterpiece)

Figure 17.1 Demonstration puzzle to illustrate agreement of location vs. agreement of quantity. Each map shows a categorical variable with two categories: dark and light. Numbers identify the individual grid cells.

can make to improve the agreement, given the quantity of the comparison map. Therefore, the disagreement of location is 2/16. The distinction between information of quantity and information of location is the foundation of this chapter's philosophy of map comparison.

It is worthwhile to consider in greater detail this concept of separation of information of quantity vs. information of location in map comparison before introducing the technical methodology of the analysis. The remainder of this introduction uses the puzzle example of Figure 17.1 to illustrate the concepts that the Methods section then formalizes in mathematical detail.

The following analogy is helpful to grasp the fundamental concept. Imagine that the reference map of Figure 17.1 is an original masterpiece that has been painted with two colors: light and dark. A forger would like to forge the masterpiece, but the only information that she knows for certain is that the masterpiece has exactly two colors: light and dark. Armed with partial information about the masterpiece (reference map), the forger must create a forgery (comparison map).

To create the forgery, the forger must answer two basic questions: What proportion of each color of paint should be used? Where should each color of paint be placed? The first question requires information of quantity and the second question requires information of location.

If the forger were to have perfect information about the quantity of each color of paint in the masterpiece, then she would use 4/16 light paint and 12/16 dark paint for the forgery, so that the proportion of each color in the forgery would match the proportion of each color in the masterpiece. The quantity of each color in the forgery must match the quantity of each color in the masterpiece in order to allow the potential agreement between the forgery and the masterpiece to be perfect. At the other extreme, if the forger were to have no information on the quantity of each color in the masterpiece, then she would select half light paint and half dark paint, since she would have no basis on which to treat either category differently from the other category. In the most likely case, the forger has a medium level of information, which is a level of information somewhere between no information and perfect information. Perhaps the forger would apply 6/16 light paint and 10/16 dark paint to the forgery, as in Figure 17.1.

Now, let us turn our attention to information of location. If the forger were to have perfect information about the location of each type of paint in the masterpiece, then she would place the paint of the forgery in the correct location as best as possible, such that the only disagreement between the forgery and the masterpiece would derive from error (if any) in the quantity of paint. If the forger were to have no information about the location of each color of paint in the masterpiece, then the she would spread each color of paint evenly across the canvas, such that each grid cell would be covered smoothly with light paint and dark paint. In the most likely case, the forger has a medium level of information of location about the masterpiece, so perhaps the forgery would have a pair of grid cells that are incorrect in terms of location, as in Figure 17.1. However, in the case of Figure 17.1, the error of location is not severe, since the error could be corrected by a swap of neighboring grid cells.

After the forger completes the forgery, we compare the forgery directly to the masterpiece in order to find the types and magnitudes of agreement between the two. There are two basic types of comparison, one based on information of quantity and another based on information of location. Each of the two types of comparisons leads to a different follow-up question.

First, we could ask, Given its medium level of information of quantity, how would the forgery appear if the forger would have had perfect information on location during the production of the forgery? For the example, in Figure 17.1, the answer is that the forger would have adjusted the forgery by swapping the location of cell #9 with cell #3, #10, or #13. As a result, the agreement between the adjusted forgery and the masterpiece would be 14/16, because perfect information on location would imply that the only error would be an error of quantity, which is 2/16.

Second, we could ask, Given its medium level of information of location, how would the forgery appear if the forger would have had perfect information of quantity during the production of the forgery? In this case, the answer is that the forger would have adjusted the forgery by using more dark paint and less light paint, but each type of paint would be in the same location as in Figure

17.1. Therefore, the adjusted forgery would appear similar to Figure 17.1; however, the light cells of Figure 17.1 would be a smooth mix of light and dark, while the dark cells would still be completely dark. Specifically, the light cells would be adjusted to be 2/3 light and 1/3 dark; hence, the total amount of light and dark paint in the forgery would equal the total amount of light and dark paint in the masterpiece. As a result, the agreement between the adjusted forgery and the masterpiece would be larger than 12/16. The exact agreement would require that we define the agreement between the light cells of the masterpiece and the partially light cells of the adjusted forgery.

The above analogy prepares the reader for the technical description of the analysis in the Methods section. In the analogy, the reference map is the masterpiece that represents the ground information, and the comparison map is the forgery that represents the classification of a remotely sensed image. The classification rule of the remotely sensed image represents the scientist's best attempt to replicate the ground information. In numerous conversations with our colleagues, we have found that it is essential to keep in mind the analogy of painting a forgery. We have derived all the equations in the Methods section based on the concepts of the analogy.

17.2 METHODS

17.2.1 Example Data

Categorical variables consisting of "forest" and "nonforest" are represented in three maps of example data (Figure 17.2). Each map is a grid of 12×12 cells. The 100 nonwhite cells represent the study area and the remaining 44 white cells are located out of the study area. We have purposely made a nonsquare study area to demonstrate the generalized properties of the methods. The methods apply to a collection of any cells within a grid, even if those cells are not contiguous, as is typically the case in accuracy assessment. Each map has the same nested stratification structure. The coarser stratification consists of two strata (i.e., north and south halves) separated by the thick solid line. The finer stratification consists of four substrata quadrates of 25 cells each, defined as the northeast (NE), northwest (NW), southeast (SE), and southwest (SW). The set of three maps illustrates the common characteristics encountered when comparing map classification rules. Imagine that Figure 17.2 represents the output maps from a standard classification rule (COM1), alternative classification rule (COM2), and the reference data (REF). Typically, a statistical test would be applied to assess the relative performance of the two classification approaches and to determine important differences with respect to the reference data. However, it would also be helpful if such a comparison would offer additional insights concerning the sources of agreement and disagreement.

Table 17.1a and Table 17.1b represent the standard confusion matrix for the comparison of COM1 and COM2 vs. REF. The agreement in Table 17.1a and Table 17.1b is 70% and 78%, respectively. Note that the classification in COM2 is identical to the reference data in the south stratum. In the north stratum, COM2 is the mirror image of REF reflected through the central vertical axis. Therefore, the proportion of forest in COM2 is identical to that in REF in both the north and south strata. For the entire study area, REF is 45% forest, as is COM2. COM1 is 47% forest. A standard accuracy assessment ends with the confusion matrices of Table 17.1.

17.2.2 Data Requirements and Notation

We have designed COM1, COM2, and REF to illustrate important statistical concepts. However, this chapter's statistical techniques apply to cases that are more general than the sample data of Figure 17.2. In fact, the techniques can compare any two maps of grid cells that are classified as any combination of soft or hard categories.

This means that each grid cell can have some membership in each category, ranging from no membership (0) to complete membership (1). The membership is the proportion of the cell that

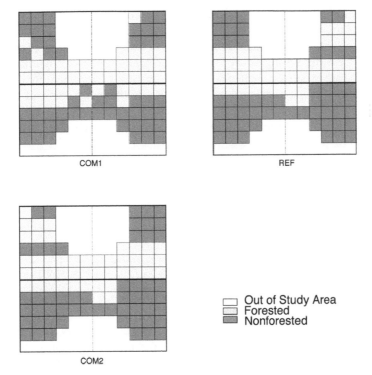

Figure 17.2 Three maps of example data.

Table 17.1a Confusion Matrix for COM1 vs. Reference

		Reference Map		
		Forest	Nonforest	Total
Comparison Map	Forest	31	16	47
	Nonforest	14	39	53
	Total	45	55	100

Table 17.1b Confusion Matrix for COM2 vs. Reference

		Reference Map		
		Forest	Nonforest	Total
Comparison Map	Forest	34	11	45
	Nonforest	11	44	55
	Total	45	55	100

belongs to a particular category; therefore, the sum of the membership values over all categories is 1. In addition, each grid cell has a weight to denote its membership in any particular stratum, where the stratum weight can also range from 0 to 1. The weights do not necessarily need to sum to 1. For example, if a cell's weights are 0 for all strata, then that cell is eliminated from the analysis. These ideas are expressed mathematically in Equation 17.1 through Equation 17.4, where j is the category index, J is the number of categories, R_{dnj} is the membership of category j in cell n of stratum d of the reference map, S_{dnj} is the membership of category j in cell n of stratum d of the comparison map, and W_{dn} is the weight for the membership of cell n in stratum d:

$$0 \le R_{dnj} \le 1 \tag{17.1}$$

$$0 \le S_{dnj} \le 1 \tag{17.2}$$

$$\sum_{j=1}^{J} R_{dnj} = \sum_{j=1}^{J} S_{dnj} = 1 \tag{17.3}$$

$$0 \le W_{dn} \le 1 \tag{17.4}$$

Just as each cell has some proportional membership to each category, each stratum has some proportional membership to each category. We define the membership of each stratum to each category as the proportion of the stratum that is covered by that category. For each stratum, we compute this membership to each category as the weighted proportion of the cells that belong to that category. Similarly, the entire landscape has membership to each particular category, where the membership is the proportion of the landscape that is covered by that category. We compute the landscape-level membership by taking the weighted proportion over all grid cells. Equation 17.5 through Equation 17.9 show how to compute these levels of membership for every category at both the stratum scale and the landscape scale. These equations utilize standard dot notation to denote summations, where N_d denotes the number of cells that have some positive membership in stratum d of the map and D denotes the number of strata. Equation 17.5 shows that $W_{d.}$ denotes the sum of the cell weights for stratum d. Equation 17.6 shows that $R_{d.j}$ denotes the proportion of category j in stratum d of the reference map. Equation 17.7 shows that $R_{..j}$ denotes the proportion of category j in the entire reference map. Equation 17.8 shows that $S_{d.j}$ denotes the proportion of category j in stratum d of the comparison map. Equation 17.9 shows that $S_{..j}$ denotes the proportion of category j in the entire comparison map:

$$W_{d.} = \sum_{n=1}^{N_d} W_{dn} \tag{17.5}$$

$$R_{d \cdot j} = \frac{\sum_{n=1}^{N_d} \left(W_{dn} * R_{dnj} \right)}{W_{d.}} \tag{17.6}$$

$$R_{\cdot \cdot j} = \frac{\sum_{d=1}^{D} \sum_{n=1}^{N_d} \left(W_{dn} * R_{dnj} \right)}{\sum_{d=1}^{D} W_{d.}} \tag{17.7}$$

$$S_{d \cdot j} = \frac{\sum\limits_{n=1}^{N_d} \left(W_{dn} * S_{dnj} \right)}{W_{d \cdot}} \tag{17.8}$$

$$S_{\cdot \cdot j} = \frac{\sum\limits_{d=1}^{D} \sum\limits_{n=1}^{N_d} \left(W_{dn} * S_{dnj} \right)}{\sum\limits_{d=1}^{D} W_{d \cdot}} \tag{17.9}$$

17.2.3 Minimum Function

The Minimum function gives the agreement between a cell of the reference map and a cell of the comparison map. Specifically, Equation 17.10 gives the agreement in terms of proportion correct between the reference map and the comparison map for cell n of stratum d. Equation 17.11 gives the landscape-scale agreement weighted appropriately with grid cell weights, where $M(\mathbf{m})$ denotes the proportion correct between the reference map and the comparison map:

$$\text{agreement in cell } n \text{ of stratum } d = \sum_{j=1}^{J} MIN \left(R_{dnj}, S_{dnj} \right) \tag{17.10}$$

$$M(\mathbf{m}) = \frac{\sum\limits_{d=1}^{D} \sum\limits_{n=1}^{N_d} W_{dn} \left[\sum\limits_{j=1}^{J} MIN(R_{dnj}, \ S_{dnj}) \right]}{\sum\limits_{d=1}^{D} \sum\limits_{n=1}^{N_d} W_{dn}} \tag{17.11}$$

The Minimum function expresses agreement between two cells in a generalized way because it works for both hard and soft classifications. In the case of hard classification, the agreement is either 0 or 1, which is consistent with the conventional definition of agreement for hard classification. In the case of soft classification, the agreement is the sum over all categories of the minimum membership in each category. The minimum operator makes sense because the agreement for each category is the smaller of the membership in the reference map and the membership in the comparison map for the given category. If the two cells are identical, then the agreement is 1.

17.2.4 Agreement Expressions and Information Components

Figure 17.3 gives the 15 mathematical expressions that lay the foundation of our philosophy of map comparison. The central expression, denoted $M(\mathbf{m})$, is the agreement between the reference map and the comparison map, given by Equation 17.11. The other 14 mathematical expressions show the agreement between the reference map and an "other" map that has a specific combination of information. The first argument in each Minimum function (e.g., R_{dnj}) denotes the cells of the reference map and the second argument in each Minimum function (e.g., S_{dnj}) denotes the cells of the other map. The components of information in the other maps are grouped into two orthogonal concepts: (1) information of quantity and (2) information of location.

There are three levels of information of quantity no, medium, and perfect, denoted, respectively, as $\mathbf{n}$, $\mathbf{m}$, and $\mathbf{p}$. For the five mathematical expressions in the "no information of quantity" column,

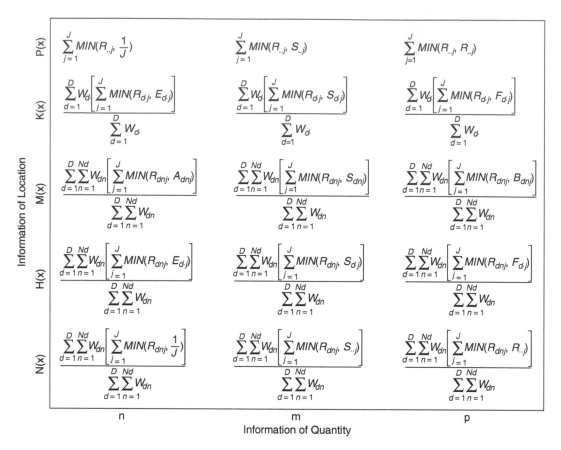

Figure 17.3 Expressions for 15 points defined by a combination of the information of quantity and location. The vertical axis shows information of location and the horizontal axis shows information of quantity. The text defines the variables.

the other maps are derived from an adjustment to the comparison map, such that the proportion of membership for each of the J categories is $1/J$ in the other maps (Foody, 1992). This adjustment is necessary to answer the question, What would be the agreement between the reference map and the comparison map, if the scientist who created the comparison map would have had no information of quantity during its production? The adjustment holds the level of information of location constant while adjusting each grid cell such that the quantity of each of the J categories in the landscape is $1/J$.

Equations 17.12 and 17.13 give the necessary adjustment to each grid cell in order to scale the comparison map to express no information of quantity:

$$A_{dnj} = S_{dnj}\left(\frac{1/J}{S_{\cdot\cdot j}}\right), \qquad \text{if } 1/J \le S_{\cdot\cdot j} \tag{17.12}$$

$$= 1 - (1 - S_{dnj})\left(\frac{S_{\cdot\cdot j}}{1/J}\right), \quad \text{else}$$

$$E_{d\cdot j} = S_{d\cdot j}\left(\frac{1/J}{S_{\cdot\cdot j}}\right), \qquad \text{if } 1/J \le S_{\cdot\cdot j} \tag{17.13}$$

$$= 1 - (1 - S_{d \cdot j})\left(\frac{S_{\cdot \cdot j}}{1/J}\right), \text{ else}$$

Equation 17.12 performs the scaling at the grid cell level, and hence creates an "other" map, denoted A_{dnj}. Equation 17.13 performs the scaling at the stratum level, and hence creates an "other" map, denoted $E_{d \cdot j}$.

The logic of the scaling is as follows, where the word "paint" can be substituted for the word "category" to continue the painting analogy. If the quantity of category j in the comparison map is less than $1/J$, then more of category j must be added to the comparison map. In this case, category j is increased in cells that are not already 100% members of category j. If the quantity of category j in the comparison map is more than $1/J$, then some of category j must be removed from the comparison map. In that case, category j is decreased in cells that have some of category j.

For expressions in the "medium information" column of Figure 17.3, the other maps have the same quantities as the comparison map. For the expressions in the "perfect information" column, the other maps are derived such that the proportion of membership for each of the J categories matches perfectly with the proportions in the reference map. This adjustment is necessary to answer the question, What would be the agreement between the reference map and the comparison map, if the scientist would have had perfect information of quantity during the production of the comparison map? The adjustment holds the level of information of location constant while adjusting each grid cell such that the quantity of each of the J categories in the landscape matches the quantities in the reference map. The logic of the adjustment is similar to the scaling procedure described for the other maps in the "no information of quantity" column of Figure 17.3.

Equation 17.14 and Equation 17.15 give the necessary mathematical adjustments to scale the comparison map to express perfect information of quantity:

$$B_{dnj} = S_{dnj}\left(\frac{R_{\cdot \cdot j}}{S_{\cdot \cdot j}}\right), \qquad \text{if } R_{\cdot \cdot j} \leq S_{\cdot \cdot j} \tag{17.14}$$

$$= 1 - (1 - S_{dnj})\left(\frac{S_{\cdot \cdot j}}{R_{\cdot \cdot j}}\right), \text{ else}$$

$$F_{d \cdot j} = S_{d \cdot j}\left(\frac{R_{\cdot j}}{S_{\cdot j}}\right), \qquad \text{if } R_{\cdot \cdot j} \leq S_{\cdot \cdot j} \tag{17.15}$$

$$= 1 - (1 - S_{d \cdot j})\left(\frac{S_{\cdot \cdot j}}{R_{\cdot \cdot j}}\right), \text{ else}$$

Equation 17.14 performs this scaling at the grid cell level, and hence creates an "other" map, denoted B_{dnj}. Equation 17.15 performs this scaling at the stratum level, and hence creates an "other" map, denoted $F_{d \cdot j}$.

There are five levels of information of location: no, stratum, medium, perfect within stratum, and perfect, denoted, respectively, as $N(x)$, $H(x)$, $M(x)$, $K(x)$ and $P(x)$. Figure 17.3 shows the differences in the 15 mathematical expressions among these various levels of information of location. In $N(x)$, $H(x)$, and $M(x)$ rows, the mathematical expressions of Figure 17.3 consider the reference map at the grid cell level, as indicated by the use of all three subscripts: d, n, and j. In the $K(x)$ row, the mathematical expressions consider the reference map at the stratum level, as indicated by the use of two subscripts: d and j. In the $P(x)$ row, the expressions consider the reference map at the study area level, as indicated by the use of one subscript: j. In the $M(x)$ row, the expressions consider the other maps at the grid cell level, as indicated by the use of all three subscripts: d, n, and j. In the H($\mathbf{x}$) and K($\mathbf{x}$) rows, the expressions consider the other maps at the stratum level, as indicated by the use of two subscripts: d and j. In the N($\mathbf{x}$) and P($\mathbf{x}$) rows, the expressions consider the other maps at the study area level, as indicated by the use of one subscript: j.

The concepts behind these combinations of components of information of location are as follows. In row N($\mathbf{x}$), the categories of the other maps are spread evenly across the landscape, such that every grid cell has an identical multinomial distribution of categories. In row H($\mathbf{x}$), the categories of the other maps are spread evenly within each stratum, such that every grid cell in each stratum has an identical multinomial distribution of categories. In row M($\mathbf{x}$), the grid cell level information of location in the other maps is the same as in the comparison map. In row K($\mathbf{x}$), the other maps derive from the comparison map, whereby the locations of the categories in the comparison map are swapped within each stratum in order to match as best as possible the reference map; however, this swapping of grid cell locations does not occur across stratum boundaries. In row P($\mathbf{x}$), the other maps derive from the comparison map, whereby the locations of the categories in the comparison map are swapped in order to match as best as possible the reference map, and this swapping of grid cell locations can occur across stratum boundaries.

Each of the 15 mathematical expressions of Figure 17.3 is denoted by its location in the table. The $\mathbf{x}$ denotes the level of information of quantity. For example, the overall agreement between the reference map and the comparison map is denoted M($\mathbf{m}$), since the comparison map has a medium level of information of quantity and a medium level of information of location, by definition. The expression P($\mathbf{p}$) is in the upper right of Figure 17.3 and is always equal to 1, because P($\mathbf{p}$) is the agreement between the reference map and the other map that has perfect information of quantity and perfect information of location.

There are seven mathematical expressions that are especially interesting and helpful. They are N($\mathbf{n}$), N($\mathbf{m}$), H($\mathbf{m}$), M($\mathbf{m}$), K($\mathbf{m}$), P($\mathbf{m}$), and P($\mathbf{p}$). For N($\mathbf{n}$), each cell of the other map is the same and has a membership in each category equal to $1/J$. For N($\mathbf{m}$), each cell of the other map is the same and has a membership in each category equal to the proportion of that category in the comparison map. For H($\mathbf{m}$), each cell within each stratum of the other map is the same and has a membership in each category equal to the proportion of that category in each stratum of the comparison map. For M($\mathbf{m}$), the other map is the comparison map. For K($\mathbf{m}$), the other map is the comparison map with the locations of the grid cells swapped within each stratum, so as to have the maximum possible agreement with the reference map within each stratum. For P($\mathbf{m}$), the other map is the comparison map with the locations of the grid cells swapped anywhere within the map, so as to have the maximum possible agreement with the reference map. For P($\mathbf{p}$), the other map is the reference map, and therefore the agreement is perfect.

17.2.5 Agreement and Disagreement

The seven mathematical expressions N($\mathbf{n}$), N($\mathbf{m}$), H($\mathbf{m}$), M($\mathbf{m}$), K($\mathbf{m}$), P($\mathbf{m}$), and P($\mathbf{p}$) constitute a sequence of measures of agreement between the reference map and other maps that have increasingly accurate information. Therefore, usually $0 <$ N($\mathbf{n}$) $<$ N($\mathbf{m}$) $<$ H($\mathbf{m}$) $<$ M($\mathbf{m}$) $<$ K($\mathbf{m}$) $<$ P($\mathbf{m}$) $<$ P($\mathbf{p}$) $= 1$. This sequence partitions the interval $[0,1]$ into components of the agreement between the reference map and the comparison map. M($\mathbf{m}$) is the total proportion correct, and $1 -$ M($\mathbf{m}$) is the total proportion error between the reference map and the comparison map. Hence, the sequence of N($\mathbf{n}$), N($\mathbf{m}$), H($\mathbf{m}$), and M($\mathbf{m}$) defines components of agreement, and the sequence of M($\mathbf{m}$), K($\mathbf{m}$), P($\mathbf{m}$), and P($\mathbf{p}$) defines components of disagreement.

Table 17.2 defines these components mathematically. Beginning at the bottom of the table and working up, the first component is agreement due to chance, which is usually N($\mathbf{n}$). However, if the agreement between the reference map and the comparison map is less than would be expected by chance, then the component of agreement due to chance may be less than N($\mathbf{n}$). Therefore, Table 17.2 defines the component of agreement due to chance as the minimum of N($\mathbf{n}$), N($\mathbf{m}$), H($\mathbf{m}$),

Table 17.2 Definition and Values of Seven Components of Agreement for COM1 vs. Reference Derived from the Mathematical Expressions of Figure 17.3

Name of Component	Definition	Percentage of Each Component	
		Stratum	Substratum
Disagreement due to quantity	$P(\mathbf{p}) - P(\mathbf{m})$	2.0	2.0
Disagreement at stratum level	$P(\mathbf{m}) - K(\mathbf{m})$	8.0	8.0
Disagreement at grid cell level	$K(\mathbf{m}) - M(\mathbf{m})$	20.0	20.0
Agreement at grid cell level	$MAX[M(\mathbf{m}) - H(\mathbf{m}), 0]$	12.2	11.5
Agreement at stratum level	If $MIN[N(\mathbf{m}), H(\mathbf{m}), M(\mathbf{m})] = N(\mathbf{m})$, then $MIN[H(\mathbf{m}) - N(\mathbf{m}), M(\mathbf{m}) - N(\mathbf{m})]$, else 0	7.5	8.2
Agreement due to quantity	If $MIN[N(\mathbf{n}), N(\mathbf{m}), H(\mathbf{m}), M(\mathbf{m})] = N(\mathbf{n})$, then $MIN[N(\mathbf{m}) - N(\mathbf{n}), H(\mathbf{m}) - N(\mathbf{n}), M(\mathbf{m}) - N(\mathbf{n})]$, else 0	0.3	0.3
Agreement due to chance	$MIN[N(\mathbf{n}), N(\mathbf{m}), H(\mathbf{m}), M(\mathbf{m})]$	50.0	50.0

and $M(\mathbf{m})$. The component of agreement due to quantity is usually $N(\mathbf{m}) - N(\mathbf{n})$; Table 17.2 gives a more general definition to account for the possibility that the comparison map's information of quantity can be worse than no information of quantity. The component of agreement at the stratum level is usually $H(\mathbf{m}) - N(\mathbf{m})$; Table 17.2 gives a more general definition to restrict this component of agreement to be nonnegative. Similarly, the component of agreement at the grid cell level is usually $M(\mathbf{m}) - H(\mathbf{m})$; Table 17.2 restricts this component of agreement to be nonnegative. Table 17.2 also defines the components of disagreement. It is a mathematical fact that $M(\mathbf{m}) \leq K(\mathbf{m}) \leq P(\mathbf{m}) \leq P(\mathbf{p})$; therefore, the components of disagreement are the simple definitions of Table 17.2.

The partition of the components of agreement can be performed for any stratification structure. Table 17.2 shows the results for the comparison of REF and COM1 at both the stratum level and the substratum level. Figure 17.4 shows this information in graphical form. The stratum bar shows the components at the stratum level and the substratum bar shows the components at the substratum level. Since the substrata are nested within the strata, it makes sense to overlay the stratum bar on

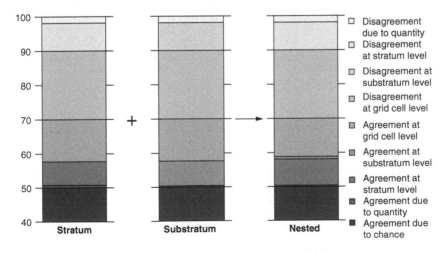

Figure 17.4 Stacked bars showing components of agreement between COM1 and REF. The vertical axis shows the cumulative percentage of cells in the study area. The nested bar is the stratum bar overlaid on top of the substratum bar to show agreement at both the stratum and substratum levels. Table 17.2 gives the numerical values for the components in the stratum and substratum bars.

top of the substratum bar to produce the nested bar. Depending on the nature of the maps, the nested bar could show nine possible components listed in the legend. In the comparison of REF and COM1, the bar shows eight nested components.

17.2.6 Multiple Resolutions

Up to this point, our analysis of the maps of Figure 17.2 has been based on a cell-by-cell analysis with hard classification. The advantage of cell-by-cell analysis with hard classification is its simplicity. The disadvantage of cell-by-cell analysis with hard classification is that if a specific cell fails to have the correct category, then it is counted as complete error, even when the correct category is found in a neighboring cell. Therefore, cell-by-cell analysis can fail to indicate general agreement of pattern because it fails to consider spatial proximity to agreement. In order to remedy this problem we perform multiple resolution analysis.

The multiple resolution analysis requires a new set of maps for each new resolution. Figure 17.2 shows maps that are hard classified, whereas Figure 17.5 shows the COM1 map at four coarser resolutions. Each cell of each map of Figure 17.5 is an average of neighboring cells of the original COM1 map of Figure 17.2. For example, for resolution 2, four neighboring cells become a single coarse cell; therefore, the 12×12 map of original cells yields a 6×6 map. At resolution three, we obtain a 4×4 map of coarse cells, in which the length of the side of each coarse cell is three times the length of the side of each original fine-resolution cell. At resolution four, we obtain a 2×2 map of coarse cells, where each coarse cell is its own substratum. At resolution 12, the entire map is in one cell. For each coarse cell, the membership in each category is the average of the memberships of the contributing cells. When using this aggregation technique, the lack of a square study area can result in an unequal number of fine-resolution cells in each of the coarser cells. This

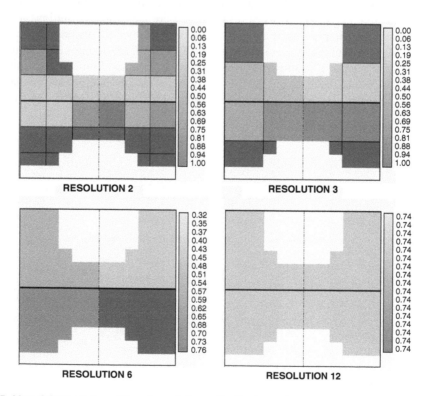

Figure 17.5 Map COM1 at four different resolutions. On the legend, 0 means completely forest, 1 means completely nonforest, and white is outside of the study area.

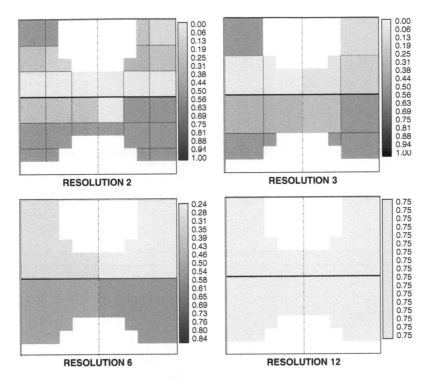

Figure 17.6 Map REF at four different resolutions. On the legend, 0 means completely forest, 1 means completely nonforest, and white is outside of the study area.

is taken into consideration by the weights that give each cell's membership in the study area. This characteristic of the technique allows the method to apply to accuracy assessment where the grid cells of interest are not contiguous.

Figure 17.5 shows the cell configuration. The darker shading shows stronger membership in the nonforested category. Figure 17.6 shows this same type of aggregation for the REF map. For each resolution, we are able to generate a bar similar to the nested bar of Figure 17.4, because the equations of Figure 17.3 allow for any cell to have partial membership in any category.

17.3 RESULTS

Figure 17.7 shows the components of agreement and disagreement between REF and COM1 at all resolutions. Figure 17.8 shows analogous results for the comparison between REF and COM2. The overall proportion correct is the top of the component of agreement at the grid cell level and the overall proportion correct at the coarser resolutions is the top of the component of agreement due to quantity. Proportion correct tends to rise as resolution becomes coarser; however, the rise is not monotonic. Proportion correct rises for each resolution that is nested within finer resolutions. That is, the proportion correct for resolution 1 < proportion correct for resolution 2 < proportion correct for resolution 6 < proportion correct for resolution 12. In addition, the proportion correct for resolution 1 < proportion correct for resolution 3 < proportion correct for resolution 6 < proportion correct for resolution 12. However, the proportion correct for resolution 2 > proportion correct for resolution 3. Note that resolution 2 is not nested within resolution 3.

The largest component is agreement due to chance, which is 50% at the finest resolution since there are two categories. Agreement due to chance rises as resolution becomes coarser. Besides the component due to chance, the largest components at the finest resolution are agreement at the grid

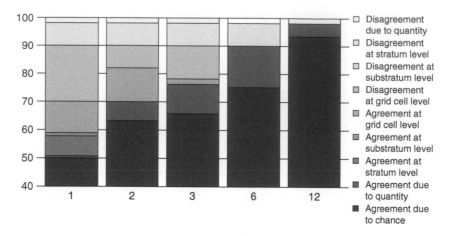

Figure 17.7 Stacked bars showing agreement between COM1 and REF. The vertical axis shows the cumulative percentage of the total study area. The numbers on the horizontal axis give the resolutions.

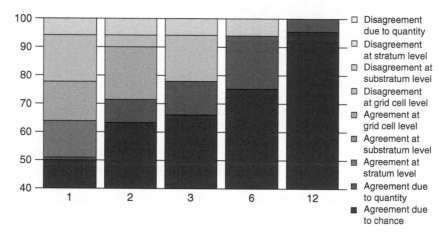

Figure 17.8 Stacked bars showing agreement between COM2 and REF. The vertical axis shows the cumulative percentage of the total study area. The numbers on the horizontal axis give the resolutions.

cell level and disagreement at the grid cell level. As resolution becomes coarser, the grid cell level information becomes less important, relative to information of quantity. At the coarsest resolution, where the entire study area is in one cell, the concept of location has no meaning; hence, the only components are agreement due to chance, agreement due to quantity, and disagreement due to quantity. COM1 has a component of disagreement due to quantity, which does not change as resolution changes, since quantity is a concept independent of resolution. COM2 has no disagreement in quantity.

Figure 17.9 shows that at a fine resolution the agreement between COM2 and REF is greater than the agreement between COM1 and REF. The components that account for the greater agreement are the agreement at the stratum level and at the grid cell level.

Table 17.3 and Table 17.4 display contingency tables that show the nested stratification structure of strata and substrata. These tables are another helpful way to present results. The information on the diagonal indicates the number of cells for each substratum that are in agreement. Therefore, the number of correct cells may be calculated for each substratum by summing the diagonal for each subset of the table. Furthermore, the row and column totals indicate stratum-level agreement. For example, Table 17.3 shows disagreement at the stratum level, since there are 31 forested cells in COM1 vs. 35 in REF for the north stratum and there are 16 forested cells in COM1 vs. 10 in

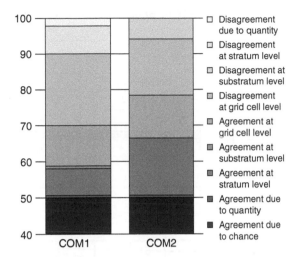

Figure 17.9 Stacked bars showing comparison of COM1 and COM2 with the REF map at the finest resolution. The vertical axis shows the cumulative percentage of the total study area.

Table 17.3 Confusion Matrix for COM1 vs. REF by Strata and Substrata; F Denotes Forest Cells and N Denotes Nonforest Cells

				REF											
				Substratum											
				NW		NE		SW		SE		Total			
				F	N	F	N	F	N	F	N				
COM1	Substratum	NW	F	13	1							14	31	F	North	Stratum
			N	3	8							11				
		NE	F			12	5					17	19	N		
			N			7	1					8				
		SW	F					5	5			10	16	F	South	
			N					1	14			15				
		SE	F							1	5	6	34	N		
			N							3	16	19				
		Total		16	9	19	6	6	19	4	21	100				
				35		15		10		40						
				F		N		F		N						
				North				South								
				Stratum											

Table 17.4 Confusion Matrix for COM2 vs. REF by Strata and Substrata; F Denotes Forest Cells and N Denotes Nonforest Cells

COM2 Substratum			REF Substratum										Stratum		
			NW		NE		SW		SE		Total				
			F	N	F	N	F	N	F	N					
NW	F		12	7							19	35	F	North	Stratum
	N		4	2							6				
NE	F				12	4					16	15	N		
	N				7	2					9				
SW	F						6	0			6	10	F	South	
	N						0	19			19				
SE	F								4	0	4	40	N		
	N								0	21	21				
Total			16	9	19	6	6	19	4	21	100				
			35		15		10		40						
			F		N		F		N						
			North				South								
			Stratum												

REF for the south stratum. This disagreement at the stratum level is reflected in the component of disagreement at the stratum level in Figure 17.7. In contrast, Table 17.4 shows perfect agreement at the stratum level; hence, Figure 17.8 shows no component of disagreement between COM2 and REF at the stratum level.

17.4 DISCUSSION

17.4.1 Common Applications

The three maps in Figure 17.2 represent a common situation in map comparison analysis. There are many applications where a scientist wants to know which of two maps is more similar to a reference map. Three likely applications are in remote sensing, simulation modeling, and land-change analysis.

In remote sensing, when a scientist develops a new classification rule, the scientist needs to compare the map generated by the new rule to the map generated by a standard rule. Two fundamental questions are (1) Did the new method perform better than the standard method concerning its estimate of the quantity of each category? and (2) Did the new method perform better than the standard method concerning its specification of the location of each category? The format of Figure 17.9 is an effective way to display the results, because it conveys the answer to both of these questions quickly. Specifically, COM1 makes some error of quantity while COM2 does not. COM2 shows a better specification of location than does COM1. When the analysis is stratified, the scientist can see whether errors of location exist at the stratum level, substratum level, or grid cell level. For example, Figure 17.9 shows that COM1 has errors at the stratum level and not the substratum level, while COM2 has errors at the substratum level and not at the stratum level. In another application from remote sensing, we could examine the influence of a hardening rule, since the techniques work for both hard and soft classification. For example, COM1 could show the soft category membership before a hardening rule is applied, and COM2 could show the hard category membership after the hardening rule is applied. The format of Figure 17.9 would then summarize the influence of the hardening rule.

In simulation modeling, a scientist commonly builds a model to predict how land changes over time (Veldkamp and Lambin, 2001). The scientist performs validation to see how the model performs and to obtain ideas for how to improve the model. When the model is run from T_1 to T_2, the scientist validates the model by comparing the simulated landscape of T_2 with a reference map T_2. A null model would predict no change between T_1 and T_2. In other words, if the scientist had no simulation model, then the best guess at the T_2 map would be the T_1 map. Therefore, to see whether the simulation model is performing better than a null model, the scientist needs to compare (1) the agreement between the T_2 simulation map and the T_2 reference map vs. (2) the agreement between the T_1 map and the T_2 reference map. In this situation, the format of Figure 17.9 is perfectly suited to address this question because the analogy is that COM1 is the T_1 map, COM2 is the T_2 simulation map, and REF is the T_2 reference map.

The methods described here are particularly helpful in this case since land-cover and land-use (LCLU) change models are typically stratified according to political units because data are typically available by political unit and because the process of land change often happens by political unit. For example, land-use activities in Brazil are planned at the regional and household scales, where the household stratification is nested within the regional stratification. Researchers are dedicating substantial effort to collecting data at a relevant scale in order to calibrate and to improve change models. Therefore, it is essential that statistical methods budget the components of agreement and disagreement at relevant scales, because researchers want to collect new data at the scale at which the most uncertainty exists.

In land-change analysis, the scientist wants to know the manner in which land categories change and persist over time. For this application, the methods of this chapter would use COM1 as the T_1 map and REF as the T_2 map. Figure 17.7 would supply a multiple-resolution analysis of LCLU change, where agreement means persistence and disagreement means change. A disagreement in quantity indicates that a category has experienced either a net gain or a net loss. Disagreement at the stratum level means that a loss of a category in one stratum is accompanied by gain in that category in another stratum. Disagreement at the grid cell level means that a loss of a category at one location is accompanied by a gain of that category at another location within the same stratum. Therefore, Figure 17.7 would show at what scales LCLU change occurs.

17.4.2 Quantity Information

We focus primarily on the center column of mathematical expressions of Figure 17.3, because those expressions give the components of agreement. However, the other two columns can be particularly helpful depending on the purpose of map comparison. In the case of remote sensing, guidance is needed to improve the classification rules. For simulation modeling, guidance is required to improve the simulation model's rules. It would be helpful to know the expected improvement if the rule's specification of quantity changes, given a specific level of information of location. The mathematical expressions in the rightmost column of Figure 17.3 show the expected results when the rule specifies the quantity of each category perfectly with respect to the reference quantities. At the other end of the spectrum, the mathematical equations of the leftmost column show the expected results when the rule uses random chance to specify the quantities of each category.

For example, M(**n**) expresses the agreement that a scientist would expect between the reference map and the other map when the other map is the adjusted comparison map that is scaled to show the quantity in each category as $1/J$. M(**p**) expresses the agreement that a scientist would expect between the reference map and the other map when the other map is the adjusted comparison map that is scaled to show the quantity in each category as matching perfectly the reference map.

The definitions of M(**n**) and M(**p**) in Table 17.2 are slightly different from the definitions of M(**n**) and M(**p**) in Pontius (2000, 2002). Table 17.2 gives expressions for M(**n**) and M(**p**) that depend on the scaling given by Equation 17.12 through Equation 17.15. The method of scaling simulates the change in quantity spread evenly across the grid cells as one moves from M(**m**) to M(**p**) or from M(**m**) to M(**n**). In contrast, Pontius (2000, 2002) does not scale the comparison map and does not represent an even spread of the change in quantity across the cells. The methods of Pontius (2000, 2002) define M(**n**) and M(**p**) in a manner that makes sense for applications of land-cover change simulation modeling and slightly confounds information of quantity with information of location. Table 17.2 defines M(**n**) and M(**p**) in a manner that is appropriate for a wider variety of applications, since it maintains complete separation of information of quantity from information of location.

17.4.3 Stratification and Multiple Resolutions

If we think of grid cells as tiny strata, then the maps of Figure 17.2 show a three-tiered, nested stratification structure. The cells are 100 tiny strata that are nested within the four substrata that are nested within the two broadest strata. The multiple-resolution procedure grows the cells such that at the resolution of 6 the four coarse grid cells constitute the four quadrants of the substrata. Another similarity between strata and cells is that they both can indicate information of location; hence, they both appear on the vertical axis of Figure 17.3.

However, there are three major conceptual differences between grid cells and strata. First, the concept of location within a grid cell does not exist because category membership within a grid cell is completely homogenous. By definition, we cannot say that a particular category is concentrated at a particular location within a cell. In contrast, the concept of location within a stratum does exist because we can say that a particular category is concentrated at a particular location within a stratum, since strata usually contain numerous cells. Second, the multiple-resolution procedure increases the lengths of the sides of the grid cells and thus reduces the number of coarse grid cells within each stratum, but the multiple-resolution analysis does not change the number of strata. Third, each cell is a square patch, whereas a stratum can be nonsquare and noncontiguous. As a consequence of these differences, analysis of multiple resolutions of cells shows how the landscape is organized in geographic space, whereas analysis of multiple strata shows how the landscape is organized with respect to the strata definitions.

17.5 CONCLUSIONS

The profession of accuracy assessment is advancing past the point where assessment consists of only a calculation of percentage correct or Kappa index of agreement (Foody, 2002). Now, measures of agreement are needed that indicate how to create more-accurate maps. Here we presented novel methods of accuracy assessment to budget the components of agreement and disagreement between any two maps that show a categorical variable. The techniques incorporate stratification, examine multiple resolutions, apply to both hard and soft classifications, and compare maps in terms of quantity and location. Perhaps most importantly, this chapter shows how to present the results of a complex analysis in a simple graphical form. We hope that this technique of accuracy assessment will soon become as common as today's use of percentage correct.

17.6 SUMMARY

This chapter presented novel methods of accuracy assessment to budget the components of agreement and disagreement between a reference map and a comparison map, where each map shows a categorical variable. The measurements of agreement can take into consideration soft classification and can analyze multiple resolutions. Ultimately, the techniques express the agreement between any two maps in terms of various components that sum to 1. The components may be agreement due to chance, agreement due to quantity, agreement due to location at one of the stratified levels, agreement due to location at the grid cell level, disagreement due to location at the grid cell level, disagreement due to location at one of the stratified levels, and/or disagreement due to quantity. These techniques can be used to compute components of agreement at all resolutions and to present the results of a complex analysis in a simple graphical form.

ACKNOWLEDGMENTS

We thank Clark University's Master of Arts program in Geographic Information Science for Development and Environment and the people of Clark Labs, especially Hao Chen, who has programmed these methods into the GIS software Idrisi32®.* We also thank the Center for Integrated Study of the Human Dimensions of Global Change at Carnegie Mellon University with which this work is increasingly tied intellectually and programmatically through the George Perkins Marsh Institute of Clark University.

REFERENCES

Congalton, R., A review of assessing the accuracy of classification of remotely sensed data, *Remote Sens. Environ.*, 37, 35–46, 1991.

Congalton, R. and K. Green, *Assessing the Accuracy of Classification of Remotely Sensed Data: Principles and Practices*, Lewis, Boca Raton, FL, 1999.

Foody, G., On the comparison of chance agreement in image classification accuracy assessment, *Photogram. Eng. Remote Sens.*, 58, 1459–1460, 1992.

Foody, G., Status of land cover classification accuracy assessment, *Remote Sens. Environ.*, 80, 185–201, 2002.

Lewis, H. and M. Brown, A generalized confusion matrix for assessing area estimates from remotely sensed data, *Int. J. Remote Sens.*, 22, 3223–3235, 2001.

Pontius, R., Quantification error versus location error in comparison of categorical maps, *Photogram. Eng. Remote Sens.*, 66, 1011–1016, 2000.

Pontius, R., Statistical methods to partition effects of quantity and location during comparison of categorical maps at multiple resolutions, *Photogram. Eng. Remote Sens.*, 68, 1041–1049, 2002.

Veldkamp, A. and E. Lambin, Predicting land-use change, *Agric. Ecosyst. Environ.*, 85, 1–6, 2001.

* Registered Trademark of Clark Labs, Worcester, Massachusetts.

CHAPTER 18

Accuracy Assessments of Airborne Hyperspectral Data for Mapping Opportunistic Plant Species in Freshwater Coastal Wetlands

Ricardo D. Lopez, Curtis M. Edmonds, Anne C. Neale, Terrence Slonecker, K. Bruce Jones, Daniel T. Heggem, John G. Lyon, Eugene Jaworski, Donald Garofalo, and David Williams

CONTENTS

18.1 INTRODUCTION

The aquatic plant communities within the coastal wetlands of the Laurentian Great Lakes (LGL) are among the most biologically diverse and productive ecosystems in the world (Mitsch and Gosselink, 1993). Coastal wetland ecosystems are also among the most fragmented and disturbed, as a result of impacts from land-use mediated conversions (Dahl, 1990; Dahl and Johnson, 1991). Many LGL coastal wetlands have undergone a steady decline in biological diversity during the 1900s, most notably within wetland plant communities (Herdendorf et al., 1986; Herdendorf, 1987; Stuckey, 1989). Losses in biological diversity can often coincide with an increase in the presence and dominance of invasive (nonnative and aggressive native) plant species (Bazzaz, 1986; Noble, 1989). Research also suggests that the establishment and expansion of such opportunistic plant species may be the result of general ecosystem stress (Elton, 1958; Odum, 1985).

Reduced biological diversity in LGL coastal wetland communities is frequently associated with disturbances such as land-cover (LC) conversion within or along wetland boundaries (Miller and Egler, 1950; Niering and Warren, 1980). Disturbance stressors may include fragmentation from road construction, urban development, or agriculture or alterations in wetland hydrology (Jones et al., 2000, 2001; Lopez et al., 2002). Specific ecological relationships between landscape disturbance and plant community composition are not well understood. Remote sensing technologies offer unique capabilities to measure the presence, extent, and composition of plant communities over large geographic regions. However, the accuracy of remote sensor-derived products can be difficult to assess, owing both to species complexity and to the inaccessibility of many wetland areas. Thus, coastal wetland field data, contemporaneous with remote sensor data collections, are essential to improve our ability to map and assess the accuracy of remote sensor-derived wetland classifications.

The purpose of this study was to assess the utility and accuracy of using airborne hyperspectral imagery to improve the capability of determining the location and composition of opportunistic wetland plant communities. Here we specifically focused on the results of detecting and mapping dense patches of the common reed (*Phragmites australis*).

18.2 BACKGROUND

Phragmites typically spreads as monospecific "stands" that predominate throughout a wetland, supplanting other plant taxa as the stand expands in area and density (Marks et al., 1994). It is a facultative-wetland plant, which implies that it usually occurs in wetlands but occasionally can be found in nonwetland environments (Reed, 1988). Thus, *Phragmites* can grow in a variety of wetland soil types, in a variety of hydrologic conditions (i.e., in both moist and dry substrate conditions). Compared to most heterogeneous plant communities, stands tend to provide low-quality habitat or forage for some animals and thus reduce the overall biological diversity of wetlands. The establishment and expansion of *Phragmites* is difficult to control because the species is persistent, produces a large amount of biomass, propagates easily, and is very difficult to eliminate with mechanical, chemical, or biological control techniques.

The differences in spectral characteristics between the common reed and cattail (*Typha* sp.) are thought to result from differences between their biological and structural characteristics. *Phragmites* has a fibrous main stem, branching leaves, and a large seed head that varies in color from reddish-brown to brownish-black; *Typha* are primarily composed of photosynthetic "shoots" that emerge from the base of the plant (at the soil surface) with a relatively small, dense, cylindrical seed head (Figure 18.1). Distinguishing between the two in mixed stands can be difficult using automated remote sensing techniques. This confusion can reduce the accuracy of vegetation maps produced using standard broadband remote sensor data.

This chapter explores the implications of the biological and structural differences, in combination with differing soil and understory conditions, on observed spectral differences within *Phragmites* stands and between *Phragmites* and *Typha* using hyperspectral data. We applied detailed ground-based wetland sampling to develop spectral signatures for the calibration of airborne hyperspectral data and to assess the accuracy of semiautomated remote sensor mapping procedures. Particular emphasis was placed on linkages between field-based data sampling and remote sensing analyses to support semiautomated mapping. Field data provided a linkage to extrapolate between airborne sensor data and the physical structure of *Phragmites* stands, soil type, soil moisture content, and the presence and extent of associated plant taxa. This chapter presents the wetland mapping techniques and results from one of the 13 coastal wetland sites currently undergoing long-term assessment by the EPA at the Pointe Mouillee wetland complex (Figure 18.2).

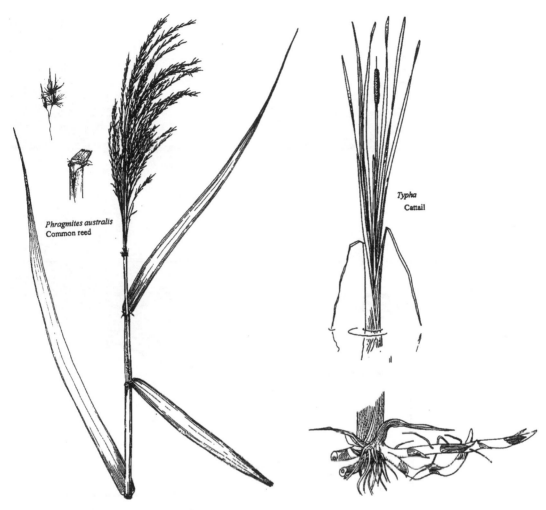

Figure 18.1 Illustrations of *Phragmites australis* and *Typha*. With permission from the Institute of Food and Agricultural Sciences, Center for Aquatic Plants, University of Florida, Gainesville.

18.3 METHODS

Thirteen coastal wetland sites were selected from a group of 65 potential coastal locations to support the EPA's wetland assessment efforts in western Lake Erie, Lake St. Clair, Lake Huron, and Lake Michigan (Lopez and Edmonds, 2001). These sites were selected after visual inspection of aerial photographs, topographic and National Wetland Inventory (NWI) maps, National Land Cover Data (NLCD) data, input from local wetland experts, and review of published accounts at each wetland (Lyon, 1979; Herdendorf et al., 1986; Herdendorf, 1987; Stuckey, 1989; Lyon and Greene, 1992). The study objectives required that each site (1) generally spanned the gradient of current LGL landscape conditions, (2) consisted of emergent wetlands, and (3) included both open lake and protected wetland systems. LC adjacent to the 13 selected study sites included active agriculture, old-field agriculture, urban areas, and forest in varying amounts (Vogelmann et al., 2001).

18.3.1 Remote Sensor Data Acquisition and Processing

Airborne imagery data were collected over the Pointe Mouillee study area using both the PROBE-1 hyperspectral data and the Airborne Data Acquisition and Registration system 5500

Figure 18.2 Thirteen wetland study sites in Ohio and Michigan coastal zone, lettered A–M. Sites were initially sampled during July–August 2001. Inset image is magnified view of Pointe Mouillee wetland complex (Site E). White arrows indicate general location of both field sampling sites for *Phragmites australis* (i.e., the northernmost stand and the southernmost stand). Field-sampled site location legend: Pa = *Phragmites australis*; Ts = *Typha* sp.; Nt = nontarget plant species; Gc = ground control point. Inset image is a grayscale reproduction of false-color infrared IKONOS data acquired in August 2001.

(ADAR). The ADAR sensor enabled remote sensing of materials at the site of < 5 m, which is the nominal spatial resolution of the PROBE-1 sensor. The ADAR system is a four-camera, multispectral airborne sensor that acquires digital images in three visible and a single near-infrared band. ADAR data acquisition occurred on August 14, 2001, at an altitude of 1900 m above ground level (AGL), providing an average pixel resolution of 75 × 75 cm. Using ENVI software, a single ADAR scene in the vicinity of the initial *Phragmites* sampling location was georeferenced corresponding to a root mean square (RMS) error of < 0.06 using digital orthorectified quarter quadrangles (DOQQs) and ground control points from field surveys.

The PROBE-1 scanner system has a rotating axe-head scan mirror that sequentially generated crosstrack scan lines on both sides of nadir to form a raster image cube. Incident radiation was dispersed onto four 32-channel detector arrays. The PROBE-1 data were calibrated to reflectance by means of a National Institute of Standards (NIS) laboratory radiometric calibration procedure, providing 128 channels of reflectance data from the visible through the short-wave infrared wavelengths (440–2490 nm). The instrument carried an on-board lamp for recording in-flight radiometric stability along with shutter-closed (dark current) measurements on alternate scan lines. Geometric integrity of recorded images was improved by mounting the PROBE-1 on a three-axis, gyro-stabilized mount, thus minimizing the effects in the imagery of changes in aircraft pitch, roll, and yaw resulting from flight instability, turbulence, and aircraft vibration. Aircraft position was assigned using a nondifferential global positioning system (GPS), tagging each scan line with the time, which was cross-referenced with the time interrupts from the GPS receiver. An inertial measurement unit added the instrument attitude data required for spatial geocorrection.

During the Pointe Mouillee overflight the PROBE-1 sensor had a 57 instantaneous field of view (IFOV) for the required mapping of vertical and subvertical surfaces within the wetland. The typical IFOV of 2.5 mrad along track and 2.0 mrad across track results in an optimal ground IFOV of 5 to 10 m, depending on altitude and ground speed. PROBE-1 data at Pointe Mouillee were collected on August 29, 2001, at an altitude of 2170 m AGL, resulting in an average pixel size of 5 m × 5 m. The data collection rate was 14 scan lines per second (i.e., pixel dwell time of 0.14 ms), and the 6.1-km flight line resulted in total ground coverage of 13 km^2. The PROBE-1 scene covering Pointe Mouillee was then georeferenced (RMS error < 0.6 pixel) using the vendor-supplied on-board GPS data, available DOQQs, and field-based GPS ground control points provided from August 2001 field surveys. Georeferencing was completed using ENVI image processing software.

The single scene of PROBE-1 data covering Pointe Mouillee was initially visually examined to remove missing or noisy bands. The resulting 104 bands of PROBE-1 data were then subjected to a minimum noise fraction (MNF) transformation to first determine the inherent dimensionality of the image data, segregate noise in the data, and reduce the computational requirements for subsequent processing (Boardman and Kruse, 1994). MNF transformations were applied as modified from Green et al. (1988). The first transformation, based on an estimated noise covariance matrix, decorrelated and rescaled the noise in the data. The second MTF step was a standard principal components transformation of the "noise-whitened" data. Subsequently, the inherent dimensionality of the data at Pointe Mouillee was determined by examining the final eigen values and the associated images from the MNF transformations. The data space was then divided into that associated with large eigen values and coherent eigen images and that associated with near-unity eigen values and noise-dominated images. By using solely the coherent portions, the noise was separated from the original PROBE-1 data, thus improving the spectral processing results of image classification (RSI, 2001).

A supervised classification of the PROBE-1 scene was performed using the ENVI Spectral Angle Mapper (SAM) algorithm. Because the PROBE-1 flights occurred 3 weeks after field sampling, there was a possibility that trampling from the field crew could have altered the physical structure of the vegetation stands. For this reason, and due to the inherent georeferencing inaccuracies, spectra were collected over a 3 × 3-pixel area centered on the single pixel with the greatest percentage of aerial cover and stem density within the vegetation stand (Figure 18.3 and Figure 18.4). The SAM algorithm was then used to determine the similarity between the spectra of homogeneous *Phragmites* and other pixels in the PROBE-1 scene by calculating the spectral angle between them (spectral angle threshold = 0.07 rad). SAM treats the spectra as vectors in an *n*-dimensional space equal to the number of bands.

The SAM classification resulted in the detection of 18 image endmembers, each with different areas mapped as potentially homogeneous regions of dense *Phragmites*. The accuracy of the 18

Figure 18.3 Field sampling activities were an important part of calibrating the hyperspectral data and assessing map accuracy. (A) dense *Phragmites* canopy and (B) dense *Phragmites* understory layer in the northernmost stand. The edges of the stand and the internal transects were mapped using a real-time differential global positioning system.

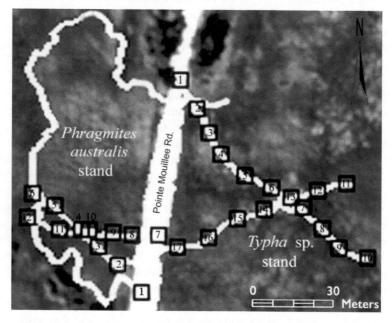

Figure 18.4 Magnified view of northernmost field-sampled vegetation stands to the east and west of Pointe Mouillee Road. Two methods were used to quadrat-sample vegetation stands: (a) edge and interior was sampled if the stand was small enough to be completely traversed (left, *Phragmites*) or (b) solely the interior was sampled if the stand was too large to be completely traversed (right, *Typha*). This example shows a *Typha* stand that extended approximately 0.75 km east of Pointe Mouillee Road. Thus, the field crew penetrated into the stand but did not completely traverse the stand. Black squares = nested quadrat sample locations. Image is a grayscale reproduction of a natural-color spatial subset of airborne ADAR data acquired August 14, 2001.

endmembers was determined based on reference data derived from the interpretation of 1999 panchromatic aerial photography and field observation data collected in 2001. Additional accuracy checking of mapped areas of *Phragmites* was accomplished using ENVI Mixture Tuned Matched Filtering (MTMF) algorithms. Visual interpretation of the MTMF "infeasibility values" (noise sigma units) vs. "matched filtering values" (relative match to spectrum) further aided in the elimination of potential endmembers. The matched filtering values provided a means of estimating the relative degree of match to the *Phragmites* patch reference spectrum and the approximate subpixel abundance. Correctly mapped pixels had a matched filter score above the background distribution and

Table 18.1 Definition(s) of Terms Used during Field Sampling Protocol at Pointe Mouillee

Term	Definition(s)
Wetland	Transitional land between terrestrial and aquatic ecosystems where the water table is usually at or near the surface, land that is covered by shallow water, or an area that supports hydrophytes, hydric soil, or shallow water at some time during the growing season (after Cowardin et al., 1979)
Target plant species	*Phragmites australis* or *Typha* spp. (per Voss, 1972; Voss, 1985)
Nontarget plant species	Any herbaceous vegetation other than target plant species
Vegetation stand	A relatively homogeneous area of target plant species with a minimum approximate size of 0.8 ha
Edge of vegetation stand	Transition point where the percentage canopy cover ratio of target:nontarget species is 50:50

Table 18.2 Nonspectral Data Parameters Collected (√) along Vegetation Sampling Transects at Pointe Mouillee

Parameter Description	1.0 m² quadrat	3.0 m² quadrat
Number of live target species stems	√	
Number of senescent target species stems	√	
Number of flowering target species stems	√	
Water depth	√	
Litter depth	√	
Mean stem diameter ($n = 5$)	√	
Percentage cover live target species in canopy		√
Percentage cover senescent target species in canopy		√
Percentage cover live nontarget species in canopy		√
Percentage cover senescent nontarget species in canopy		√
Percentage cover live nontarget species in understory		√
Percentage cover senescent nontarget species in understory		√
Percentage cover senescent target species in understory (i.e., senescent material that is not litter)		√
Percentage cover exposed moist soil		√
Percentage cover exposed dry soil		√
Percentage cover litter		√
Percentage cover water		√
General dominant substrate type (i.e., sand, silt, or clay)		√
Distance to woody shrubs or trees within 15 m		√
Direction to woody shrubs or trees within 15 m		√
Total canopy cover (area) of woody shrubs		√

a low infeasibility value. Pixels with a high matched filter result and high infeasibility were "false positive" pixels that did not match the *Phragmites* target.

18.3.2 Field Reference Data Collection

To minimize ambiguous site identifications, specific definitions of wetland features were provided to field investigators (Table 18.1). Vegetation was sampled on August 7–8, 2001, to provide training data for the semiautomated vegetation mapping (Table 18.2) and subsequent accuracy assessment effort. Prior to field deployment, aerial photographs were used along with on-site assessments to locate six large stands of vegetation at the site. They included (1) two stands of *Phragmites,* (2) two stands of *Typha,* and (3) two nontarget vegetation stands for comparison to the target species (Figure 18.2). Digital video of each vegetation stand was recorded to fully characterize the site for reference during image processing and accuracy assessment. Additional field data used to support accuracy assessment efforts included vegetation stand sketches, notes of the general location and shape of the vegetation stand, notes of landmarks that might be recognizable in the imagery, and miscellaneous site characterization information.

Figure 18.5 Field spectroradiometry sampling conducted August 14–17, 2001, at 4 of 13 wetland sites for comparison to the PROBE-1 reflectance spectra. The procedure involved recording (A) reference spectra and (B) vegetation reflectance spectra during midday solar illumination. Vegetation spectra were recorded from 1 m above the vegetation canopy.

Transects along the edges of target-species stands were recorded using a real-time differential GPS for sampled target species (Figure 18.3). Each of the two nontarget stands of vegetation was delineated with a minimum of four GPS points, evenly spaced around the perimeter. Five GPS ground control points (GCPs) were collected at Pointe Mouillee, generally triangulating on the sampled areas of the wetland (Figure 18.2). GPS location points were recorded along with multiple digital photographs, as necessary, to provide multiple angle views of each sample location. The edge polygons, GPS points, GCPs, field notes, and field-based images (camera) were used to provide details about ground data for imagery georeferencing, classification, and accuracy assessments.

A quadrat sampling method was used within each target-species stand to sample herbaceous plants, shrubs, trees, and other characteristics of the stand (Mueller-Dombois and Ellenberg, 1974; Barbour, 1987). Depending on stand size, 12 to 20 (nested) 1.0-m^2 and 3.0-m^2 quadrats were evenly spaced along intersecting transects (Figure 18.4). The approximate percentage of cover and taxonomic identity of trees and shrubs within a 15-m radius were also recorded at each quadrat. Where appropriate, the terminal quadrat was placed outside of the target-species stand perimeter to characterize the immediately adjacent area. This placement convention improved the accurate determination of vegetation patch edge locations. The location of SAM classification output was accomplished partly by identifying a uniform corner of each quadrat with the real-time differential GPS to provide a nominal spatial accuracy of 1 m. Field data were collected to characterize both canopy and understory in targeted wetland plant communities (Table 18.2).

Reflectance spectra were measured in the field for each of the target species at four selected wetland sites (Site A, Site B, Site F, and Site J; Figure 18.2) on August 14–17, 2001, using a field spectroradiometer (Figure 18.5). Field spectra collected from 1 m above the top of the *Phragmites* canopy were compared to PROBE-1 to confirm target species spectra at Pointe Mouillee and were archived in a wetland plant spectral library.

18.3.3 Accuracy Assessment of Vegetation Maps

A three-tiered approach was used to assess the accuracy of PROBE-1 vegetation maps. This approach included unit area comparisons with (1) photointerpreted stereo panchromatic (1999) aerial photography (1:15,840 scale), (2) GPS vector overlays and field transect data from 2001 (Congalton and Mead, 1983), and (3) field measurement data (2002).

Pointe Mouillee 2002 sampling locations were based on a stratified random sampling grid and provided to a field sampling team as a list of latitude and longitude coordinates along with a site orientation image, which included a digital grayscale image of the site with the listed coordinate points displayed as an ArcView point coverage. Stratification of samples was based on Universal Transverse Mercator (1000 m) grid cells ($n = 17$), from which the total number of potential sampling points were selected ($n = 86$). The supplied points represented the center point of mapped areas of dense *Phragmites* (> 25 stems/m^2 and > 75% cover). Accordingly, the 86 sampling points selected to support the validation and accuracy assessment effort contained no "false positive" control locations. At each field validation sampling location, both 1-m^2 and 3-m^2 quadrats were used. Five differentially corrected GPS ground control points were collected to verify the spatial accuracy of field validation locations.

18.4 RESULTS

18.4.1 Field Reference Data Measurements

The northernmost *Phragmites* stand sampled at Pointe Mouillee was bounded on the eastern edge by an unpaved road with two small patches of dogwood and willow in the north and a single small patch of willow in the south (Figure 18.4). A mixture of purple loostrife (*Lythrum salicaria*) and *Typha* bounded the eastern edge of the stand. Soil in the *Phragmites* stand was dry and varied across the stand from clayey-sand to sandy-clay, to a mixture of gravel and sandy-clay near the road. Litter cover was a constant 100% across the sampled stand; nontarget plants in the understory included smartweed (*Polygonum* spp.), jewel weed (*Impatiens* spp.), mint (*Mentha* spp.), Canada thistle (*Cirsium arvense*), and an unidentifiable grass. Cattail was the sole additional plant species in the *Phragmites* canopy.

The southernmost Pointe Mouillee *Phragmites* stand was completely bounded by manicured grass or herbaceous vegetation, with dry and clayey soil throughout. Litter cover was 100% and nontarget plants in the understory included smartweed, mint, purple loosestrife, and an unidentifiable grass. Nontarget plants were not observed in the canopy. Comparisons of the two field-sampled stands indicated that quadrat-10 region of the northernmost stand was the most homogeneous of all sampled quadrats. Accordingly, field transect data were used to determine which pixel(s) in the PROBE-1 data had the greatest percentage of cover of nonflowering *Phragmites* and the greatest stem density (Figure 18.6).

18.4.2 Distinguishing between *Phragmites* and *Typha*

Phragmites and *Typha* are often interspersed within the same wetland, making it difficult to distinguish between the two species. Because plant assemblage uniformity was measured in the field (Figure 18.6), we could compare the PROBE-1 reflectance spectra of *Phragmites* within a single stand of *Phragmites* (Plate 18.1) and with *Typha* (Figure 18.7). There was substantial spectral variability among pixels within the northernmost stand of *Phragmites* (Plate 18.1). The greatest variability for *Phragmites* corresponded to the spectral range associated with plant pigments (470 to 850 nm) and structure (740 to 840 nm). Comparison of reflectance characteristics in the most homogeneous and dense regions of *Phragmites* (quadrat-10) and *Typha* (quadrat-8) (Figure 18.4) indicated that *Phragmites* was reflecting substantially less energy than *Typha* in the near-infrared (NIR) wavelengths and reflecting substantially more energy than *Typha* in the visible wavelengths (Figure 18.7).

18.4.3 Semiautomated *Phragmites* Mapping

Based on the analyses of field measurement data, digital still photographs, digital video images, field sketches, and field notes, we selected nine relatively pure pixels of *Phragmites* centered on quadrat-10 in the northernmost stand (Figure 18.4). A supervised SAM classification of the PROBE-1 imagery, using precision-located field characteristics, resulted in a vegetation map indicating the

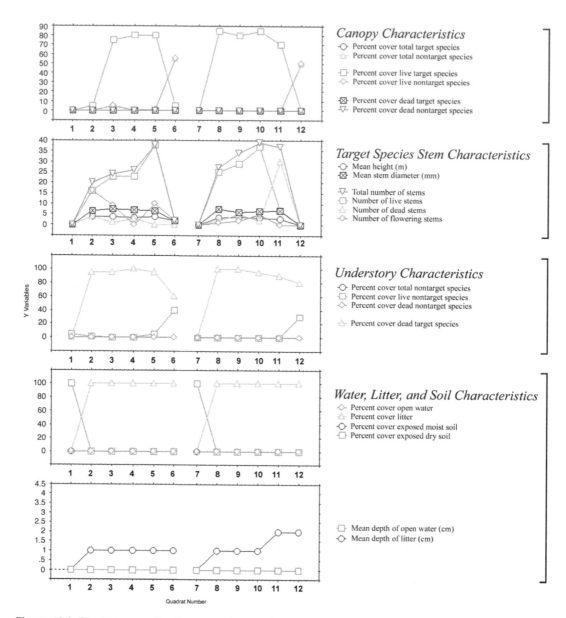

Figure 18.6 The heterogeneity of canopy, stem, understory, water, litter, and soil characteristics in the northernmost *Phragmites* stand was used to calibrate the PROBE-1 data for the purpose of detecting relatively homogeneous areas of *Phragmites* throughout the Pointe Mouillee wetland complex. The most homogeneous area of *Phragmites* in the northernmost stand was in the vicinity of quadrat-10. These pixels were used in the Spectral Angle Mapper (supervised) classification of PROBE-1 reflectance data.

likely locations of homogeneous *Phragmites* stands (Plate 18.2). Several of the mapped areas were within the drier areas of the Pointe Mouillee wetland complex, which was typical of *Phragmites* observed in other diked Lake Erie coastal wetlands.

18.4.4 Accuracy Assessment

Tier-1 accuracy assessments that compared *Phragmites* maps to photointerpreted reference data supplemented with field notes resulted in an estimated accuracy of 80% ($n = 11$) for the presence

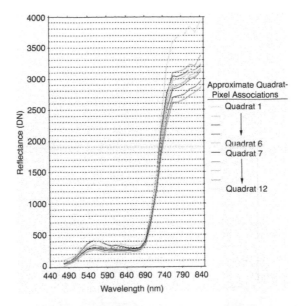

Plate 18.1 (See color insert following page 114.) Comparison of *Phragmites australis* among 10 field-sampled quadrats using spectral reflectance of PROBE-1 data (480 nm–840 nm). Pixel locations were in the approximate location of quadrats in the northernmost *Phragmites* stand at Pointe Mouillee.

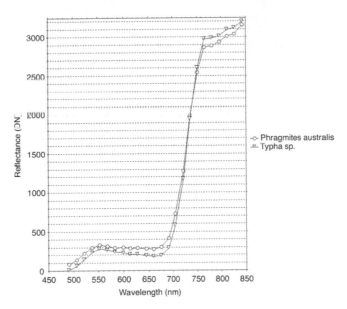

Figure 18.7 Comparison of *Phragmites australis* and *Typha* sp. spectral reflectance in separate relatively homogeneous stands (5 m × 5 m). Pixel locations were in the northernmost *Phragmites* (quadrat-10) and *Typha* (quadrat-8) field sites.

or absence of *Phragmites*. Tier-2 assessments resulted in an approximate ± 1-pixel accuracy relative to the actual location of *Phragmites* on the ground. Tier-3 field-based accuracy assessments resulted in 91% accuracy ($n = 86$). Eight of the sampling points were located in vegetated areas other than *Phragmites* (i.e., either *Typha* or other mixed wetland species), resulting in an omission error rate of 9%. Because the analyses presented here solely pertain to locations of relatively dense *Phragmites* (> 25 stems/m^2 and > 75% cover), errors of commission were not calculated.

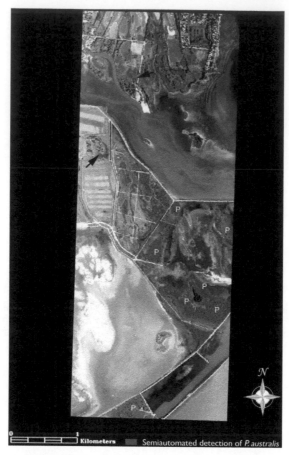

Plate 18.2   (See color insert following page 114.) Results of a Spectral Angle Mapper (supervised) classification, indicating likely areas of relatively homogeneous stands of *Phragmites australis* (solid blue)
and field-based ecological data. Black arrows show field-sampled patches of *Phragmites*. Areas
of mapped *Phragmites* are overlaid on a natural-color PROBE-1 image of Pointe Mouillee wetland
complex (August 29, 2001). Yellow "P" indicates location of generally known areas of *Phragmites*,
as determined from 1999 aerial photographs.

18.5 DISCUSSION

The nominal spatial resolution associated with the ADAR data acquired at Pointe Mouillee was
ideal for viewing field GPS overlays and for ensuring the accuracy of coarser-resolution PROBE-
1 data. The ADAR data were also easily georeferenced using DOQQ image-based warping techniques. However, these four-band data were limited in their usefulness for developing *Phragmites*
spectral signatures.

Field data from quadrat sampling was an essential part of effectively assessing the accuracy of
PROBE-1 *Phragmites* maps. The nominal 1-m spatial accuracy associated with field data collections
at vegetation sampling sites provided essential information to support the accuracy assessment at
Pointe Mouillee. The observed heterogeneity in *Phragmites* stands was likely the result of variability
within underlying vegetation, litter, and soil conditions, as evidenced by field data and PROBE-1
spectral variability within stands. The use of precision-located field data enabled the selection of
specific pixels within the imagery that contained the highest densities of *Phragmites*. Additionally,
the ground imagery data (i.e., video and digital still images) corresponding to individual quadrats
improved the decision-making processes for identifying which specific stand locations were dominated by high-density plant assemblages.

Pointe Mouillee field results demonstrated that a major impediment to the automated detection of wetland vegetation can be the inaccurate assessment of mixtures of biotic and abiotic wetland characteristics, even when wetland vegetation is predominated by a single taxon, such as *Phragmites* (Figure 18.6). For example, those bands observed in the near-infrared wavelengths for *Phragmites* may have caused image classification confusion (Plate 18.1). Heterogeneity and interspersion of different wetland species are also thought to contribute to a relatively wide range of reflectance values observed within wetland stands. Although water was not present at the selected Pointe Mouillee sample locations in 2001, changes in hydrology and variability in soil moisture could also contribute to inaccurate wetland classification. Thus, the biological and physical characteristics of wetland plant communities at the time of imagery collection must be factored into the analysis.

To improve the accuracy of PROBE-1–derived maps we accounted for plant community heterogeneity by: (1) selecting plant taxa that were least likely to exist in diverse, heterogeneous plant communities; (2) using GPS points with a nominal spatial accuracy that exceeds that of the imagery data for locating sampled quadrats, stand edges, and ground control points; (3) acquiring a variety of remote sensing data types to provide a range of spectral and spatial characteristics; (4) collecting relevant ecological field data most likely to explain the differences in spectral reflectance characteristics among pixels; (5) using archived aerial photography to assess and understand site history; and (6) collaborating with local wetland experts to better understand the ecological processes at the site and the historical context of changes.

18.6 CONCLUSIONS

The use of hyperspectral data at Pointe Mouillee demonstrated the spectral differences between *Phragmites* and *Typha*. Spectral differences between taxa are likely attributable to differences in chlorophyll content, plant physical structure, and water relations of the two taxa. The combined use of detailed ecological field data, field spectrometry data, and multiscalar accuracy assessment approaches were instrumental to our ability to validate mapping results for *Phragmites* and provide important information to assess the future coastal mapping efforts in the LGL. Additional classification and accuracy assessment procedures are ongoing at 12 other wetland study sites to determine the broader applicability of these techniques and results (Lopez and Edmonds, 2001; Figure 18.2). Other important ongoing research related to advanced hyperspectral wetland remote sensing includes: (1) improving techniques for separating noise from signal in hyperspectral data, (2) determining the relevant relationships between imagery data and field data for other plant species and assemblages, (3) calibrating sensor data with field spectral data, (4) merging cross-platform data to improve detection of plant taxa; and (5) employing additional assessment techniques using field reference data.

The results of this study describe the initial steps required to investigate the correlations between local landscape disturbance and the presence of opportunistic plant species in coastal wetlands. These results support general goals to develop techniques for mapping vegetation in ecosystem types other than wetlands, such as upland herbaceous plant communities. The results of this and other similar research may help to better quantify the cost-effectiveness of semiautomated vegetation mapping and accuracy assessments so that local, state, federal, and tribal agencies in the LGL can decide whether such techniques are useful for their monitoring programs.

18.7 SUMMARY

The accuracy of airborne hyperspectral PROBE-1 data was assessed for detecting dense patches of *Phragmites australis* in LGL coastal wetlands. This chapter presents initial research results from a wetland complex located at Pointe Mouillee, Michigan. This site is one of 13 coastal wetland field sites currently undergoing long-term assessment by the EPA. Assessment results from wetland field sampling indicated that semiautomated mapping of dense stands of *Phragmites* were 91% accurate using a supervised classification approach. Results at Pointe Mouillee are discussed in the larger context of the long-term goal of determining the ecological relationships between landscape disturbance in the vicinity of wetlands and the presence of *Phragmites*.

ACKNOWLEDGMENTS

We thank Ross Lunetta and an anonymous reviewer for their comments regarding this manuscript. We thank Joe D'Lugosz, Arthur Lubin, John Schneider, and EPA's Great Lakes National Program Office for their support of this project. We thank Marco Capodivacca, Karl Leavitt, Joe Robison, Matt Hamilton, and Susan Braun for their help with the field sampling work. The EPA's Office of Research and Development (ORD) and Region 5 Office jointly funded this project. This publication has been subjected to the EPA's programmatic review and has been approved for publication. Mention of any trade names or commercial products does not constitute endorsement or recommendation for use.

REFERENCES

Barbour, M.G., J.H. Burk, and W.D. Pitts, *Terrestrial Plant Ecology*, Benjamin/Cummings, Menlo Park, CA, 1987.

Bazzaz, F.A., Life history of colonizing plants, in *Ecology of Biological Invasions of North America and Hawaii*, Mooney, H.A. and J.A. Drake, Eds., Springer-Verlag, New York, 1986.

Boardman, J.W. and F.A. Kruse, Automated Spectral Analysis: A Geological Example Using AVIRIS Data, North Grapevine Mountains, Nevada, in Proceedings of the ERIM Tenth Thematic Conference on Geologic Remote Sensing, Environmental Research Institute of Michigan, Ann Arbor, MI, 1994.

Congalton, R. and R. Mead, A quantitative method to test for consistency and accuracy in photointerpretation, *Photogram. Eng. Remote Sens.*, 49, 69–74, 1983.

Cowardin, L.M., V. Carter, F.C. Gollet, and E.T. LaRoe, Classification of Wetlands and Deepwater Habitats of the United States, FWS/OBS-79/31, U.S. Fish and Wildlife Service, Washington, DC, 1979.

Dahl, T.E., Wetlands Losses in the United States, 1780s to 1980s, U.S. Fish and Wildlife Service, Washington, DC, 1990.

Dahl, T.E. and C.E. Johnson, Status and Trends of Wetlands in the Conterminous United States, Mid-1970s to Mid-1980s, U.S. Fish and Wildlife Service, Washington, DC, 1991.

Elton, C.S., *The Ecology of Invasions by Animals and Plants*, Methuen, London, 1958.

Green, A.A., M. Berman, P. Switzer, and M.D. Craig, A transformation for ordering multispectral data in terms of image quality with implications for noise removal, *IEEE Trans. Geosci. Remote Sens.*, 26, 65–74, 1988.

Herdendorf, C.E., The Ecology of the Coastal Marshes of Western Lake Erie: A Community Profile, Biological Report 85(7.9), U.S. Fish and Wildlife Service, Washington, DC, 1987.

Herdendorf, C.E., C.N. Raphael, E. Jaworski, and W.G. Duffy, The Ecology of Lake St. Clair Wetlands: A Community Profile, Biological Report 85(7.7), U.S. Fish and Wildlife Service, Washington, DC, 1986.

Jones, K.B., A.C. Neale, M.S. Nash, K.H. Riitters, J.D. Wickham, R.V. O'Neill, and R.D. Van Remortel, Landscape correlates of breeding bird richness across the United States Mid-Atlantic Region, *Environ. Monit. Assess.*, 63, 159–174, 2000.

Jones, K.B., A.C. Neale, M.S. Nash, R.D. Van Remortel, J.D. Wickham, K.H. Riitters, and R.V. O'Neill, Predicting nutrient and sediment loadings to streams from landscape metrics: a multiple watershed study from the United States Mid-Atlantic Region, *Landsc. Ecol.*, 16, 301–312, 2001.

Lopez, R.D., C.B. Davis, and M.S. Fennessy, Ecological relationships between landscape change and plant guilds in depressional wetlands, *Landsc. Ecol.*, 17, 43–56, 2002.

Lopez, R.D. and C.M. Edmonds, An Ecological Assessment of Invasive and Aggressive Plant Species in Coastal Wetlands of the Laurentian Great Lakes: A Combined Field Based and Remote Sensing Approach, EPA/600/R-01/018, U.S. Environmental Protection Agency, Washington, DC, 2001.

Lyon, J.G., Remote sensing analyses of coastal wetland characteristics: The St. Clair Flats, Michigan, in Proceedings of the Thirteenth International Symposium on Remote Sensing of Environment, April 23–27, Ann Arbor, MI, 1979.

Lyon, J.G. and R.G. Greene, Use of aerial photographs to measure the historical areal extent of Lake Erie coastal wetlands, *Photogram. Eng. Remote Sens.*, 58, 1355–1360, 1992.

Marks, M., B. Lapin, and J. Randall, *Phragmites australis (P. communis)*: threats, management, and monitoring, *Nat. Areas J.*, 14, 285–294, 1994.

Miller, W. and F. Egler, Vegetation of the Wequetequock-Pawcatuck tidal marshes, CT, *Ecol. Monogr.*, 20, 147–171, 1950.

Mitsch, W.J. and J.G. Gosselink, *Wetlands*, Van Nostrand Reinhold, New York, 1993.

Mueller-Dombois, D. and H. Ellenberg, *Aims and Methods of Vegetation Ecology*, John Wiley & Sons, London, 1974.

Niering, W.A. and R.S. Warren, Vegetation patterns and processes in New England salt marshes, *BioScience*, 30, 301–307, 1980.

Noble, I.R., Attributes of invaders and the invading process: terrestrial and vascular plants, in *Biological Invasions: A Global Perspective*, Drake, J.A., H.A. Mooney, F. di Castri, R.H. Groves, F.J. Kruger, M. Rejmanek, and M. Williamson, Eds., John Wiley, Chichester, UK, 1989.

Odum, E.P., Trends expected in stressed ecosystems, *Bioscience*, 35, 419–422, 1985.

Reed, P.B., Jr., National List of Plant Species that Occur in Wetlands, U.S. Fish and Wildlife Service, Biological Report 88(26.3), Washington, DC, 1988.

RSI (Research Systems, Inc.), ENVI User's Guide, ENVI version 3.5, RSI, Boulder, CO, 2001.

Stuckey, R.L., Western Lake Erie aquatic and wetland vascular-plant flora: its origin and change, in *Lake Erie Estuarine Systems: Issues, Resources, Status, and Management*, Krieger, K.A., Ed., National Oceanographic and Atmospheric Administration, Washington, DC, 1989.

Vogelmann, J.E., S.E. Howard, L. Yang, C.R. Larson, B.K. Wylie, and N. Van Driel, Completion of the 1990s National Land Cover Data Set for the conterminous United States from Landsat Thematic Mapper Data and ancillary data sources, *Photogram. Eng. Remote Sens.*, 67, 650–662, 2001.

Voss, E.G., *Michigan Flora (Part I, Gymnosperms and Monocots)*, Cranbrook Institute of Science and University of Michigan Herbarium, Ann Arbor, MI, 1972.

Voss, E.G., *Michigan Flora (Part II, Dicots, Saururaceae - Cornaceae)*, Cranbrook Institute of Science and University of Michigan Herbarium, Ann Arbor, MI, 1985.

CHAPTER 19

A Technique for Assessing the Accuracy of Subpixel Impervious Surface Estimates Derived from Landsat TM Imagery

S. Taylor Jarnagin, David B. Jennings, and Donald W. Ebert

CONTENTS

19.1 INTRODUCTION

An emerging area in remote sensing science is subpixel image processing (Ichoku and Karnieli, 1996). Subpixel algorithms allow the characterization of spatial components at resolutions smaller than the size of the pixel. Recent studies have shown the general effectiveness of these techniques (Huguenin, 1994; Huguenin et al., 1997). The importance of subpixel methods is particularly relevant to the field of impervious surface mapping where the predominance of the "mixed pixel" in medium-resolution imagery forces the aggregation of urban features such as roadways and rooftops into general "developed" categories (Civco and Hurd, 1997; Ji and Jensen, 1999; Smith, 2001). The amount of impervious surface in a watershed is a landscape indicator integrating a number of concurrent interactions that influence a watershed's hydrology, stream chemical quality, and ecology and has emerged as an important landscape element in the study of nonpoint source pollution (NPS) (USEPA, 1994). As such, Schueler (1994) proposed that impervious surfaces should be the single unifying environmental theme for the analysis of urbanizing watersheds. Effectively extracting the percentage of impervious surface from medium-resolution imagery would provide a time and cost savings as well as allowing the assessment of these landscape features over extensive geographic areas such as the Chesapeake Bay. As part of the Multi-Resolution Landscape Characterization 2000 program (MRLC 2000, 2002), the United States Geological Survey (USGS) has embarked on an effort to map impervious surfaces across the conterminous U.S. utilizing subpixel techniques. This study proposes to produce a spatial and statistical framework from within which we can investigate subpixel-derived estimates of a material of interest (MOI) utilizing multiple accuracy assessment strategies.

Traditional map accuracy assessment has utilized a contingency table approach for assessing the per-pixel accuracy of classified maps. The contingency table is referred to as a confusion matrix or error matrix (Story and Congalton, 1986). This type of assessment is a "hit or miss" technique and produces a binary output in that a pixel is either "correct" or "not correct." The generally accepted overall accuracy level for land-use (LU) maps has been 85% with approximately equal accuracy for most categories (Jensen, 1986). While alternative techniques to assess the accuracy of land-cover (LC) maps using measurement statistics such as the Kappa coefficient of agreement have been proposed, most methods still rely on the contingency table and use per-pixel assessments of the thematic map class compared to "truth" sample points (Congalton and Green, 1999). However, as noted by Ji and Jensen (1999), this classic "hit or miss" approach is problematic with respect to assessing the accuracy of a subpixel-derived classification. A subpixel algorithm allows the pixel to be classified based on the percentage of a given MOI such that for any given pixel the "fit" to truth can be assessed. A level of accuracy can be still be obtained from a pixel that "misses" the truth. The derivation of a percentage of a MOI per-pixel allows for alternative accuracy assessment approaches such as aggregate whole-area assessments (i.e., watershed) and correlations (Ji and Jensen, 1999). These alternative approaches may produce adequate accuracies despite the fact that a lower per-pixel accuracy is derived from the standard error matrix.

An accuracy assessment of subpixel data is largely dependent upon high-resolution planimetric maps or images to provide reference data. Concurrent with the emergence of subpixel techniques has been a trend in the production of high-resolution data sets, including high-resolution multispectral satellite imagery, GIS planimetric data, and USGS Digital Ortho Quarter Quads (DOQQs). All these data sources can be readily processed within standard GIS software packages and used to assess the accuracy of subpixel estimates, as derived from Landsat data, over large geographic regions.

In this study we compared classified subpixel impervious surface data derived from Landsat TM imagery and planimetric impervious surface maps produced from photogrammetric mapping processes. Comparisons were performed on the classified subpixel (30 m) using planimetric reference data in a raster GIS overlay environment. Our goal was to produce a spatial framework in which to test the accuracy of subpixel-derived estimates of impervious surface coverage. In addition to a traditional per-pixel assessment of accuracy, our technique allowed for a correlation assessment and an assessment of the whole-area accuracy of the impervious surface estimate per unit area (i.e., watershed). The latter is important for ecological and water quality models that have percentage of impervious surface as a variable input.

19.2 METHODS

19.2.1 Study Area

Our study area was the Dead Run watershed, a small, 14-km^2 subwatershed located 9 km west of Baltimore, Maryland (Figure 19.1). The Dead Run subwatershed is a portion of the greater Baltimore Long Term Ecological Research (LTER) area located in Baltimore County, Maryland, and resides within the coastal plain and piedmont geologic areas of the Mid-Atlantic physiographic region. Previously produced planimetric and subpixel data sets were available for the area.

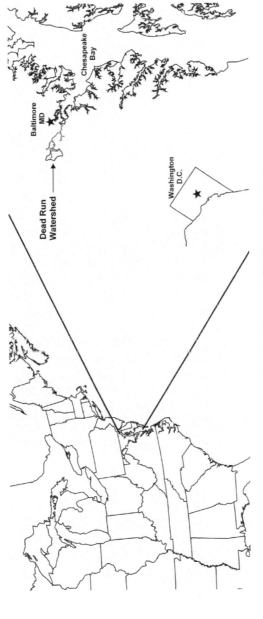

Figure 19.1 Location of the 14-km² Dead Run subwatershed approximately 9 km west of Baltimore, Maryland.

**Table 19.1 The University of Maryland
Mid-Atlantic RESAC Impervious
Surface Percentage per Pixel Classes**

Impervious Class
0
1–10
11–20
21–30
31–40
41–50
51–60
61–70
71–80
81–90
91–100

Note: Classes are represented in the raster data as 10, 20, etc., such that class 1–10 = 10, 11–20 = 20, etc.

19.2.2 Data

Subpixel impervious surface cover data derived from TM imagery were provided by the University of Maryland's Mid-Atlantic Regional Earth Sciences Application Center (RESAC) impervious surface mapping effort (Mid-Atlantic RESAC, 2002). The Mid-Atlantic RESAC process utilized a decision tree classification system to map 11 different levels of impervious surface percentage per 30-m pixel (Table 19.1) (Smith, 2001). Reference data were obtained using photogrammetrically derived GIS planimetric vector data provided by Baltimore County, Maryland. The vector data included anthropogenic features such as roads, parking lots, and rooftops but did not include driveways associated with single-family homes. The lack of compiled driveways was a limitation of the truth set and has the potential to be a source of error.

The Dead Run subwatershed was delineated using USGS Digital Raster Graphics (DRG) and "heads-up" digital collection methods. The compiled Dead Run subwatershed was subsequently utilized to clip both the Mid-Atlantic RESAC raster data and the Baltimore county impervious surface planimetric data. This produced a spatially coincident Dead Run 30-m subpixel estimate GRID and a Dead Run impervious surface truth vector file (Figure 19.2). All data were processed in the UTM Zone 18, NAD83 projection. The respective data sets were independently registered (prior to our study) and no attempt was made to coregister the data via image-to-image methods.

19.2.3 Spatial Processing

GIS raster overlay techniques were utilized to compute the reference values for percentage of impervious surface for each 30-m grid cell within the Dead Run subwatershed. The process was a modified form of zonal analysis. Here, however, the zones are the individual 30-m classified pixels as opposed to individual land LU/LC zones. This method was a variation of the overlay processes reported by Prisloe et al. (2000) and Smith (2001) and included the following analysis procedures:

- A vector-to-raster conversion of the Dead Run impervious surface reference data was performed to produce a high-resolution (3-m) impervious surface grid cell (0 = nonimpervious, 1 = impervious).
- A comparison of the classified 30-m Dead Run data with the 3-m impervious surface reference data was performed using an overlay process, which calculated the number of reference data cells spatially coincident with the classified data (Plate 19.1). The count of coincident reference data cell percentage for each Dead Run grid cell was tallied.

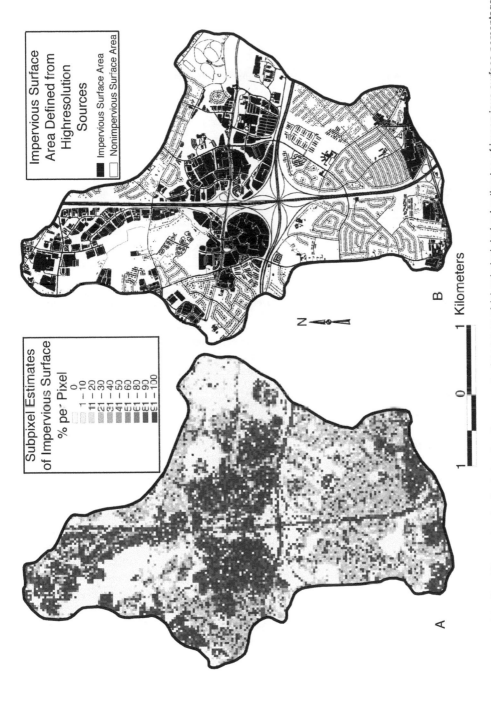

Figure 19.2 Two graphics of the Dead Run subwatershed showing the separate data types of (a) subpixel derived estimates of impervious surface percentage and (b) truth impervious surface vector file.

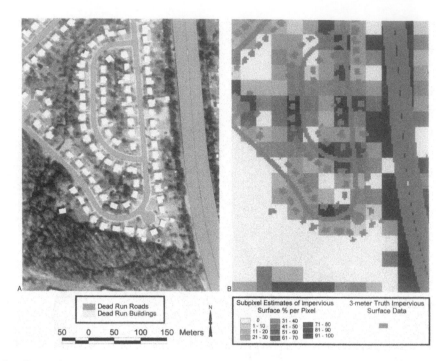

Plate 19.1   (See color insert following page 114.) An approximately 15-ha portion of the Dead Run subwatershed showing (a) truth vector file (roads and rooftops) overlain on a USGS DOQQ and (b) rasterized 3-m reference GRID overlain on the 30-m subpixel estimate grid.

- A vector point file was created based on the Dead Run cell centroids. Table 19.2 summarizes the percentage of impervious reference data (REFERENCE_IS%) and associated subpixel impervious surface estimate (SUBPIXEL_IS%) for each Dead Run cell record. Reference data were "rounded up" to coincide with the subpixel estimate class structure implemented by the Mid-Atlantic RESAC (1–10 = 10, 11–20 = 20, etc.). Table 19.2 was used to derive per-pixel or aggregate watershed error assessment statistics.

19.2.4 Statistical Processing

We tested the overall classification accuracy of the subpixel derived impervious surface estimates by comparing "per-pixel" measures of accuracy with whole-area measures of accuracy for a series of simple random samples (with replacement). A range of sample sizes corresponding to various unit areas were utilized to determine whether the calculated accuracies were dependent on sample size. Sample sizes ranged from the entire Dead Run watershed (15,651 pixels) to simple random samples of 225 pixels. We wanted to achieve an absolute ± 95% confidence interval of < 5% and found that six replicates per unit area provided that level of accuracy for every sample size except the smallest. Given the incomplete nature of the planimetric truth data set, a more rigorous sampling scheme was considered to be unnecessary. To explore the issue of spatial autocorrelation of subpixel classified imagery, we sampled a series of discrete pixel blocks without replacement and compared "per-pixel" measures of accuracy with whole-area measures of accuracy. Six replicates per block sizes 3 × 3, 5 × 5, 9 × 9, 15 × 15, and 25 × 25 were used.

To assess pixel accuracy, we processed the reference and subpixel classified data within an 11-category contingency table and calculated a per-pixel overall accuracy value and a Kappa coefficient of agreement (Khat) value. To assess the whole-area accuracy, we compared the subpixel derived impervious estimates with the reference data estimates per unit area and calculated the absolute value of the relative error ([abs(REFERENCE_IS% – SUBPIXEL _IS%)]/(REFERENCE_IS%))

**Table 19.2 Attribute Table Produced from the
Overlay of the Truth and Subpixel
Classified Data Sets (Plate 19.1)**

POINT_ID	SUBPIXEL_IS%	REFERENCE_IS%
24944	100	7
24945	20	11
24946	20	0
24947	20	0
24948	40	39
24949	40	30
24950	40	8
24951	40	43
24952	40	27
24953	30	0
24954	90	63
24955	90	82
25093	40	28
25094	90	36
25095	40	16
25096	50	0
25097	100	45
25098	50	31
25099	30	4
25100	100	34
25101	90	31
25102	50	5
25103	90	56
25104	90	93

Note: The continuous REFERENCE_IS% field data
are subsequently "rounded up" to coincide with
the Mid-Atlantic RESAC classification structure
(Table 19.1). Data from the SUBPIXEL_IS%
and REFERENCE_IS% fields are utilized in the
accuracy assessment.

for each of the six sample replicates. From the six replicates we computed the mean and coefficient of variation (defined as the standard deviation divided by the mean, expressed as a percentage) at both the pixel and whole-area measures of accuracy for each of the unit areas.

Additionally, per Ji and Jensen (1999), we performed a per-pixel rank Spearman correlation test between the subpixel estimates and the reference data for all cells in the Dead Run watershed (no smaller unit areas were processed for the rank test).

19.3 RESULTS AND DISCUSSION

We wish to stress that this study is not an accuracy assessment of the Mid-Atlantic RESAC subpixel classification, but rather a discussion of alternative assessment methodologies that may be more compatible with the characteristics of subpixel classified data.

The classified data utilized for this study were in preliminary form and were not meant for external distribution for use in watershed assessments. Although we did not quantitatively assess the registration displacement of the two datasets, a manual review showed an approximate and unsystematic 1-m difference throughout the Dead Run subwatershed.

Our results indicated that the pixel-based methods of determining the accuracy of the subpixel estimates yielded results that were consistently lower than the whole-area method of determining accuracy (Table 19.3 and Table 19.4). The whole-area estimate of impervious surface percentage

Table 19.3 A Comparison of Accuracy Assessment Statistics Derived at Different Spatial Scales of Analysis Using Per-Pixel and Whole-Area Assessment Comparisons of the Classified and Planimetric Reference Data Sets Based on Simple Random Samples of the Data

Portion of Watershed Sampled (with Replacement)	Pixels Sampled per Run	Area Analyzed (km²)	Relative Percentage Correct (± 95% CI)	Relative Percentage Correct CV (%)	Error Matrix Overall Accuracy (± 95% CI)	Error Matrix Overall Accuracy CV (%)	K_{hat} (± 95% CI)	K_{hat} CV (%)
Full	15,651	14.086	70.85 (–)	—	28.41 (–)	—	0.1853 (–)	—
Half	7,825	7.0425	70.61 (0.39)	1.65	28.31 (0.30)	1.31	0.1846 (0.0027)	1.85
Quarter	3,913	3.5217	71.18 (0.63)	2.74	28.72 (0.71)	3.08	0.1871 (0.0060)	3.98
Eighth	1,956	1.7604	72.33 (2.47)	11.16	29.12 (0.44)	1.87	0.1928 (0.0033)	2.11
Sixteenth	978	0.8802	71.67 (2.33)	10.26	27.74 (0.91)	4.12	0.1774 (0.0078)	5.48
1/25th	625	0.5625	67.20 (4.38)	16.67	27.79 (0.81)	3.65	0.1787 (0.0085)	5.93
1/40th	400	0.36	73.92 (3.06)	14.65	28.50 (2.13)	9.35	0.1899 (0.0197)	12.99
1/70th	225	0.2025	71.43 (5.74)	25.13	27.31 (2.43)	11.13	0.1780 (0.0244)	17.15

Note: The relative percentage correct column lists the whole-area accuracy and the error matrix overall accuracy and the Khat columns list pixel-based accuracy estimate values. CI = confidence interval; CV = coefficient of variation.

for the entire Dead Run watershed was approximately 71% accurate. Also, the whole-area estimates were robust with respect to the size of the sample subset, although the variability of the estimate increased with smaller sample sizes. The per-pixel assessments of accuracy for the same unit-area data sets were approximately 28% (Kappa 0.19) for the error matrix overall accuracy measurement. For simple random sampling, the per-pixel assessments of accuracy showed less variability with smaller sample sizes than the whole-area method, with the error matrix overall accuracy measurement being particularly stable in this regard (Table 19.3). For pixel block sampling, measured accuracy declined with smaller block sizes when considering both the whole area and per-pixel methods of accuracy assessment (Table 19.4).

Table 19.4 A Comparison of Accuracy Assessment Statistics Derived at Different Spatial Scales of Analysis Using Per-Pixel and Whole-Area Assessment Comparisons of the Classified and Planimetric Reference Data Sets Based on Pixel Blocks Sampled without Replacement

Block Size Sampled (without Replacement)	Pixels Sampled per Run	Area Analyzed (km²) per Block	Relative Percentage Correct (± 95% CI)	Error Matrix Overall Accuracy (± 95% CI)	Error Matrix CV	K_{hat} (± 95% CI)	K_{hat} CV
25 × 25 blocks	625	0.5625	63.68 (13.43)	26.75 (4.48)	20.91	0.1593 (0.0439)	34.47
15 × 15 blocks	225	0.2025	54.67 (13.33)	23.78 (8.42)	44.24	0.1214 (0.0751)	77.30
9 × 9 blocks	81	0.0729	51.47 (17.21)	21.81 (8.04)	46.08	0.1018 (0.0730)	89.66
5 × 5 blocks	25	0.0225	48.05 (26.33)	17.33 (7.75)	55.90	0.0455 (0.0576)	158.06
3 × 3 blocks	9	0.0081	−17.81 (114.25)	18.52 (7.26)	48.99	0.0732 (0.0742)	126.71

Note: The relative percentage correct column lists the whole-area accuracy and the error matrix overall accuracy and the Khat columns list pixel-based accuracy estimate values. CI = confidence interval; CV = coefficient of variation.

The Spearman correlation results were 0.609, suggesting an increased estimate of accuracy compared to the result from the contingency table assessment. Ji and Jensen (1999) also noted an increase in accuracy when utilizing the rank correlation test. Of particular benefit would be subpixel classifications that yield continuous data estimates as opposed to rank order data. These data would allow for regression modeling that could be applied to the individual per-pixel errors.

The results presented here would have greatly benefited from a more accurate reference data set. The lack of driveways in the planimetric data set affected a large proportion of the pixels and probably served to underreport the actual truth for any given pixel. Intuitively, we feel that this "lack of truth" probably had a greater effect on the per-pixel assessments (error matrix) than on either the Spearman correlation or the whole-area approaches. This probably explains a portion of the low per-pixel accuracy. However, all three approaches have been affected by the inaccuracies in the truth set. For example, using nonrandom techniques, we sampled 50 driveway areas to derive a total driveway area in the subwatershed. Summing the driveway area to the previously compiled planimetric impervious surface area increased the accuracy of the whole-area approach to approximately 85%. This underscores the need for high-quality reference data when assessing subpixel estimates. However, reference data sets in the "real world" will always contain a certain proportion of error. The GIS overlay framework effectively extrapolated the reference impervious surface to correspond to the classified 30-m Dead Run cell. These spatial overlay methods provided here may be repeated over any region to assess the accuracy of any pixel-based product.

The overlay framework also allows for the analysis of the spatial distribution of errors. Figure 19.3 is an error grid showing the absolute error per 30-m cell for the entire Dead Run subwatershed. A cursory review of the error grid reveals that approximately 66% of the errors exist within the 1–20% (absolute error) range, signifying that a majority of cells (within the two data sets) are in close agreement. This explains why the correlation assessment outperformed the assessment from the contingency matrix. We can also discern that the contiguous blocks of error in the 90–100% range are primarily due to areas not compiled in the truth data but present in the Landsat data. This would include anthropogenic areas of interest such as parking lots as well as bare soil areas not included in the truth data. Generally, in these contiguous areas of large error, the subpixel classification outperformed the truth data. We feel that this is in part due to a temporal disconnection between the two data sets. Areas that were not included in the truth data set as of the date of the imagery acquisition were actually present and imaged by the sensor. Misregistration between the two data sets can also be observed in the error grid. Linear patterns appear to be associated with the large roadways that traverse the area and are generally associated with the middle range of error (30–80%). Spatial aggregation of error also contributes to decreased measures of accuracy when using sample blocks of pixels compared to a simple random sampling scheme.

19.4 CONCLUSIONS

Results indicate that accuracy assessments of subpixel derived estimates based on per-pixel sampling strategies may underestimate the overall accuracy of the map product. We believe this is because per-pixel assessments of subpixel estimates are sensitive to registration accuracies, the accuracy of the truth data, and classification variability at the pixel scale. A more robust subpixel assessment may be achieved by applying whole-area (aggregate) or correlation-based approaches. These approaches are less sensitive to differences in image registrations as well as errors in the truth set and, in certain large-area applications such as watersheds, are probably a more realistic indicator of the subpixel classification map accuracy.

With respect to impervious surfaces, we believe this technique has considerable merit when considering water quality and watershed runoff models that require, as an input, the percentage of impervious surface area above a given gauge or "pour point." Our analysis shows that whole-area scale estimates of a subpixel derived MOI can be relatively accurate even when the per-pixel

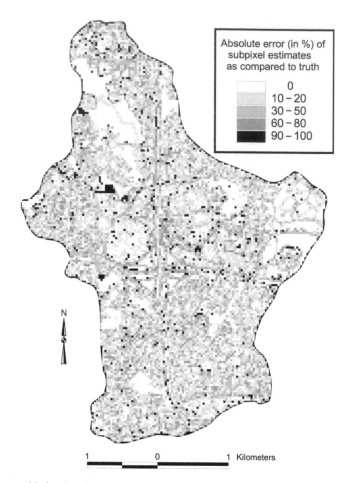

Figure 19.3 An error grid showing the absolute per-pixel error between the truth and subpixel estimates.

measurement of accuracy, derived from the contingency table, is very low. Furthermore, the assessment of accuracy of subpixel estimates over large areas using a sampling scheme based on sampled unit areas (i.e., 5×5 or 9×9 windows) may not be as accurate as one based on the simple random sampling of individual pixels to derive a whole-area estimation of an MOI. For applications over large geographic regions, high-resolution, multispectral satellite data could provide an optimal data source for these sampling situations. Further investigation is necessary to corroborate these results over multiple watershed areas using more accurate reference data.

The spatial and statistical techniques reported here provide an analytical tool that can be used to make per-pixel, unit-area, or correlation-based accuracy assessments of subpixel derived classification estimates easily. In addition, these techniques allow the spatial relationship of the per-pixel error to be explored. The raster overlay technique easily extracted the data necessary to derive these assessments. The ArcView Avenue format script used here, although primarily suited for the assessment of data at the subpixel level, can be utilized (or altered) to derive the accuracy of any classified data set in which higher-resolution truth data are available.

19.5 SUMMARY

This chapter presents a technique for assessing the accuracy of subpixel derived estimates of impervious surface extracted from Landsat TM imagery. We utilized spatially coincident subpixel derived impervious surface estimates, high-resolution planimetric GIS data, vector-to-raster conversion methods, and raster GIS overlay methods to derive a level of agreement between the subpixel classified estimates and the planimetric truth in the Dead Run watershed, a small (14-km^2) subwatershed in the Mid-Atlantic physiographic region. From the planimetric data we produced a per-pixel reference data estimate of impervious surface percentage as a means for assessing the accuracy of preliminary subpixel estimates of impervious surface cover derived from TM imagery. The spatial technique allows for multiple accuracy assessment approaches. Results indicated that even though per-pixel-based estimates of the accuracy of the subpixel data were poor (28.4%, Kappa = 0.19), the accuracy of the impervious surface percentage estimated using whole-area and rank correlation approaches was much improved (70.9%, Spearman correlation = 0.608). Our findings suggest that per-pixel-based approaches to the accuracy assessment of subpixel classified data need to be approached with some caution. Per-pixel-based approaches may underestimate the actual whole-area accuracy of the MOI map, as derived from subpixel methods, when applied over large geographic areas. The raster overlay technique easily extracted the data necessary to derive these assessments. Although the ArcView Avenue script used here was primarily suited for the assessment of data at the subpixel level, it can be utilized to derive the accuracy of any classified data set in which higher-resolution digital truth data are available.

ACKNOWLEDGMENTS

We gratefully acknowledge the United States Geological Survey and the Mid-Atlantic RESAC for providing the preliminary subpixel derived impervious surfaces estimates and Baltimore County, Maryland for providing the planimetric GIS data. The U.S. Environmental Protection Agency (EPA), through its Office of Research and Development (ORD), funded and performed the research described. This manuscript has been subjected to the EPA's peer and programmatic review and has been approved for publication. Mention of any trade names or commercial products does not constitute endorsement or recommendation for use.

REFERENCES

Civco, D.L. and J.D. Hurd, Impervious Surface Mapping for the State of Connecticut, in Proceedings of the American Society of Photogrammetry and Remote Sensing (ASPRS) 1997, American Society of Photogrammetry and Remote Sensing, Bethesda, MD, 1997.

Congalton, R.G. and K. Green, *Assessing the Accuracy of Remotely Sensed Data: Principles and Practices*, Lewis, Boca Raton, FL, 1999.

Huguenin, R.L., Subpixel analysis process improves accuracy of multispectral classifications, *Earth Observation Magazine*, 3, 37–40, 1994.

Huguenin, R.L., M.A. Karaska, D.V. Blaricom, and J.R. Jensen, Classification of bald cypress and tupelo gum trees in Thematic Mapper imagery, *Photogam. Eng. Remote Sens.*, 63, 717–727, 1997.

Ichoku, C. and A. Karnieli, A review of mixture modeling techniques for sub-pixel land cover estimation, *Remote Sens. Rev.*, 13, 161–186, 1996.

Jensen, J.R., *Introductory Digital Image Processing: A Remote Sensing Perspective*, Prentice-Hall, Englewood Cliffs, NJ, 1986.

Ji, M. and J.R. Jensen, Effectiveness of subpixel analysis in detecting and quantifying urban imperviousness from Landsat Thematic Mapper, *Geocarto Int.*, 14, 31–39, 1999.

Mid-Atlantic RESAC, Impervious Surface Mapping Using Multi-Resolution Imagery, available at http://www.geog.umd.edu/resac/impervious/htm, 2002.

MRLC 2000 (Multi-Resolution Landscape Characteristics 2000), available at http://landcover.usgs.gov/natlandcover_2000.html, 2002.

Prisloe, M., L. Gianotti, and W. Sleavin, Determining Impervious Surfaces for Watershed Modeling Applications, in the 8th National Non-point Monitoring Workshop, Hartford, available at http://www.canr.uconn.edu/ces/nemo/gis/pdfs/nps_paper.pdf, 2000.

Schueler, T.R., The importance of imperviousness, *Watershed Protection Techniques,* 1, 100–111, 1994.

Smith, A., Subpixel estimates of impervious cover from Landsat TM image, available at http://www.geog.umd.edu/resac/impervious2.htm, 2001.

Story, M. and R.G. Congalton, Accuracy assessment: a user's perspective, *Photogram. Eng. Remote Sens.*, 52, 397–399, 1986.

USEPA, The Quality of Our Nation's Water: 1992, United States Environmental Protection Agency, USEPA Office of Water, Report # EPA-841-S-94-002, Washington, DC, 1994.

CHAPTER 20

Area and Positional Accuracy of DMSP Nighttime Lights Data

Christopher D. Elvidge, Jeffrey Safran, Ingrid L. Nelson, Benjamin T. Tuttle, Vinita Ruth Hobson, Kimberly E. Baugh, John B. Dietz, and Edward H. Erwin

CONTENTS

20.1 INTRODUCTION

The Operational Linescan System (OLS) is an oscillating scan radiometer designed for cloud imaging with two spectral bands (visible and thermal) and a swath of approximately 3000 km. The OLS is the primary imager flown on the polar orbiting Defense Meteorological Satellite Program (DMSP) satellites. The OLS nighttime visible band straddles the visible and near-infrared (NIR) portion of the spectrum from 0.5–0.9 μm and has six-bit quantitization, with digital numbers (DNs) ranging from 0 to 63. The thermal band has eight-bit quantitization and a broad band-pass from 10–12 μm. The wide swath widths provide for global coverage four times a day: dawn, daytime, dusk, and nighttime. DMSP platforms are stabilized using four gyroscopes (three-axis stabilization) and platform orientation is adjusted using a star mapper, Earth limb sensor, and a solar detector.

At night the OLS visible band is intensified using a photomultiplier tube (PMT), enabling the detection of clouds illuminated by moonlight. With sunlight eliminated, the light intensification results in a unique data set in which city lights, gas flares, lightning-illuminated clouds, and fires can be

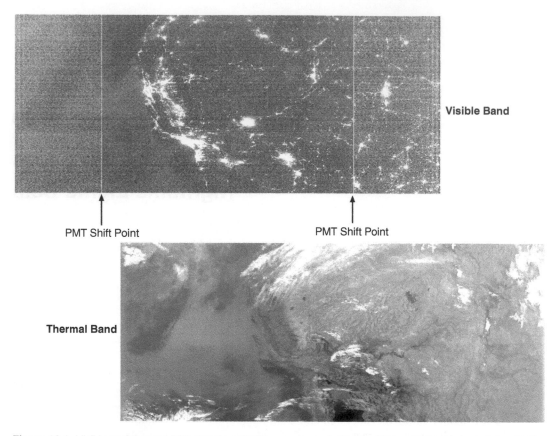

Figure 20.1 Visible and thermal NIR nighttime OLS images over California. With sunlight eliminated, the OLS's light intensification results in the detection of lights present at the Earth's surface.

observed (Figure 20.1). The OLS visible band sensor system is designed to produce visually consistent imagery of clouds at all scan angles for use by U.S. Air Force meteorologists with a minimal amount of ground processing. The visible band base gain is computed on-board based on scene source illumination predicted from solar elevation and lunar phase and elevation. This automatic gain setting can be overridden or modified by commands transmitted from the ground. The automatic gain is lowest when lunar illuminance is high. As lunar illuminance wanes, the gain gradually rises. The highest visible gain settings occur when lunar illumination is absent. The combination of high gain settings and low lunar illuminance provides for the best detection of faint light sources present at the Earth's surface. The drawback of these high gain setting observations is that the visible band data of city centers are typically saturated. Data acquired under a full moon when the gain is turned to a lower level are generally not as useful for nighttime lights product generation since they exhibit fewer lights and have the added complication of bright clouds and terrain features.

In addition to tracking lunar illuminance, gain changes occur within scan lines with the objective of making visually consistent cloud imagery, regardless of scan angle. The base gain is modified every 0.4 ms by an on-board along-scan-gain algorithm. A bidirectional reflectance distribution function (BRDF) algorithm further adjusts the gain to reduce the appearance of specular reflectance in the scan segment where the solar or lunar illumination angle approaches the observation angle.

The OLS design provides imagery with a constant ground-sample distance (GSD) both along- and across-track. The along-track GSD is kept constant through a sinusoidal scan motion, which keeps the track of the scan lines on the ground parallel. The analog-to-digital conversion within individual scan lines is timed to keep the GSD constant from the nadir to the edge of scan. OLS data can be acquired in two spatial resolution modes corresponding to fine-resolution data (0.5-km

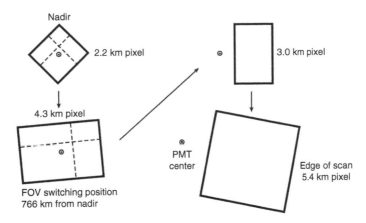

Figure 20.2 The OLS fine-resolution nighttime visible band instantaneous field of view (IFOV) data starts at 2.2 km at the nadir and expands to 4.5 km at 766 km out from the nadir. After the PMT electron beam is switched the IFOV is reduced to 3 km and expands to 4.8 km at the far edges of the scan.

GSD) and smoothed data (2.7-km GSD). All data are acquired in fine resolution mode, but in most cases the recorded data are converted to the smoothed resolution by averaging of 5×5 pixel blocks.

While the GSD of OLS data is kept constant, the instantaneous field of view (IFOV) gradually expands from the nadir to the edge of the scan (Figure 20.2). At nadir the low-light imaging IFOV of the fine resolution data is 2.2 km and it expands to 4.3 km 800 km out from the nadir. At this point in the scan the electron beam within the OLS PMT automatically shifts to constrain the enlargement of pixel dimensions, which normally occurs as a result of cross-track scanning (Lieske , 1981). This reduces the IFOV to 3 km. The IFOV then expands to 5.4 km at the edge of the scan, 1500 km out from the nadir. Thus, the IFOV is substantially larger than the GSD in both the along-track and along-scan directions. At the nadir the smoothed OLS low-light imaging pixel has an IFOV of 5 km and at the edge of the scan the IFOV is approximately 7 km.

In order to build cloud free global maps of nighttime lights and to separate ephemeral lights (e.g., fires) from persistent lights from cities, towns, and villages, a compositing procedure is used to aggregate lights from cloud-free portions of large numbers of orbits, spanning months or even multiple years (Elvidge et al., 1997, 1999, 2001). To avoid the inclusion of moonlit clouds in the products, only data from the dark half of the lunar cycle are composited. The lights in the resulting composites are known to overestimate the actual size of lighting on the ground.

The objective of this chapter is to document the area and positional accuracy of OLS nighttime lights and to examine the causes for the area overestimation of OLS lighting. We have done this using light from isolated sources located in southern California. The analyses were conducted using data from four OLS sensors spanning a 10-year time period.

20.2 METHODS

20.2.1 Modeling a Smoothed OLS Pixel Footprint

A scaled model of an OLS PMT smoothed pixel IFOV at nadir was built by placing 25 fine-resolution pixel footprints onto a 5×5 grid, each displaced by a 0.5-km GSD. The number of times a light would get averaged into a smoothed pixel was tallied for each of the resulting polygon outlines (Figure 20.3). A similar model was built to show the IFOV overlap between adjacent PMT smoothed pixels. This model was constructed by placing nine of the smoothed pixel footprints from Figure 20.5 onto a 3×3 grid using a scaled GSD of 2.7 km. The number of smoothed pixel detection opportunities was then tallied for each polygon zone (Figure 20.4).

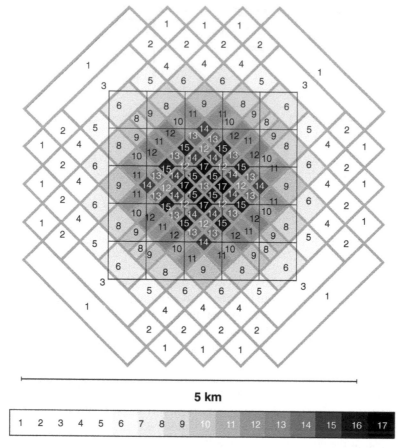

5 km

| 1 | 2 | 3 | 4 | 5 | 6 | 7 | 8 | 9 | 10 | 11 | 12 | 13 | 14 | 15 | 16 | 17 |

Figure 20.3 Scaled model of a PMT smoothed pixel at the nadir composed of 25 fine-resolution pixel footprints. The overlap between IFOVs of adjacent fine-resolution pixels results in the possibility that lights present on the Earth's surface will be averaged into the smoothed pixels multiple times. The number labels marked on the polygons indicate the number of times lights present in the polygons would be averaged into the resulting smoothed pixel.

20.2.2 OLS Data Preparation

Nighttime DMSP-OLS data from 2210 orbits acquired between April 26, 1992, and April 4, 2001, were processed to produce georeferenced images of lights and clouds of the southern California region. The data were initially processed for the NOAA National Marine Fisheries Service to determine the locations and temporal patterns of squid fishing activities conducted using heavily lit boats offshore from the Channel Islands. Data were included from four day–night DMSP satellites: F-10, F-12, F-14, and F-15. DMSP data deliveries to the archive were irregular during 1992, resulting in gaps in the early part of our time series.

Orbits were selected from the archive based on their acquisition time to include nighttime data over California. The orbits were automatically suborbited based on the nadir track to 32°–42° north latitude. Lights and clouds were identified using the basic algorithms described in Elvidge et al. (1997). The next step in the processing was to geographically locate (geolocate) the suborbits. The geolocated images covered the area from 32°–36° north latitude and 117°–122° west longitude. The OLS geolocation algorithm uses satellite ephemeris (latitude, longitude, and altitude at nadir) generated by the SPEPH (Special Ephemeris) orbital model developed by the U.S. Air Force specifically for the DMSP platforms. The orbital model was parameterized by bevel vectors derived from daily RADAR sightings of each DMSP satellite. Ephemeris data were calculated for each

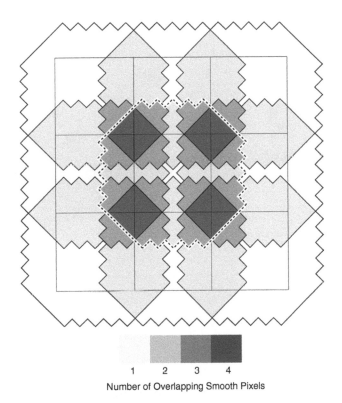

Number of Overlapping Smooth Pixels

Figure 20.4 Scaled model of a three-by-three block of smoothed OLS PMT pixels. The dashed line indicates the boundary of a single smoothed pixel IFOV at the nadir, as modeled in Figure 20.3. Because of the substantial overlap between adjacent smoothed pixel IFOVs it is possible for point sources of light to show up in more than one smoothed pixel. The number of overlapping smoothed OLS pixel IFOVs for each polygon is indicated using the grayscale. Additional levels of overlap are encountered in actual OLS imagery as the pattern is extended beyond this three-by-three example and as the IFOV expands at off-nadir scan angle conditions.

scan line. The geolocation algorithm calculates the position of each OLS pixel center using the satellite ephemeris, a calculation of the scan angle, an earth geode model, and a terrain correction using GTOPO30. The pixel center positions were used to locate the corresponding 30 arc second grid cells, which are filled with the OLS DN values. This generates a sparse grid, having DN data only in cells containing OLS pixel centers. The complete 30 arc second grids were then filled to form a continuous image using nearest-neighbor resampling of the sparse grids.

20.2.3 Target Selection and Measurement

A composite of cloud-free light detections was produced using data from the entire time series. The composite values indicated the number of times lights were detected for each 30 arc second grid cell. These were then filtered to remove single-pixel light detections, a set that contains most of the system noise (Figure 20.5). The cloud-free composites were then used to identify persistent light sources (present through the entire time series) for potential use in the study. Two types of persistent lights were selected: (1) isolated point sources with lighting ground areas much smaller than the OLS pixel, such as oil and gas platforms in the Santa Barbara Channel (Figure 20.6) and (2) isolated lights with more extensive areas of ground lighting. We identified five point sources: four oil and gas platforms (Channel Islands 1–3 and Gaviota 1) and a solitary light present at an airfield on San Nicolas Island. Calibration targets with more extensive area of lighting included a series of cities, towns, and facilities found on land (Table 20.1).

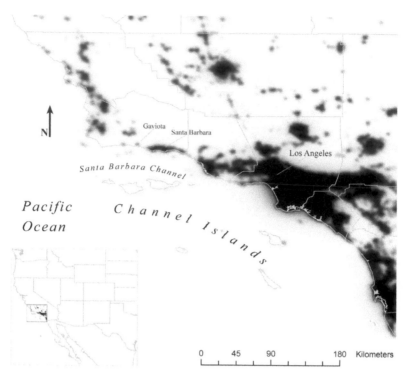

Figure 20.5 Image indicating the number of time lights were detected for each 30 arc second grid cell.

The area of each of the lighting sources was estimated using Landsat Enhanced Thematic Mapper plus (ETM+) data from 2000 by manually drawing a polygon around each of the targets using ENVI software. The number of ETM+ pixels in the polygon was then multiplied by the pixel area to estimate the total target size (km²). For the point sources lights, locations and area estimates were performed using the 15-m panchromatic data. Area extraction for the onshore targets was based on visual interpretation of an ETM+ color composite formed using bands 2, 4, and 5 as blue, green, and red. The ETM+ color composite was individually contrast-enhanced for each target prior to the manual polygon generation.

20.3 RESULTS

20.3.1 Geolocation Accuracy

Light detections of the point sources from individual suborbits in the time series were examined to determine the geolocation accuracy of OLS nighttime visible band data. The latitude/longitude locations of the five point sources of light were extracted for the center of each feature using the 15-m panchromatic ETM+ data. Vector shorelines were overlain on the panchromatic data to confirm that the geolocation accuracy of the ETM+ data was in the range of ± one or two pixels. This was deemed fully adequate for use as a geolocation accuracy reference source for the 2.7-km GSD OLS data.

We followed the geolocation accuracy assessment procedures outlined by the U.S. Federal Geographic Data Committee (FGDC, 1998). This procedure used the root-mean-square error (RMSE) to estimate positional accuracy. RMSE was calculated as the square root of the average of the set of squared differences between data set coordinate values and coordinate values from an independent source of higher accuracy for identical points.

Figure 20.6 Oil and gas platforms detected by the OLS are approximately 0.01 km². This represented approximately 0.2% of the IFOV of a smoothed PMT pixel from the OLS.

Table 20.1 List of Calibration Target Characteristics

Name	Latitude	Longitude	ETM+ Area (km²)	OLS Area (km²)
Gaviota 1	34.3506	−120.2806	0.00765	20.445
Channel Island 3	34.1253	−119.401	0.008325	15.51
Gaviota 2 West	34.3768	−120.1688	0.00855	18.33
Gaviota 2 East	34.3907	−120.1218	0.009	17.625
Structure East of CA City	35.1583	−117.8585	0.318375	41.595
Gaviota Plant	34.4751	−120.2084	0.348075	39.48
Avalon	33.3501	−118.3251	1.133	40.185
San Nicolas	33.2584	−119.4918	1.417	21.15
California Correctional Institute	35.1169	−118.5718	1.834	38.775
Johannesburg	35.3667	−117.6502	2.13	23.265
Helendale	34.7584	−117.3419	8.614	34.545
Lake Los Angeles	34.6224	−117.8329	20.961	51.465
California City	35.1333	−117.9668	36.126	67.68
Edwards AFB	34.925	−117.9002	37.597	174.135
Ridgecrest	35.6418	−117.6751	73.644	169.2
Santa Barbara	34.4334	−119.7084	147.188	425.82
Santa Clarita	34.4521	−118.5418	197.802	457.545
Bakersfield	35.3667	−119.0418	333.866	763.515
Lancaster/Palmdale	34.625	−118.1252	357.979	661.995
Victorville	34.4917	−117.3085	369.411	719.1

For the OLS geolocation accuracy assessment we compared the latitude/longitude position of the centroid of OLS detected lights against the latitude/longitude position extracted from the ETM[+] panchromatic band. We tested the geolocation accuracy of lights detected for the five point sources. The analysis was performed for the data from the individual satellites.

For each image in the time series an automated process searches for a light near the specified latitude/longitude position from the ETM[+] data. The algorithm looks for the presence of cloud-free lights in an 11×11 box of 30 arc second grid cells centered on the ETM[+] latitude/longitude. When a light was found the algorithm identified the full the extent of the light (extending beyond the initial 11×11 box as needed). Valid lights for the analysis were limited to those no larger than 50 grid cells in extent. A bounding rectangle of 30 arc second grid cells was established for each of the valid lights. The 30 arc second grid cell representing the centroid of the light was identified through a separate analysis of the DN values in both the x and y directions inside the bounding rectangles. Two arrays were generated containing the average DN of the grid cells for the lines and columns. The centroid x,y was identified based on the average DN peak found in the two arrays. The centroid x,y was then converted to a latitude/longitude for the center of the identified 30 arc second grid cell. The algorithm then calculated the positional offset between the centroid and the ETM[+] derived latitude/longitude of the light. The process was repeated for each of five point sources and each of the images in the time series. The resulting lists of offsets were used to calculate RMSEx, RMSEy and accuracy in accordance with the FGDC procedure.

The geolocation accuracy assessment results for lights detected in F-10, F-12, F-14, and F-15 satellite data are shown in Figure 20.7. The white triangles indicate the 30 arc second grid cell containing the ETM[+] latitude/longitude of the light sources. The RMSEx and RMSEy values ranged from 0.74 to 1.13 km. RMSEx was lower than RMSEy for each satellite. This indicates that there is more dispersion in the along-track geolocation accuracy than in the cross-track direction. The satellite F-14 data yielded the highest geolocation accuracy (1.55 km). The satellite F-12 and F-15 data had nearly identical RMSEx, RMSEy, and geolocation accuracy results. Satellite F-10 data had the lowest geolocation accuracy (2.36 km). Data from all four satellites produced geolocation accuracies of less than one pixel.

20.3.2 Comparison of OLS Lighting Areas and ETM[+] Areas

The area of OLS lighting was extracted for each of the targets from the F-14 cloud-free composite from 2000. This composite was selected because it had large numbers of cloud-free observations and was most contemporaneous with the ETM[+] data from 2000. The composite was filtered to remove light detections that occurred only once. Figure 20.8 shows the area of OLS lighting vs. ETM[+] area for 20 light sources, indicating that the OLS overestimated the area of lighting. However, the OLS lighting area was highly correlated to the area of lighting estimated from the daytime ETM[+] data. Regression analysis indicated that the OLS lighting areas were approximately twice the size of the area of ground lighting for lights ranging from 20–400 km[2]. This overestimation was substantially higher for lighting sources that were smaller than the OLS IFOV. The OLS was able to detect lights as small as 0.01 km[2], representing approximately 0.01% of an OLS smoothed pixel IFOV (Table 20.1).

20.3.3 Multiplicity of OLS Light Detections

The multiplicity of OLS light detections was examined by tallying the number of OLS pixels detected as lights for the point sources. For this test we pooled the data from all four sensors and five point sources of light. For each point source we tallied the number of OLS light pixels present (1, 2, 3, etc.) on nights with light detection and zero lunar illuminance. From this we calculated the percentage of observations resulting in single OLS light detections, double, triple, and higher.

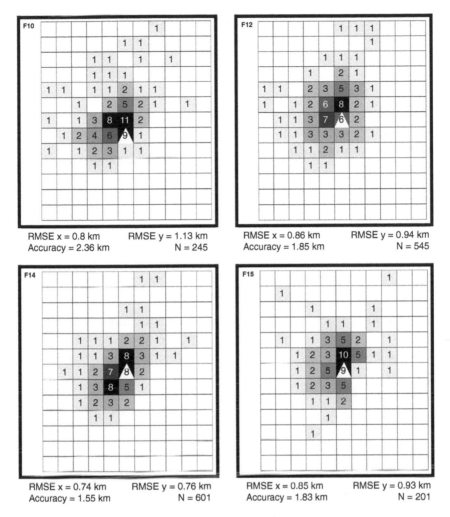

Figure 20.7 Geolocation accuracy of the centroid positions of OLS lights from DMSP satellites F-10, F-12, F-14, and F-15. The numbers printed on the grid cells indicate the percentage of observations in which the OLS light centroids were found in that grid cell position relative to the actual location of the light.

The results (Figure 20.9) show that point sources of light were detected in solitary OLS pixels 38% of the time, in two OLS pixels in 28% of the detections, and in three OLS pixels for 13% of the observations. This phenomenon was caused by the substantial overlap in the footprints of adjacent OLS pixels (Figure 20.3 and Figure 20.4).

20.4 CONCLUSIONS

The DMSP-OLS provides a global capability to detect lights present at the Earth's surface. This chapter provides the first quantitative assessment of the area and positional accuracy of DMSP-OLS–observed nighttime lights.

Light sources from isolated oil and gas platforms with areas as small as 0.1 km² were detected in this study. Since these platforms are heavily lit, the 0.1-km² area approximates the detection limits of the OLS for other heavily lit sources. For detection, the aggregated radiances within an OLS pixel must produce a DN value that exceeds the background noise present in the PMT data. A larger area of lighting would be required for OLS detection of more dimly lit features than the

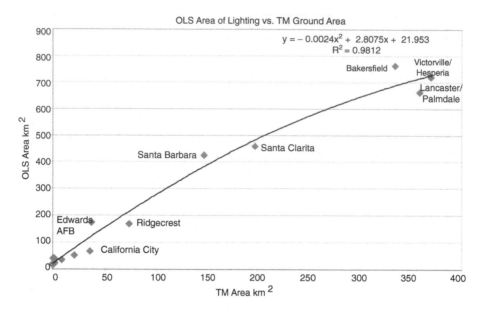

Figure 20.8 Area of OLS lighting vs. area of lighting estimated from Landsat 7 ETM+ imagery.

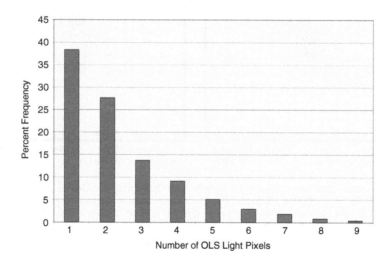

Figure 20.9 Multiplicity of OLS light detections from point sources of surface lighting.

oil and gas platforms. Based on preflight calibrations of the OLS, the oil and gas platforms produce a top-of-atmosphere brightness of approximately 10^{-9} watts/cm²/sr.

Using isolated point sources of light we have tested the geolocation accuracy of nighttime lights data from the four day–night DMSP satellites for which there is a digital archive. This includes satellites F-10, F-12, F-14, and F-15. The OLS lights from single orbits have geolocation accuracies ranging from 1.55 to 2.36 km. This was less than the GSD of the raw data (2.7 km). This subpixel geolocation accuracy is achieved without use of ground control points. Being able to position lights with comparable geolocation accuracy from the multiple satellites will be crucial to the analysis of changes in the extent of development from the DMSP-OLS time series. While further testing will be required, the geolocation accuracy results reported here are encouraging in terms of the prospects for using nighttime OLS data to analyze changes in the extent of lighting over time.

In examining the relationship between the area of OLS lighting and the area of lighting present on the ground our study confirms that cloud-free composited DMSP nighttime lights overestimated the area of lighting on the ground. This overestimation was the result of a combination of factors, including (1) the large OLS pixel size; (2) the OLS's capability to detect subpixel light sources; (3) overlap in the IFOV footprints of adjacent pixels, resulting in multiple pixel detections from subpixel-sized lights; and (4) geolocation errors. These effects, present in data from single observations, were accumulated during the time-series analysis.

Three other mechanisms that may have enlarged OLS lights beyond the extent of surface lighting under certain conditions were not explicitly explored in the current study. One was the scattering of light in the atmosphere as it was transmitted from the Earth's surface to space. The second was the reflection of lights off surface waves in cases where bright city lights were adjacent to water bodies. The third possible mechanism was the detection of terrain illuminated by downward-scattered light arising from very bright urban centers or gas flares.

Imhoff et al. (1997) developed thresholding techniques to accurately map urban areas. The disadvantage of these techniques is that they eliminate lights from small towns owing to their low frequency of detection. We believe that it would be possible to reduce the overestimation of the area of lighting based on an empirical calibration to the extent of surface lighting or via the modulation transfer function (MTF) of the OLS nighttime visible band imagery.

20.5 SUMMARY

The Defense Meteorological Satellite Program (DMSP) Operational Linescan System (OLS) has a unique low-light imaging capability developed for the detection of moonlit clouds. In addition to moonlit clouds, the OLS also detects lights from human settlements, fires, gas flares, heavily lit fishing boats, lightning, and the aurora. Because all these lights are detected in a single spectral band, and to remove the effects of cloud cover, time-series compositing is used to make stable light products that depict the location and area of persistent light sources. This compositing is done using data collected on nights with low lunar illumination to avoid the detection of moonlit clouds and the lower number of lights detected due to the OLS gain settings during periods of high lunar illumination. A number of studies have found that these stable lights products overestimate the size of light sources present on the Earth's surface. This overestimation is due to a combination of factors: the large OLS pixel size, the OLS's capability to detect subpixel light sources, and geolocation errors. These effects, present in data from single observations, are accumulated during the compositing process.

ACKNOWLEDGMENTS

The NOAA NESDIS, Ocean Remote Sensing Research Program, funded this project. Information Integration and Imaging LLC, Fort Collins, Colorado provided the Landsat imagery used in this study imagery.

REFERENCES

Elvidge, C.D., K.E. Baugh, J.B. Dietz, T. Bland, P.C. Sutton, and H.W. Kroehl, Radiance calibration of DMSP-OLS low-light imaging data of human settlements, *Remote Sens. Environ.*, 68, 77–88, 1999.

Elvidge, C.D., K.E. Baugh, E.A. Kihn, H.W. Kroehl, and E.R. Davis, Mapping of city lights using DMSP Operational Linescan System data, *Photogram. Eng. Remote Sens.*, 63, 727–734, 1997.

Elvidge, C.D., M.L. Imhoff, K.E. Baugh, V.R. Hobson, I. Nelson, J. Safran, J.B. Dietz, and B.T. Tuttle,
    Nighttime lights of the world: 1994–95, *ISPRS J. Photogram. Remote Sens.*, 56, 81–99, 2001.
FGDC (U.S. Federal Geographic Data Committee), Geospatial Positioning Accuracy Standards, Part 3:
    National Standards for Spatial Data Accuracy (NSSDA), FGDC-STD-007.3-1998, 1998.
Imhoff, M.L., W.T. Lawrence, D.C. Stutzer, and C.D. Elvidge, A technique for using composite DMSP/OLS
    "city lights" satellite data to accurately map urban areas, *Remote Sens. Environ.*, 61, 361–370, 1997.
Lieske, R.W., DMSP primary sensor data acquisition, *Proc. Int. Telemetering Conf.*, 17, 1013–1020, 1981.

Index

Milton Keynes UK
Ingram Content Group UK Ltd.
UKHW052019071024
449327UK00027B/2337